Beyond the MNCs: Exploring Knowledge Transfer in Diverse Organizations

perry

TABLE OF CONTENTS

1. Introduction

The introduction will present to the reader the background of the study, the problem discussion and research gap and conclude this section with the purpose of the book along with our proposed research question.

1.1 Background

Knowledge transfer is a vital source of a firm sustainable competitive advantage (Easterby-Smith et al., 2008; Cardoni et al., 2019) because merely knowledge creation and retention within the organization are not sufficient. Knowledge must be transferable and shareable both within and across an organization for leveraging competitive advantage and improving business performance (Minbaeva et al., 2003; Levine & Prietula, 2012).

Knowledge transfer (KT) is recognized as an important research area within the knowledge management (KM) field (Tangaraja et al., 2016). KM consists of the five key processes: knowledge identification, knowledge creation, knowledge storage and retention, knowledge transfer, knowledge sharing, and knowledge utilization (Cerchione et al., 2020). However, KT is one of the most investigated research dimensions among academics (Durst & Runar Edvardsson, 2012) due to its complexity and importance for enhancing innovation capabilities that are considered as one of the key aspects of the competitive advantage for companies (Easterby-Smith et al., 2008; Milagres & Burcharth, 2019) and individual career success (Tangaraja et al., 2016).

To establish aspects that affect KT, it is essential to introduce the KT concept's outgoing boundaries. KT is defined as the process of knowledge transmission between a source (sender who transfers knowledge) and the recipient (who acquires the knowledge) that can appear in both individual and higher levels such as group, division, and organizations (Minbaeva et al., 2003; Paulin & Suneson, 2012). The KT process occurs in two ways, codification strategy or personalization strategy (Tangaraj et al., 2016). The former is based on explicit knowledge and the latter is tacit knowledge. Furthermore, knowledge types distinction in terms of tacit and explicit knowledge are essential due to their impact on easiness to transfer the knowledge (Durst & Runar Edvardsson, 2012). Particularly, explicit knowledge comes from extracting information from a source that is stored, commonly in the form of formalized instructions. Tacit knowledge includes embedded knowledge from an individual that is transferred, in the form of skill set or experiences which is described as know-how knowledge regarding a topic and is delivered directly from an individual (Minbaeva & Michailova, 2004; Hutzschenreuter & Horstkotte, 2010; Korbi & Chouki, 2017; Ishihara & Zolkiewski, 2017).

Notably, knowledge sharing (KS) is often considered interchangeable with the KT concept (Paulin & Suneson 2012). However, KS is instead a critical process of KT (Tangaraja et al., 2016). Particularly, it occurs at the individual level and possesses behavioral characteristics such as providing and exchanging knowledge and referring to the "people-to-people" process. In contrast, KT is more complex; it includes behavioral and nonbehavioral features such as knowledge search, access, sharing, utilization, and embeddedness (Filieri & Alguezaui, 2014; Tangaraja et al., 2016).

Equally important, the bulk of literature investigates KT in large companies through the lens of multinational companies (MNC) and their subsidiaries (Foss & Pedersen, 2002; Minbaeva & Michailova, 2004), because large companies in terms of MNCs considered as *"network of capital, product and knowledge transactions among units located in different countries"*

(Ishihara & Zolkiewski, 2017). Moreover, as MNCs possess greater capabilities of KT, this yields them with an additional source of competitive advantage (Minbaeva et al., 2003).

At the same time, another branch of the literature concentrates on knowledge management, transfer and sharing in small and medium-sized enterprises context (SMEs) due to their importance for national economies. For example, according to the Organization for Economic Co-operation and Development (OECD), SMEs represent 99% of all enterprises and provide about 60% of all jobs in the OECD area (OECD, 2019). Furthermore, SMEs are increasing their presence in the technology and software industries (Zapata Cantú et al., 2009), contributing to further economic growth. As a result, the literature supports that knowledge management, transfer, and sharing in small and medium-sized firms are key ingredients of SMEs' performance and long-term survival (Desouza & Awazu, 2006; Wee & Chua, 2013).

1.2 Problem discussion and Research question

Successful companies are differentiated from others if they can effectively leverage the transfer of knowledge both within and across the organizations (Easterby-Smith et al., 2008). Firms, both LC/MNCs and SMEs, who are successful, normally experience higher innovation performance rates that are linked to the ability in managing and transferring knowledge (Desouza & Awazu 2006; Cerchione et al. 2017; Korbi & Chouki 2017; Milagres & Burcharth 2019; Cerchione et al. 2020). However, SMEs do not manage the knowledge similarly to larger counterparts and it is incorrect to assume that distinction is associated with only different magnitude or scales (Desouza & Awazu, 2006). First, knowledge transfer might be significantly distinguished because partners with different sizes face the following constraints: (1) divergent dynamic capabilities; (2) distinct corporate governance and culture; (3) various extent of specialization (Korbi & Chouki, 2017; Wee & Chua, 2013). In addition, a company's size and power can impact the knowledge transfer process and the requisites for further coordination and control (Loebbecke et al., 2016).

Second, the critical difference is that larger organizations manage knowledge in a systematic and formalized way, while SMEs, due to lack of resources, relies more on the informal way of organizing knowledge (Cerchione et al., 2020, 2017; Wee & Chua, 2013; Durst & Runar Edvardsson, 2012). In detail, LC/MNC codify tacit knowledge into external through formalized and standardized ways such as reports to reduce uncertainty and ambiguity associated with tacit knowledge transfer (Korbi & Chouki, 2017). On the contrary, the knowledge created by SMEs is tacit in nature (Cerchione et al., 2020) because SMEs more rely on knowledge migration between individuals through socialization activities and utilize the "common knowledge" that referred to knowledge known to all employees like "common sense", which leads to a faster speed of knowledge transfer in SMEs (Desouza & Awazu, 2006). Likewise, SMEs face two main tensions: adhering to formality in knowledge transfer to improve efficiency and informality for facilitating the decision-making process (Fletcher & Prashantham, 2011).

Third, actors engaged in KT should consider various aspects related to different firm sizes, since they affect the KT process differently alongside each other. Particularly, LC/MNC generate more knowledge internally because of their increased capacity of resources allocation availability and higher access to knowledge resources, thus providing larger firms with a higher absorptive capacity over SMEs (Wijk et al., 2008). In contrast, literature agrees that SMEs need external knowledge from their network due to scarcity of resources that eventually help to generate knowledge internally (Chen et al., 2006; Desouza & Awazu, 2006; Corral de Zubielqui et al., 2015). Furthermore, when LC/MNCs and SMEs operate in one alliance structure, where the LC/MNC has more power and overinfluence, this can be harmful to SMEs

' knowledge sharing and learning processes and further the firm's performance (O'Dwyer & O'Flynn, 2005). In particular, SMEs may allocate too many resources for meeting the needs of MNC that eventually slows down SMEs' ability for knowledge transfer and decision making, independently that might lead to a business failure and become costly for the MNC in the future.

Finally, as companies increase in size, LC/MNC are focused more on developing incremental innovation and might miss opportunities for breakthrough discoveries that can be seized by smaller firms (Christensen, 2013). This might be due to larger organizations have a more complex and bureaucratic organizational structure that hinders knowledge transfer, and consequently innovation performance (Qian & Xu, 1998; Thompson, 1965). On the other hand, SMEs, as compared to LCs/MNC, is characterized by a flat organizational structure, unique management style, flexibility, informal communication, non-bureaucratic relationships that facilitate knowledge flow leading to fostering innovation capabilities and competitive advantage of SMEs (Harel et al., 2020; Corral de Zubielqui et al., 2019; Durst & Runar Edvardsson, 2012).

Accordingly, the literature on knowledge transfer in the context of LC/MNCs and SMEs is still fragmented (Milagres & Burcharth, 2019; Cerchione et al., 2017) and this book aims to examine in-depth aspects that affect KT by considering perspective in terms of different firm sizes. In particular, one branch of the literature explores aspects that affect KT within organizations. However, this study will conduct an exploration of aspects that affect inter-organizational knowledge transfer for different firm sizes such as LC/MNCs and SMEs on the organizational level. This is crucial because by overlooking knowledge transfer in the different firm size contexts, researchers and practitioners might lose a holistic picture of aspects that impact knowledge flow on inter and intra-organizational levels that eventually, might significantly influence a firm's innovation performance and further success.

Given the foregoing, this study will focus on the following research question:

What aspects affect inter-organizational knowledge transfer in different company's sizes?

2. Systematic literature review

This systematic literature review consists of three chapters. The beginning of each chapter will briefly summarize what will be included in the corresponding chapter.

Chapter I. The literature surrounding knowledge transfer

This chapter will introduce how different schools of thought and theoretical contributions have affected the development of KT. The second part of this chapter will introduce the topic of knowledge and knowledge transfer (KT), how these concepts are of great importance for firms for their innovation capabilities and sustainable competitive advantage. Third, the concept of knowledge itself along with how it transpires through different stages, along with knowledge management (KM), as KT belongs to this field, this section will also explain how the concepts of KT are affected by KM, as they are interconnected. Fourth, the process of KT will be analysed to comprehend how different authors' perspectives towards it is, with the purpose to gain an understanding of how the process might occur differently in different contexts. This part will also discuss the perspective of knowledge transfer used for this book, to separate the concept to our main topic of KT aspects. Lastly, depending on which perspective to KT is adopted, this section will provide an elaboration to the thesis's perspective on inter and intra-organizational KT.

As the context of perspectives to how the KT aspects are researched might yield different results, it is important to establish boundaries to the thesis's perspective due to the outcome might appear different in another context. To provide clarity of what will be discussed in the concept of KT during this chapter, figure 2.1 presents the inter and intra-perspectives towards KT. The literature of intra-and inter-knowledge transfer will be discussed here to provide a holistic picture and formalize boundaries to the perspective of KT that will be researched in this book

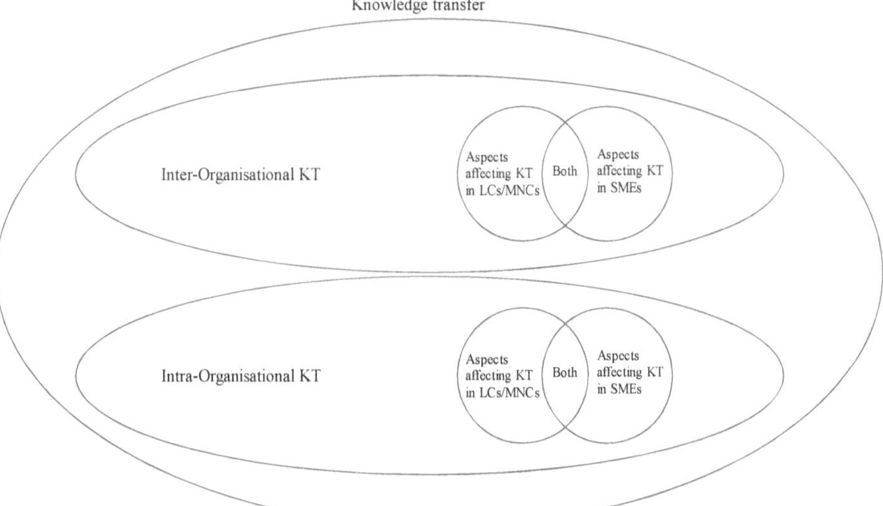

Figure 2.1. The Knowledge transfer concepts and perspectives to it.

2.1 Development of Knowledge transfer and its origins

According to Lucas & Ogilvie (2006) KT is viewed by primarily in two schools of thoughts. First school argues that successful KT is achieved by controlling resources and managing them, with focus on aspects within KT such as absorptive and retentive (disseminative) capabilities. They argue that this school focus on the ability to institutionalize knowledge, making it available and usable for firms (Szulanski, 1996), implying that from the individual level of acquiring said knowledge, to disseminate it into the organization. The second school's perspective of KT is seen as a social activity, on the individual level. Whereas they state to achieve successful KT is to understand the process of how individuals develop and manage relationships by aspects such as culture, networks and team compositions. From these schools of thought, according to Paulin & Suneson (2012) a pair of research streams have been developed in the field of knowledge transfer and sharing. They argue that the first school focus on product innovation and technology transfer, which researches the relationship and communication between firms, ergo transferring knowledge between them, the second stream developed a focus on tacit and explicit knowledge.

Moreover, Penrose (1959) concept of a resource-based view of a firm is the underlaying foundation of the theories (Eisenhardt & Santos, 2006). The theories contribute towards the KT field by identifying its processes and the determinants. Categorized in groups of four; characteristics of knowledge, knowledge senders, knowledge receivers and the relationships between them, as such, there are four major theories contributing towards the literature of KT among them are the resource-based view, dynamic capabilities approach, organizational learning theory and knowledge-based view (Minbaeva, 2007). However, the knowledge-based view conceptualized by Dierickx & Cool (1989) it is not yet recognized as a finished strategy or theory of the firm (Eisenhardt & Santos, 2006). What all theories have in common is that they share the initial premises in the foundation of the theory, which is based upon firms' structure, behaviour and performance of firms (Grant, 1996).

First, Wernerfelt (1984) argue that resources which are uncommon, difficult to replicate and do not possess substitutes, along with routines and technological skills are strong sustainable advantages for firms. Whereas the resource-based view, conceptualized by Penrose (1959), view knowledge as one of the primary resources for firms which contributes to competitive advantage, along with structure and behaviour for firms in regard to their performance (Nonaka & von Krogh, 2009; Conner & Prahalad, 1996; Grant, 1996). In the resource-based view, the view on knowledge is of a generic nature (explicit knowledge) and acknowledges that the transfer of knowledge is unambiguous and easy to transfer (Eisenhardt & Santos, 2006). Also utilize KT to measure how dissemination of knowledge within the firm (Barney, 1991). Moreover, optimization of managerial decisions to identify how firms achieve maximum benefits from their resources and capabilities to obtain a certain level of sustainable competitive advantage (Grant, 1996), thus relating to the KT process.

Secondly, the theory of organizational learning is about how learning could be described as a process, attempting to codify (explicit knowledge) it with a perspective of exploitation of knowledge (Newell, 2005). Then how information is treated, incorporated into behaviour of agents and how they adapt to the received information (Eisenhardt & Santos, 2006). The literature of organizational learning consists of several contributions. Such as by organizational routines (Cyert & March, 1963), which identifies capabilities of a firm and how routines could be utilized to get consistent performance as the organization learns from their processes (Eisenhardt & Santos, 2006). At the organizational level, this is accomplished by the transfer of knowledge as best practices, while on an individual level, employees attempt to codify their tacit knowledge in how-to documents and lessons learnt (Newell, 2005). Organizational memory (Nonaka & Takeuchi, 1995) builds on the routines established within a firm and acts as a guide on how to improve routines by past experience (Eisenhardt & Santos, 2006) by utilizing the explicit knowledge acquired from best practices and how-to documents or lessons learnt (Newell, 2005).

Third, the dynamic capabilities approach developed by Teece et al., (2009) focus on managerial and organizational processes to achieve competitive advantage. It is based upon the ability of the firm to deploy, acquire and reconfiguration internal and external competencies with knowledge transfer to adapt and adjust to a changing environment (Eisenhardt & Santos, 2006). With this dynamic process, the transfer of knowledge determines the competitive advantage for firms by combining their assets and coordinating it (Nonaka & von Krogh, 2009). Moreover, the dynamic capabilities approach perspective is based on equilibrium and do not consider that knowledge changes or how its created within firms, disregarding that firms' routines need alterations regarding the environment (Eisenhardt & Santos, 2006), which may result in inadequate adjustment in modern industries.

Fourth, depending on the theory researched, perspectives that are not optimized towards how firms manage and transfer their knowledge. As a result, combining aspects of the multiple theories, the knowledge-based view emerged and acts as an extension depending on perspective towards each theory, where the knowledge-based view of a firm conceptualizes internal and external knowledge flows as knowledge streams and knowledge assets as stocks of knowledge (Eisenhardt & Santos, 2006). As modern organizations are increasingly striving towards being knowledge-based (Sikombe & Phiri, 2019; Martins & Terblanche, 2003) and instead of the previous view of knowledge in an explicit matter, the knowledge-based view considers tacit and explicit knowledge as a competitive resource (Nonaka & von Krogh, 2009). Moreover, it attempts to integrate tacit knowledge into a firm's products or services (Grant, 1996) by strategies concerning managing knowledge assets and characteristics of knowledge types, which affects the process of KT (Nonaka & von Krogh, 2009), since tacit and explicit knowledge possess different transfer difficulties regarding the firm's organizational structure (Grant, 1996) and characteristics of knowledge types (Giannakis, 2008; Nonaka, 1991).

2.2 Importance of Knowledge and Knowledge transfer

Knowledge comes in various definitions and perspectives seen in a spectrum of individual to industry (Paulin & Suneson, 2012), as such, according to different authors knowledge is viewed as; knowledge of the firm is observable in operating rules, technologies and data banks (Kogut & Zander, 1992), knowledge is seen as a firm's stock (Szulanski, 1996), knowledge is defined as justified true beliefs of individuals (Nonaka & von Krogh, 2009), knowledge comes from the individual's ability to make judgements in social contexts (Fernie et al., 2003). Thus, as the broad-spectrum place knowledge differently, it reveals the importance of differentiating knowledge in its context, so that it could be understood depending in which context its applied to.

In a dynamic market, one of the most reliable sources of competitive advantage for firms is knowledge (Argote & Ingram, 2000; Giannakis, 2008; Nonaka, 1991; Sikombe & Phiri, 2019), implying that the knowledge itself is acquirable, transferrable and integrable while considering it as a strategy to sustain competitive advantage (Eisenhardt & Santos, 2006). Moreover, knowledge contributes to a firm's competitive advantage only when it can be used to provide a form of value, in the sense of a strategy or a process to which increases its efficiency (Barney, 1991). As such, the containment of knowledge consists in members (employees), tools (hardware or software) and tasks (intentions, purpose or goals) within an organization and embedded in networks (Argote & Ingram, 2000). Thus, stating the importance of being aware of knowledge itself, knowledge types, characteristics and how to manage knowledge, to adequately assimilate and utilize acquired knowledge (Easterby-Smith et al., 2008).

Every activity conducted within a firm that holds value is affected by technological change and industry structure, which is a primary driver for competition that reflect on how companies need to constantly adapt and change to achieve competitive advantage (Porter, 1985). Knowledge creation in turn could be seen as the creation of organizational resources, which is generated by new knowledge (Argote et al., 2003) that occurs during the changes and adaptations, which forms competitive advantage for firms (Argote & Ingram, 2000). As such, organizations assimilate and accumulates new knowledge via individuals who generate and transfer new solutions (Nonaka & Takeuchi, 1995), which is how KT is connected to sustainable competitive advantage (Rogers, 2004). Moreover, KT occurs when an organization is affected by experiences or learning outcomes of others, which is achieved by per example interactions between individuals, publications and disclosures with the intention of providing a new competence or skill-set for the organization (Giannakis, 2008). Thus, the changes and adaptations of organizational resources occur through knowledge transfer activities, revealing the importance for firms to possess adequate KT processes, as it contributes to firm's sustainable competitive advantage and long-term survival in terms of higher firm performance over industry competitors (Argote & Ingram, 2000). Which leads the literature toward how firms can sustain their competitive advantage with innovation.

Such as innovation according to Schumpeter (1947), is defined as doing something new or doing something in a new way. Which could be seen aligned with Nonaka (1991) view of innovation, as the author defines innovation as a knowledge-creating companies are those who strive for creating new knowledge, disseminating it and embed it in firms' technologies and products. As firms who follows the logic of developing new products or services gain better insight in how to utilize new knowledge in a financial matter (Cohen & Levinthal, 1990). Similarly, innovation could be seen as "the ability of the firm to generate new ideas" (Rogers, 2004) and the process of innovation includes three steps acquisition (to obtain), dissemination (distribute) and deployment (implementation) of old or new knowledge (Hamdoun et al., 2018). Moreover, one of the critical factors to achieve successful innovation in firms is by the ability of transferring knowledge and managing knowledge, as firms need to be available for the external environment and adapt accordingly to the change (Gilbert & Cordey-Hayes, 1996). As such, knowledge transfer and knowledge management are competitive resources for firms which foster innovation while generating sustainable competitive advantage within firms (Joia & Lemos, 2010), which indicates the need for firms to be aware of KT aspects and how to manage, to create value from the knowledge acquired (Newell, 2005).

As such, innovation capabilities depend on the organizations ability to transfer knowledge (Hamdoun et al., 2018) internally and externally between the firm and customers, from external ideas or research and trends (Rogers, 2004). Firms who are successful tend to reach their success by different means, some examples are by utilizing external partnerships (Milagres & Burcharth, 2019) or knowledge within firms (Durst & Runar Edvardsson, 2012). These firms normally experience higher innovation performance rates that are linked to the ability in managing and transferring knowledge (Desouza & Awazu 2006; Cerchione et al., 2017; Korbi & Chouki 2017; Milagres & Burcharth 2019; Cerchione et al., 2020). Moreover, on the individual level, discussions and interactions between an employee and second party, is the process which generates new ideas and acts as a conduit for innovation in firms as certain are skill-sets experience based (Nonaka, 1991). Whereas the importance of KT between actors is to possess adequate transmission and ability to absorb and integrate to firm's routines

(Szulanski, 1996). As they reflect on the ability of firms to explore and exploit knowledge, which determines the level of innovation and the innovation performance of firms (Hamdoun et al., 2018).

2.3 Knowledge and Knowledge management

The literature of knowledge illustrates the importance of understanding it, in regard to its context of; knowledge types, its characteristics, how it's managed and how it's transferred. Since without these concepts, knowledge, by itself does not provide or generate value for firms (Milagres & Burcharth, 2019). Moreover, as stated before, in the field of knowledge management, knowledge transfer is one of the more researched areas (Tangaraja et al., 2016). Thus, emphasizing on why the transfer process of knowledge itself is of great importance to understand, since with an absence of understanding the process, it will not be properly transferred. As so, this section will research which characteristics knowledge possess, how it contributes to a firm and how used to sustain competitive advantage. Hence, to provide clarity of how the concepts are connected, in figure 2.3.1 below, an illustration of the conceptual boundaries to the fields which are researched in this book is presented.

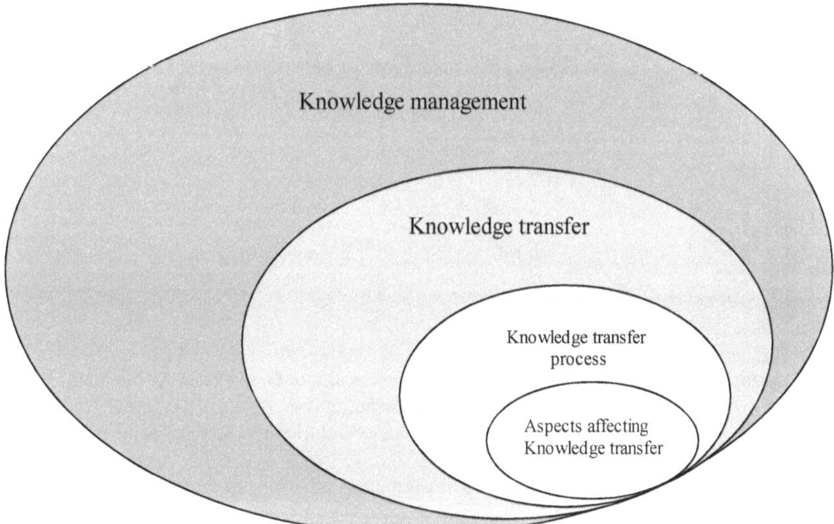

Figure 2.3.1 The field of Knowledge Management, Knowledge transfer, its process and aspects.

In the literature, elements about how knowledge evolves through different stages described by Bender & Fish (2000) provides a good overview of knowledge in different settings. As mention in the previous section, knowledge consists of several definitions depending on which context its applied to, this author argues that the hierarchy of knowledge (figure 2.3.2) transpires through different stages such as by expertise, knowledge, information and data. At the lowest stage, data is mere objectified facts, acting as a material of information creation, when data is correlated with understanding and comprehension it transforms to information. As such, information transforms into knowledge by adding personal experiences, values and beliefs.

Furthermore, expertise in an area of knowledge is seen as a deeper understand of it, being untransferable between individual since it builds on the individuals training, personal experience and education. Thus, it is important to be aware that knowledge originates from the individual and is *"the mental state of ideas, facts, concepts, data and techniques, recorded in an individual's memory."* and transform through the hierarchy of knowledge. Hence, how knowledge transpires through these stages provides clarity towards understanding the categorization of knowledge characteristics and how knowledge transforms, which assist in the understanding of knowledge itself and how knowledge transfer occurs through different stages.

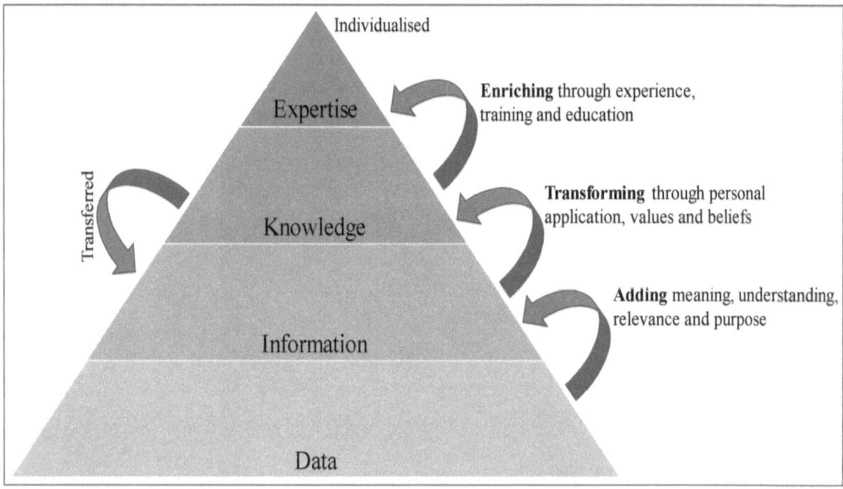

Figure 2.3.2 Knowledge hierarchy by Bender & Fish, (2000).

According to the literature provided in the previous paragraph, it relates to knowledge types, as they distinct in terms of tacit and explicit knowledge (Nonaka & Takeuchi, 1995; Nonaka & von Krogh, 2009; Spender, 1996). Since it is essential to view them differently due to their impact on the transfer the knowledge (Durst & Runar Edvardsson, 2012). As such, codified knowledge (explicit knowledge) refers to information which is specifiable and formalizable and is storable for later reusage (Cavusgil et al., 2003; Loebbecke et al., 2016; Nonaka, 1991; Sikombe & Phiri, 2019), also known as indirect KT that indicated that the knowledge is being transferred by a medium (Sveiby, 1996). Explicit knowledge is created by two ways; tacit to explicit and explicit to explicit. The former, is achieved by articulating the foundation of a tacit knowledge and storing it, which then is accessible and sharable with others, the latter involves reconceptualizing of explicit knowledge (Nonaka, 1991), a personalized way of an individual's interpretation of an acquired explicit knowledge suitable for the situation. Tacit knowledge includes embedded knowledge from an individual that is transferred, in the form of skill set or experiences which is described as know-how knowledge regarding a topic and is delivered directly from an individual (Minbaeva & Michailova, 2004; Hutzschenreuter & Horstkotte, 2010; Korbi & Chouki, 2017; Ishihara & Zolkiewski, 2017), which is seen as a direct transfer of knowledge (Sveiby, 1996). Creation of tacit knowledge is achieved by two ways; tacit to tacit and explicit to tacit formerly explained. The former is an acquirable by observation, imitation, practice (Nonaka, 1991) and experience (Giannakis, 2008). Lastly, there is a noted disagreement between authors (Nonaka, 1991, Sikombe & Phiri, 2019) which regards the observability and teachability of tacit knowledge. Moreover, Sikombe & Phiri (2019) illustrates the differences of tacit and explicit knowledge (see table 2.3.3).

Scope	Tacit knowledge	Explicit knowledge
Definition	Know-how, know-why, skills expressed through performance	Know-about: comprise facts, theories and instructions
Quality, speed, cost of transfer	Slow, costly and uncertain (high stickiness)	Fast, may be costly, accurate (low stickiness)
Diffusion	Difficult to convey	Easier to convey
Residence	General information, experiences and memories	Books, documents, databases, policy manuals
Complexity	Relative complex	Relatively simple
Teachability	Not teachable	Teachable
Observability	Not observable	Observable
Codifiability	Difficult	Easy

Table 2.3.3. Differences between tacit and explicit knowledge (Sikombe & Phiri, 2019).

In turn, by acknowledging that the knowledge type is either; tacit or explicit. Knowledge characteristics in turn differentiate (Sikombe & Phiri, 2019; Ishihara & Zolkiewski 2017), by these two parameters. The importance towards understanding knowledge lays in comprehending its characteristics. The characteristic of knowledge consists of tacitness, ambiguity and complexity (Easterby-Smith et al., 2008) and is generated from actualisation of a skilful action or ideation of a situation (Nonaka & von Krogh, 2009). Milagres & Burcharth (2019) argues that knowledge characteristics such as ambiguity, dependency and tacitness makes knowledge "sticky" (Szulanski, 1996), by embedding the knowledge in a knowledge repositor which increases its persistence over time (Argote et al., 2003). Moreover, knowledge disruptiveness impacts on willingness to share knowledge or being undermined regarding one's own knowledge (Newell, 2005). Tacit knowledge contains higher complexity, tacitness and stickiness (Ishihara & Zolkiewski, 2017) compared to explicit (Milagres & Burcharth, 2019), since it is difficult to formalize (Hutzschenreuter & Horstkotte, 2010) and codify (Ishihara & Zolkiewski, 2017).

However, as mentioned before, according to Durst & Runar Edvardsson (2012) if knowledge itself remains dormant it does not yield any benefit or value towards a firm, it is only when utilized it creates value for firms, which reflect on the importance of knowledge management (KM), as it contributes to the organization innovation performance and competitive advantage (Joia & Lemos, 2010) by increasing firms' efficiency in management of knowledge, utilization and dissemination of knowledge (Garavelli et al., 2002). Thus, by integrating these concepts with knowledge within firms, it creates value (Argote et al., 2003).

KM is defined as an organizational or technical tool (Cerchione et al., 2020). While stated by Durst & Runar Edvardsson that these tools are used in KM of organizations processes, culture, learning, knowledge transfer, sharing and storing illustrated in figure 2.3.4. They argue that the first step is that knowledge needs to be identified by activities such as research or experiments, either by existing or new knowledge, which could be beneficial for the firm. It its only when knowledge has been identified it can be created, in means of how it will provide value to the firm. The thirds step involves retaining the knowledge within the firm, so it is not lost (Nonaka, 1991). This is achieved by codifying acquired knowledge and dissemination of knowledge within the firm (Garavelli et al., 2002; Hamdoun et al., 2018). Acquired knowledge is then transferred into a firm´s product or service which is how the knowledge is utilized and how knowledge itself creates value for firms, as by integrating the knowledge with a product or service (Durst & Runar Edvardsson, 2012). This process occurs in the context of; properties of units (such as groups, individually or in the organization), affecting the relationship between these units and the knowledge properties (Argote et al., 2003). As the process could be seen as a circular one, parallel to the process of KM, new insights and personal experiences emerges which could provide a basis for new knowledge identifications. Thus, repeating the process.

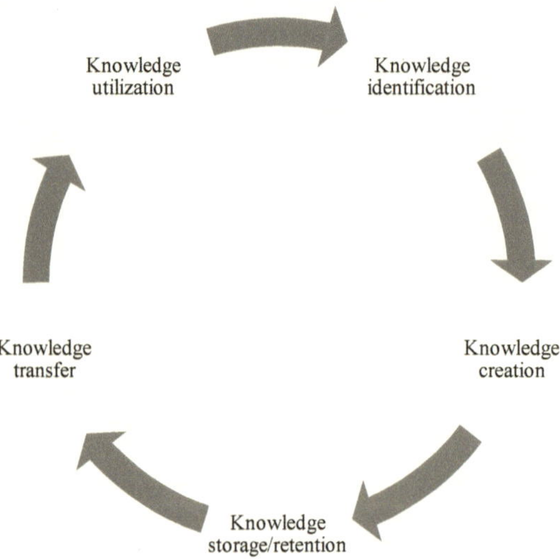

Figure 2.3.4 Knowledge management process according to Durst & Runar Edvardsson (2012).

Concludingly, this section covered the process of knowledge and how its managed, the following section will research the process of transferring knowledge, as it is by this process which knowledge generates its value.

2.4 Knowledge transfer as a process

This section will describe the process of knowledge transfer by considering different literature perspectives on: (1) KT concept and basic definitions; (2) actors involved in KT; and (3) the description of KT sub-processes.

First, the bulk of the literature agrees that knowledge transfer is a process and not just a one-time act (De Luca & Cano Rubio, 2019; Milagres & Burcharth, 2019; Battistella, 2016; Ajith Kumar & Ganesh, 2009; Argote & Ingram, 2000; Szulanski, 2000; Gilbert & Cordey-Hayes, 1996). According to Zairi (1997) the process is defined as "an approach for converting inputs into outputs. It is the way in which all the resources of an organization are used in a reliable, repeatable and consistent way to achieve its goals". At the same time, according to the Cambridge dictionary, the term "act" as a noun refers to something done or performed with a short performance duration. Furthermore, acts are considered as one of the critical components of the process (Zairi, 1997). Thus, this implies that the act should be regarded as one-time short action rather repeatable and consistent activity, which is assumed as a process.

Since knowledge is considered as an essential resource for firms that should effectively be converted and transferred to benefit the company (Nonaka & von Krogh, 2009; Conner & Prahalad, 1996; Grant, 1996), hence, knowledge transfer is the process rather than the act (Szulanski, 2000). Furthermore, according to Liyanage et al., (2009), who developed the process model for KT, argue that KT is a process because it consists of two features such as (1) the act of collaboration between involved actors and (2) transformation of the knowledge into applicable form.

Besides, a process view on knowledge transfer instead of modelling KT as mere acts facilitates a deep understanding of the factors that enhance or hinder knowledge movement along with KT processes (Szulanski, 2000). For instance, effective managing ambiguous set of routines, related to R&D capabilities, know-how, managerial practices, and so on, require examining knowledge transfer from the process perspective that provides an in-depth investigation of barriers for effective KT within a firm. Moreover, Gilbert & Cordey-Hayes (1996) supports that KT is dynamic rather than static implying that the KT process is repeatable, thus KT should be regarded as a process.

However, the concept of knowledge transfer is defined in various ways (see table 2.4.1). This overview of KT concepts from different scholars allows considering the KT process as a part of continuous learning process of the learning organization (Newell, 2005; Goh, 2002; Gilbert & Cordey-Hayes, 1996). Particularly, the learning organization is distinguished from other organizations due to its ability to continuously acquiring a number of skills in the following areas such as uncertainty and conflict management, negotiation, efficient communication in order to respond to the fast-changing internal and external environment, which can be achieved via effective KT process at the company (Gilbert & Cordey-Hayes, 1996).

Moreover, other literature also supports a learning process perspective of the KT concept and that is why entities engage in knowledge transfer (Milagres & Burcharth, 2019; Ajith Kumar & Ganesh, 2009; Argote & Ingram, 2000), where the learning process appears in two ways: organization learns from its own experience or experiences of others and the term "experience" helps to distinguish the transfer of knowledge from just transfer of data or information (Ajith Kumar & Ganesh, 2009).

At the same time, according to Foss & Pedersen (2002), who examined KT from the intra-organizational level, argue that KT is defined as a transformation of existing knowledge and it does not entail knowledge replication in a new place, instead prior knowledge modifies and fits the new contexts to solve specific issues.

Author (Year)	Definition	Perspective
Gilbert & Cordey-Hayes (1996)	*"The process of knowledge transfer is not a static one, it is dynamic, and is part of a process of continuous learning".*	Knowledge is a resource and KT enhances the learning capabilities of the learning organizations that lead to technological change and innovation.

Argote & Ingram (2000)	*"Knowledge transfer in organizations is the process through which one unit (e.g., group, department, or division) is affected by the experience of another".*	Knowledge is regarded as a key resource that is called knowledge reservoirs that contribute to the firms' competitive advantage. But KT can be challenged due to differences in subnetworks and adapting knowledge in different contexts.
Szulanski (2000)	*"Knowledge transfer is seen as a process (not a one-time act) in which an organization recreates a complex, causally ambiguous set of routines in new settings and keeps it functioning".*	Knowledge is viewed as best practice in terms of a set of routines but complex and difficulties with transfer cannot bring benefits to the organizations due to various barriers (stickiness) that exist at different substages of KT.
Goh (2002)	*"Knowledge transfer is also a key dimension of a learning organization. Learning occurs when knowledge in one part of an organization is transferred effectively to other parts and used to solve problems there or to provide new and creative insights".*	Knowledge is viewed as intellectual capital and companies should effectively manage it via knowledge transfer that is a key dimension of a learning organization.
Newell (2005)	*"Knowledge transfer implies that each individual/group/organizational unit need not learn from scratch but can rather learn from the experiences of others".*	Knowledge is viewed as previous experience and KT is examined via the learning cycle model. The knowledge characteristics such as distribution, ambiguity, disruptiveness of knowledge, can impede the KT process.
Ajith Kumar. & Ganesh (2009)	*''a process of exchange of explicit or tacit knowledge between two agents, during which one agent purposefully receives and uses the knowledge provided by another''.*	Knowledge is regarded as specialized existing knowledge that should be exploited and applied for organizational goals. KT is considered as the exchange process that *"involves two complementary acts: the act of giving or delivering knowledge by one agent (the source), complemented by the act of receiving and using knowledge by another (the recipient). Without either, the process of transfer is incomplete.".*
Foss & Pedersen (2002)	*"Indeed, transfer of knowledge is often associated with modification of the existing knowledge to the specific context. Therefore, what is transferred is not the underlying knowledge but rather applications of this knowledge in the form of solutions to specific problems.".*	Knowledge is a strategic resource for KT between companies. Internal, network, and cluster knowledge positively affect the extent of KT.
Milagres & Burcharth (2019)	KT defined as *"a process through which an organization purposefully learns from another".*	Knowledge is an asset from the external partner source. Knowledge transfer factors examined from the partnership perspective that contributes to the firm innovativeness.

Table 2.4.1 Knowledge transfer definitions from the scientific papers.

Second, the literature agrees that there are two sides involved in the knowledge transfer process (see figure 2.4.2): a source (sender who transfers knowledge) and the recipient (who acquires the knowledge) that can appear in both individual and higher levels such as group, division, and organizations (Tangaraja et al., 2016; Battistella, 2016; Ajith Kumar & Ganesh, 2009; Newell, 2005; Argote & Ingram, 2000; Szulanski, 1996).

All things considered, KT is defined as the process of knowledge transmission between a source (sender who transfers knowledge) and the recipient (who acquires the knowledge) that can occur at individual, team, unit, division, and organizations levels (Ajith Kumar & Ganesh, 2009; Argote & Ingram, 2000; Paulin & Suneson, 2012; Tangaraja et al., 2016).

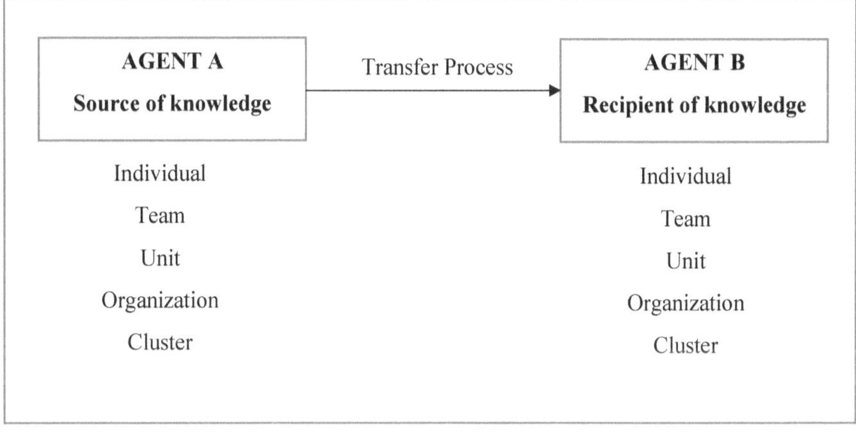

Figure 2.4.2 A simple model of knowledge transfer with key actors according to Ajith Kumar & Ganesh (2009).

Third, scholars agree that knowledge transfer is a complex process because KT includes behavioral features (knowledge sharing and application) and non-behavioral features such as knowledge search, acquisition, access, utilization, and embeddedness (Tangaraja et al., 2016). Furthermore, various factors and contextual issues affect KT such as the knowledge characteristics, knowledge barriers, motivational factors, and so on (Shen et al., 2015; Paulin & Suneson, 2012; Lee & Choi, 2003; Szulanski, 2000, 1996). Overall, the KT process consists of different core sub-processes and different academics suggest them in various perspectives that are summarized in table 2.4.3.

Author (Year)	Knowledge transfer process stages	Perspective
Gilbert & Cordey-Hayes (1996)	*Acquisition, Communication, Application, Acceptance, and Assimilation.*	KT process as a key factor for successful technological innovation.
Szulanski (2000, 1996)	*"Initiation, Implementation, Ramp-up, and Integration".*	KT process has barriers that can impede the best practices transfer within a firm.
Millie & Cheung (2006)	*Motivation, Matching, Implementation, Retention.*	KT success by identifying KT stages needs including knowledge

		management tools and technologies.
Maurer et al., (2011)	*"Mobilization (Search), Assimilation, Utilization".*	KT at the intra-organizational level as a mediator for transferring social capital (i.e. number of ties, tie strength, and trust) to boost the firm growth and innovation performance.
Filieri & Alguezaui (2014)	*"Knowledge search, Knowledge access (or acquisition), knowledge assimilation (or absorption), knowledge integration (or combination)".*	The role that structural social capital (SC) plays for knowledge transfer and innovation at the interpersonal, inter-unit and inter-firm levels.

Table 2.4.3 Knowledge transfer process from the scientific papers.

From the table 2.4.3, it can be argued that the scholars suggest that KT process divided into main sub-phases such as (1) knowledge identification (i.e., initiation/search); (2) acquisition (i.e., access); (3) assimilation (i.e., absorption/rump-up); (4) integration (i.e., embeddedness/combination).

Knowledge identification stage: The majority of academics agree that the knowledge process begins from the knowledge identification stage but different authors call this phase in various terms such as initiation (Szulanski, 1996), knowledge search (Filieri & Alguezaui, 2014), mobilization (Maurer et al., 2011). Despite these differences in titles, authors agree that the knowledge identification stage refers to meeting the needs for undiscovered knowledge, identification knowledge gaps and knowledge needs to fulfill specific tasks or reach a certain performance level at the company, as a result, the suitability for transfer of that knowledge is explored at this stage (Filieri & Alguezaui, 2014; Maurer et al., 2011; Szulanski, 1996). Moreover, the motivation why firm actors engage in knowledge search is because they are looking for useful knowledge from other firm members (1) to avoid duplications in efforts to fulfill the assigned tasks; and (2) to receive better technical expertise that can consequently shorten the task time or project completion time (Hansen, 1999).

Knowledge acquisition stage: The literature agrees that after identification, the knowledge should be acquired in terms of accessing the external superior knowledge that is critical to the firm activities (Tangaraja et al., 2016; Filieri & Alguezaui, 2014; Gilbert & Cordey-Hayes, 1996). Furthermore, Millie & Cheung (2006) argue that the need for knowledge acquisition triggers searching for an appropriate transfer partner(s) to access better knowledge. The main goal for knowledge acquisition is learning due to the increased complexity of various projects including R&D in terms of higher uncertainty and associated costs that eventually improve the performance of the company in terms of business growth and productivity (Lyles & Salk, 1996).

Knowledge assimilation stage: The knowledge assimilation stage is related to the process of analysing, processing, understanding, and absorbing the knowledge transferred (Filieri & Alguezaui, 2014). Moreover, the knowledge assimilation starts from the "decision to proceed" and apply by involved actors in KT, and during this stage knowledge exchanges between the sender and the recipient via establishing the "social-specific ties" (Szulanski, 2000), and this stage is considered completed when the recipient side begins to apply obtained knowledge resource from the source (Millie & Cheung, 2006). Importantly, Gilbert & Cordey-Hayes (1996) found that when the knowledge is applied and before it is routinized to the company's processes should be accepted by individuals that follow specific beliefs and behavior patterns, as a result, one more step in the KT process is acceptance stage that occurs after knowledge application but before an assimilation phase.

Knowledge integration stage: The Integration stage comes as the final phase in the KT process, which is equivalent to the embedding of absorbed and satisfying knowledge into the daily routines of the recipient, (Tangaraja et al., 2016; Filieri & Alguezaui, 2014; Maurer et al., 2011). The embedding of the acquired external knowledge and combining it with the internal one requires knowledge to achieve satisfactory outcomes by solving unexpected problems by the recipient of the knowledge transferred (Szulanski, 2000).

2.5 Knowledge sharing

The literature supports that knowledge sharing is a critical subset of the knowledge transfer process (Paulin & Suneson, 2012; Nonaka & Takeuchi, 1995) since it occurs at the individual level and possesses behavioral characteristics such as providing and exchanging knowledge and referring to the "people-to-people" process (Tangaraja et al., 2016). Thus, knowledge sharing (KS) is defined as "the process where individuals mutually exchange their (implicit and explicit) knowledge and jointly create new knowledge" (van den Hooff & de Ridder, 2004). This definition of KS corresponds to the bidirectional view for sharing of the knowledge capital (Lin, 2007; Tohidinia & Mosakhani, 2010) In more detail, the bidirectional KS refers to the knowledge exchange between individuals through the knowledge denotation (e.g., when person communicating own intellectual capital to others) and knowledge collection which occurs when person consulting others to acquire their intellectual (Tangaraja et al., 2015). Additionally, from the bidirectional perspective, the term "knowledge exchange" has been applied interchangeably with "knowledge sharing" by implying both providing knowledge to others and searching for knowledge from others (Wang & Noe, 2010).

However, KS can also take a unidirectional mode that refers to the prior agreed knowledge flow in a one-way direction from the source to the recipient (Bock & Kim, 2002; Loebbecke et al., 2016). For instance, training activities are directed in one way of KS wherein a teacher delivers knowledge to the students (see table 2.5.1).

Author (Year)	Definition	Perspective
Bock & Kim (2002)	*"The degree to which one actually shares one's knowledge"*	Unidirectional
Loebbecke et al. (2016)	*"Unilateral knowledge sharing is of a pooled or sequential nature; it comprises steps of identifying and transferring, in a single direction, prior agreed-upon knowledge and information"*	Unidirectional
Lin (2007)	*"KS is a social interaction culture, involving the exchange of employee knowledge, experiences, and skills through the whole department or organization"*	Bidirectional
Tohidinia & Mosakhani (2010)	*"KS occurs when organizational members exchange organization-related information, ideas, suggestions and expertise with each other. It is an actual KS behaviour involving knowledge donating and knowledge collecting processes"*	Bidirectional

Wang & Noe (2010)	*"Knowledge sharing refers to the provision of task information and know-how to help others and to collaborate with others to solve problems, develop new ideas, or implement policies or procedures"*	Bidirectional
van den Hooff & de Ridder (2004)	*"KS is the process where individuals mutually exchange their (implicit and explicit) knowledge" and it includes two processes, namely, knowledge donating and knowledge collecting"*	Bidirectional

Table 2.5.1 Knowledge sharing definitions according to Tangaraja et al. (2016).

To summarize, knowledge sharing is a key subprocess of KT, hence, it is essential to cover aspects related to KS that are interconnected to knowledge transfer (Tangaraja et al., 2016).

2.6 Knowledge transfer perspectives

This section of the chapter is intended to examine KT perspectives because KT within or between organizations can have various factors and contextual aspects, which consequently influence the KT process (Nakauchi et al., 2017; Wijk et al., 2008). Particularly, knowledge recipient's ability to identify, assimilate and apply new knowledge varies and depends on transfers of knowledge across or within organizations (Cohen & Levinthal, 1990). As a result, the literature considers KT at different levels: (1) intra-organizational or within a firm (Goh, 2002; Argote & Ingram, 2000; Gilbert & Cordey-Hayes, 1996; Szulanski, 1996); (2) inter-organizational perspective or knowledge transfer between organizations and even clusters (Milagres & Burcharth, 2019; Battistella, 2016; Ajith Kumar & Ganesh, 2009; Easterby-Smith et al., 2008).

Hence, it is important to delimitate the knowledge transfer perspective boundaries between intra-organizational KT, inter-organizational KT, and KT at the individual level for this research.

2.7 Intra-and-Inter organizational KT delimitation

To define the boundaries of KT perspective in terms of intra or inter-firm KT, it is important to understand that knowledge transfer can take two modes such as internal knowledge transfer and external knowledge transfer (Wijk et al., 2008; Argote & Ingram, 2000; Gupta & Govindarajan, 2000). However, firstly, it is essential to distinguish the internal and external knowledge. Internal knowledge refers to the knowledge generated inside of the company and the firm already possess the rights for using this knowledge treated as the firm's resource, whereas external knowledge is the knowledge that the company does not have previously owned and transfers from external sources into the company (Becker & Knudsen, 2006). For instance, knowledge acquired and transferred from firms after merger and acquisitions, suppliers, customers related to the external knowledge (Sikombe & Phiri, 2019; Blome et al., 2014; Segarra-Ciprés et al., 2014; Wijk et al., 2008; Giannakis, 2008; Bresman et al., 1999), while knowledge transfer between different units but within a company like multinational corporations (subsidiaries to headquarter and vice versa) is recognized as the internal knowledge transfer (Chen & Lovvorn, 2011; Gupta & Govindarajan, 2000).

After defining external and internal knowledge, then external knowledge transfer can be defined as the ability of the firm to utilize external knowledge and expertise from external sources including suppliers, customers, alliances, or mergers and acquisitions (Milagres &

Burcharth, 2019; Blome et al., 2014; Giannakis, 2008; Sikombe & Phiri, 2019). In contrast, internal knowledge transfer is the ability of the company to share knowledge with other units within the organization (Tsai, 2001; Gupta & Govindarajan, 2000). Furthermore, internal knowledge transfer facilitates knowledge flow within the firm and motivates coordination among members and the integration of external knowledge into the organization (Schulz & Jobe, 2001).

At the same time, intra-organizational KT refers to the knowledge flow within the organization (Gupta & Govindarajan, 2000; Szulanski, 1996; Tsai, 2001), while inter-organizational KT is associated with acquiring, learning and transfer of the knowledge from the outside organizational borders (Martinkenaite, 2011). The levels of KT are shown in figure 2.7.1.

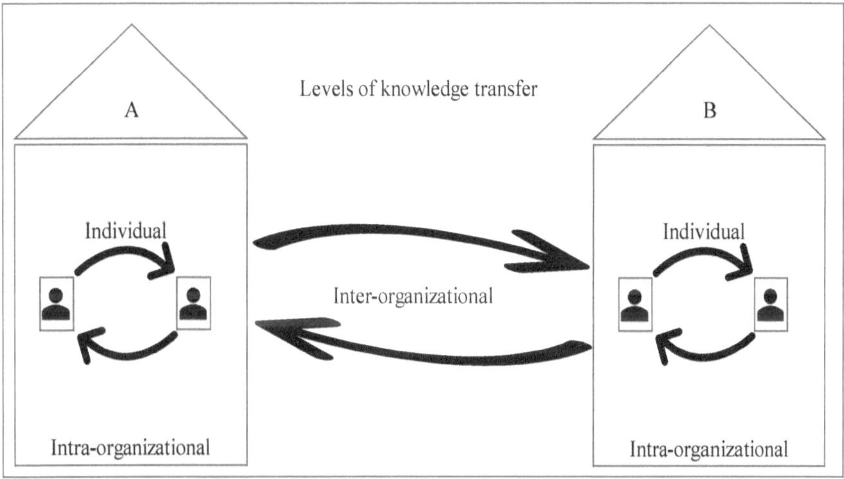

Figure 2.7.1. The levels of knowledge transfer according to Wilkesmann et al. (2009).

Overall, internal knowledge transfer and intrafirm KT are used interchangeably as well as for external knowledge transfer and inter-KT for this book.

2.8 Intra-organizational knowledge transfer

As discussed in the previous section, intra-organizational knowledge transfer implies the internal knowledge movement within the firm (Szulanski, 1996). To determine more precisely, the intrafirm KT is defined as the process through which the units, departments, or divisions are affected by the experience of another unit (Argote & Ingram, 2000). In other words, different business units within the organization purposefully learn and exchange their knowledge; for instance, one production unit may need to learn from another division on product assembling or other issues. This knowledge exchange process at the intra-firm level is conducted through various KT activities such as training programs, teaching, meetings, best practice documents, mentoring, coaching, cross-functional teams, dynamic and virtual communities (Spraggon & Bodolica, 2012), communication and cooperation (e.g. project cooperation) between employees within the organization (Tang, 2011).

Equally important, intrafirm knowledge transfer not only appears within a single organization but also can take place in the multinational company network between the parent organization and its subsidiaries (Gaur et al., 2019; Blomkvist, 2012) due to (1) *"intracorporate network consists of a group of organizations operating under a unified corporate identity, with the headquarters of the network having controlling ownership interest in its subsidiaries"* (Inkpen & Tsang, 2005); and (2) internal knowledge generated within MNC is not shared with an independent actor (Becker & Knudsen, 2006). Moreover, knowledge transfer in MNC is associated with complexity because of the KT challenges arising in different levels such as (1) national culture and geographical distance at the country level (Vlajcic et al., 2019; Fong Boh et al., 2013); (2) organizational boundaries that exist within each entity: the headquarters and between subsidiaries that make knowledge exchange more problematic (Zeng et al., 2018); and (3) motivation, ability, trustworthy relations, reputation and incentives for KT at the individual level (Andersson et al., 2015; Huang et al., 2013; Martín Cruz et al., 2009; Lucas & Ogilvie, 2006)

Besides, a growing scholar's attention within intra-firm KT in multinational enterprises received a reverse knowledge transfer concept that manifests the knowledge transfer process in MNC network as bidirectional: from the subsidiary to the headquarter and vice versa (Jeong et al., 2017; Driffield et al., 2016; Najafi-Tavani et al., 2015; Kumar, 2013; Yang et al., 2008; Ambos et al., 2006) rather than traditional single directional way: from the parent to its foreign subsidiaries (Minbaeva et al., 2003; Foss & Pedersen, 2002; Gupta & Govindarajan, 2000). Particularly, a conventional KT is associated with "forward" transfer of the knowledge from the headquarters to the subsidiary in the multinational companies and less traditional "reverse" from the subsidiary to its parent firm (Yang et al., 2008; Ambos et al., 2006). Furthermore, the importance of the reverse KT is determined by scholars because (1) subsidiaries possess better local knowledge about technologies, market, products and, so on that is essential for the parent to coordinate its strategy for enhancing the competitive advantage (Jeong et al., 2017); (2) subordinates performance influence the parent productivity in terms of turnover and profitability (Driffield et al., 2016). To summarize, intrafirm KT is the essential dimension of the KT field.

2.9 Inter-organizational knowledge transfer

Knowledge transfer across different organizations has become a popular research sub-field of knowledge transfer over the last decade because knowledge transfer between different entities is more complicated as compared to the intra-firm knowledge transfer (Milagres & Burcharth, 2019; Battiestella, 2016; Easterby-Smith et al., 2008). The complexity of interfirm KT is explained in terms of distinctions of organizations in technological capabilities, geographical distance (Battistella, 2016), processes, cultural issues (Easterby-Smith et al., 2008), and social capital (Filieri & Alguezaui, 2014). Furthermore, knowledge transfer among different firms entails multi-mode learning processes that happen simultaneously (i.e., learning about the partner, with the partner, from the partner, and about alliance management) (Milagres & Burcharth, 2019). Besides, knowledge leakage is another *"dark side"* of knowledge transfer between different enterprises that raises complexity, and ambiguity during organizational collaboration (Frishammar et al., 2015).

However, the enterprises engage in inter-organizational KT with a purpose for learning from another enterprise (Milagres & Burcharth, 2019) and it is possible if at least two organizations are involved in the knowledge transmission process (Easterby-Smith et al., 2008). In addition, inter-organizational knowledge transfer appears in the form of international joint ventures (Minbaeva et al., 2018), alliances and partnerships (Korbi & Chouki, 2017; Mazloomi Khamseh & Jolly, 2008; Mowery et al., 1996), mergers and acquisitions (Bresman et al., 1999), supply and customer collaborations (Sikombe & Phiri, 2019). Thus, indeed KT at the inter-organizational level is complex and recognized as the large sub-field of KT that should be considered while examining the knowledge transfer domain.

2.10 Importance of different firm sizes for KT

This section is aimed to examine the significance of the firm size for the knowledge transfer through consideration of various academic studies that investigate the key factors that influence KT and ultimately can lead to an unsuccessful KT between different firm-sized companies.

First, the role of firm size in knowledge transfer is considered by scholars as one of the essential organizational characteristics that can have positive or negative effects on the extent of knowledge transferred (Dhanaraj et al., 2004; Gupta & Govindarajan, 2000; Makino & Delios, 1996). As a result, some branches of literature focused on investigating KT in large companies such as multinational corporations due to their diversified network and higher ability to exploit knowledge flow benefits (Gaur et al., 2019; Ishihara & Zolkiewski, 2017; Chen & Lovvorn, 2011; Yang et al., 2008; Minbaeva, 2007; Minbaeva et al., 2003; Foss & Pedersen, 2002). At the same time, other scholars examine KT in small and medium-sized enterprises (Corral de Zubielqui et al., 2015; Durst & Runar Edvardsson, 2012; Chen et al., 2006) due to their firm specifics that can affect KT.

Second, the literature recognizes that managing and transferring knowledge in LC/MNCs is different as compared to smaller firms due to an asymmetry in power, corporate structure, resources, and capabilities (Korbi & Chouki, 2017; Wee & Chua, 2013; Durst & Runar Edvardsson, 2012; Desouza & Awazu, 2006). Particularly, the literature suggests that SMEs are characterized by a flat organizational structure, unique management style, flexibility, informal communication, non-bureaucratic relationships that facilitate knowledge flow leading to fostering innovation capabilities of SMEs (Harel et al., 2020; Corral de Zubielqui et al., 2015; Durst & Runar Edvardsson, 2012). However, SMEs often face a resource scarcity that leads to having the knowledge in critical employees' minds rather than formally sharing and saving in physical artifacts (Cardoni et al., 2019; Corral de Zubielqui et al., 2015; Durst & Runar Edvardsson, 2012; Durst & Wilhelm, 2012). Furthermore, the knowledge created by SMEs is tacit in nature (Cerchione et al., 2020). On the contrary, large companies in terms of MNCs are complex, multi-layer, and geographically diversified entities that incorporate a network of capital, product, and knowledge assets (Zakova Talpova, 2019; Ishihara & Zolkiewski, 2017; Minbaeva, 2007). Moreover, knowledge flows within MNCs occur in multiple dimensions and multiple directions (Minbaeva, 2007) and the ability of MNCs to manage knowledge transfer across the dispersed cross-border subsidiaries has become crucial for reaching long-term competitive advantage and survival (Yang et al., 2008).

Third, both stream of the research agrees that knowledge transfer in different firm size can bring different barriers: SMEs experience with the lack of resources that might hinder KT and, as a result, it can distort the firm performance, whereas a larger company might act as bureaucratic machines that can lead to knowledge flow jams that eventually diminish innovation capabilities (Durst & Runar Edvardsson, 2012; Desouza & Awazu, 2006; Qian & Xu, 1998; Thompson, 1965).

Given the foregoing, companies that have diverse firm-sized subsidiaries, partners, or other forms of collaborations can be challenged due to the asymmetry in size. Hence, it is necessary to incorporate aspects that affect KT in the context of different firm sizes into policies and procedures.

2.11 Concluding remarks for the KT concept section

To conclude this part of the literature review, in the figure 2.11.1 below, the concept of KT is illustrated. To provide clarity moving to next chapter of the systematic literature review, this figure presents the perspective taken to KT. Moving forward, this book will investigate aspects that affect inter-organizational knowledge transfer for SMEs and LC/MNCs, as mentioned before, the literature of intra-knowledge transfer has been discussed here to provide a holistic picture and formalize boundaries to the perspective of KT that will be researched in this book.

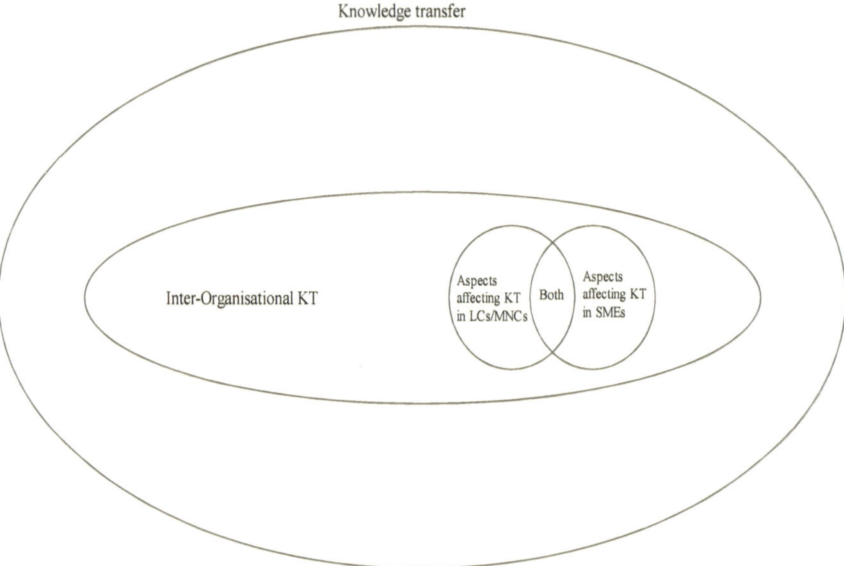

Figure 2.11.1 KT concept and the perspective moving forward.

Chapter II. Aspects that affect inter-organizational knowledge transfer in the context of different firm sizes

This chapter will provide inter-organizational aspects regarding knowledge transfer, elaborating on how and why these concepts are important for successful KT between firms. As previously discussed in Chapter 1, KT between organizations is a critical process for LC/MNCs as well as for SMEs to enhance their competitive advantage and innovation capabilities (Milagres & Burcharth, 2019; Battistella, 2016; Easterby-Smith et al., 2008). Particularly, LC/MNCs engage in inter collaborations for varied reasons not always related to the knowledge type acquired or shared, but for the experience of the collaboration itself (Narteh, 2008). The larger the size of companies the higher positive effect to successful KT on an inter-organizational level, due to (1) higher resource allocation available; (2) more knowledge resources which increase their absorptivity capacity; (3) a large pool of employees contains diversified experience and knowledge (Wijk et al., 2008).

At the same time, approximately 99% of SMEs show the inter-organizational KT needs, and they involved in various form of inter-organizational KT including knowledge exchange with customers, suppliers, friends and even competitors (Chen et al., 2006). These findings were also supported by more recent research of Wei Chong et al. (2011), which found that SMEs participate in KT for acquiring external knowledge from their network such as suppliers, customers that is more important than internal knowledge. Furthermore, Corral de Zubielqui et al. (2015) suggest that SMEs take part in inter-organizational research collaboration in terms of receiving access and adoption of external knowledge resources that help to overcome the resource constraints and decrease the effects of being "small".

Thus, to achieve successful inter-organizational KT, the aspects related to KT must be identified in-depth. As such, this section will study aspects that affect different firm sizes from the perspective of; (1) both small, medium enterprises (SME) and large or multinational companies (LC/MNCs), (2) LC/MNCs, and (3) SMEs, in regard to an inter-organizational perspective.

Therefore, the next sections describe the most reoccurring critical aspects, which have been identified in the literature that impact inter-organizational knowledge transfer in LC/MNCs and SMEs and are organized as follows: cultural distance, geographical distance/proximity, knowledge governance mechanisms, trust, motivation, social ties, knowledge characteristics, absorptive and disseminative capacities.

2.12 Knowledge governance mechanisms

One of the key barriers that can hinder the knowledge transfer between different partners is the opportunistic behaviour of parties in terms of evading their obligations and responsibilities related to the improper allocation of knowledge, efforts, and other knowledge-based assets as a result, partners apply various knowledge governance mechanisms to avoid opportunistic behaviours of their partners (Yang et al., 2015; Mesquita et al., 2008). Furthermore, the interfirm KT depends not only on the acquisition and absorption of the external knowledge but also needs to be internalized and integrated within the organization by using knowledge governance techniques (Fang et al., 2013). Thus, knowledge governance mechanisms are the critical aspects that affect KT between different organizations.

The term "governance" is derived from the Latin word implying the notion of "steering" and it describes the processes that need to be in place for a successful business project outcome (Robinson et al., 2010). Moreover, governance mechanisms between partners are defined as *"the underlying and concrete management and control activities, which describe in detail how the required behaviour of the partner will become motivated, influenced, and established, or more generally, in which ways the desirable or predetermined gains are to be fulfilled"* (Hoetker & Mellewigt, 2009). Hence, the knowledge governance mechanisms refer to the selecting organizational structures and mechanisms that can impact the processes of application, sharing, integrating, and creating knowledge in preferred directions and towards preferred levels (Foss & Michailova, 2009).

Knowledge governance mechanisms can take two modes such as formal and informal: (1) the formal governance mechanisms consist of management policies and procedures, organizational structure, incentive schemes, information systems, and other control and coordination systems, whereas (2) informal (also referred to as relational) includes corporate culture, trust, social networks and communities (Foss, 2007).

First, formal governance mechanisms in terms of contracts through the specification of obligations to perform certain activities can facilitate knowledge transfer between alliance partners, which remotely collaborate and are geographically distant and can counterbalance the absence of face-to-face relationships and informal communication (Berchicci et al., 2016). Moreover, systematic governance mechanisms by means of *"summaries, weekly reports, monthly reports, milestone reports in project routines"* have greatly facilitated the KT between cross-project teams (Zhao et al., 2015).

However, the fully-contract partnership where all cooperation is based on an extensive formal agreement between independent companies can block knowledge transfer and diminish innovation capabilities, especially for SMEs, whereas a better alternative can be a fully-trust based partnership where contracts are substituted by mutual trust between partners or by applying mutual trust and contract complementary (Bosch-Sijtsema & Postma, 2010). This is also supported by findings of Al-Salti & Hackney (2011), who explored interfirm KT from vendors to customers in information system outsourcing relations and concluded that inter-organizational knowledge exchange was facilitated to a greater extent due to social relationships rather than formal contracts.

Second, Fang et al. (2013) argue that inter-organizational knowledge transfer can be fostered through four knowledge governance mechanisms, see table 2.12.1.

Knowledge governance mechanisms	Description
Market-based mode	*"This market form relies on the price mechanism to coordinate inter-organizational relationship. In inter-organizational knowledge transfer context, with standard knowledge assets and strong property rights, marginal pricing assurances to optimize knowledge transfer"* Measured through *"the frequency of using price mechanism (e.g. auction) for knowledge transfer between organizations"*. For example, the higher the frequency of using simple auction mechanisms in terms of establishing standardized rules for communication and interaction between partners, the better can be managed risks related to conflict of interest during KT process between different parties.
Trust-based mode	"Trust-based mechanism is a way to control and coordinate to build trust between partners for facilitating knowledge transfer through long term and frequently interaction between parties" Can be measured via assessing *"the degree of sharing value between organizations when transferring knowledge"*. For instance, the higher degree of interaction for

	building a common value, the more efficient KT and leads to faster decision-making process between partners.
Reciprocity mode	*"A reciprocity-based mechanism is a way to control and coordinate to build reciprocal relationships between partners for knowledge transfer"* Assessed in terms of "the degree of responding to a positive action with another positive action, rewarding kind actions between organizations when transferring knowledge". The higher the level of reciprocity the more successful KT between parties in terms of better collaboration and sharing know-hows and expertise.
Norm-based mode	*"The norm-based mechanism as the construction of inter-organizational knowledge sharing standard is a way of control and coordination for building social bonds between partners for facilitating knowledge transfer. This mechanism allows inter-organizational members to act together in a coordinated way for knowledge transfer purposes"* Evaluated in terms of *"the frequency of providing knowledge transfer guidance for members between organizations in their behaviour, thinking, judgment-making and perceptions of the world"*. The high frequency of supplying guidance among partners it enhances common understanding between different organizations, which in turn reduces barriers for interorganizational knowledge transfer.

Table 2.12.1 Description and measurement of knowledge governance mechanisms according to Fang et al. (2013).

Similarly, Mesquita et al. (2008), who studied buyer-supplier alliances, found that relational or informal governance mechanisms in terms of (1) collaboration commitments for an extensive exchange of market demand information; (2) commitments based on mutual assistance for managing supply-chain interruptions or handling losses; (3) commitments for reciprocity by promoting fairness in splitting the cost savings and benefits; have significant influence knowledge transfer between partners in turn leads to higher firm performance.

Furthermore, Cheung et al. (2011) suggest that cultivating and investing in relation-specific mechanisms between partners such as knowledge integration, information exchange, and joint problem solving can support the development of shared memory where common values and beliefs are stored and routinized in different formal and informal processes of involved parties. Moreover, Eiriz et al. (2017) found that the adaptation capacity is a vital governance mechanism for managing and transferring knowledge, which is defined as "the ability to promote organizational change in order to meet the other parties' requirements", that can be fostered by cooperation and frequent interactions between different firms. Overall, knowledge governance mechanisms, both formal and informal can enhance KT between different partners but can be different in terms of application for LC/MNCs and SMEs (Bosch-Sijtsema & Postma, 2010). Lastly, a summary of knowledge governance mechanisms has been created in figure 2.12.2.

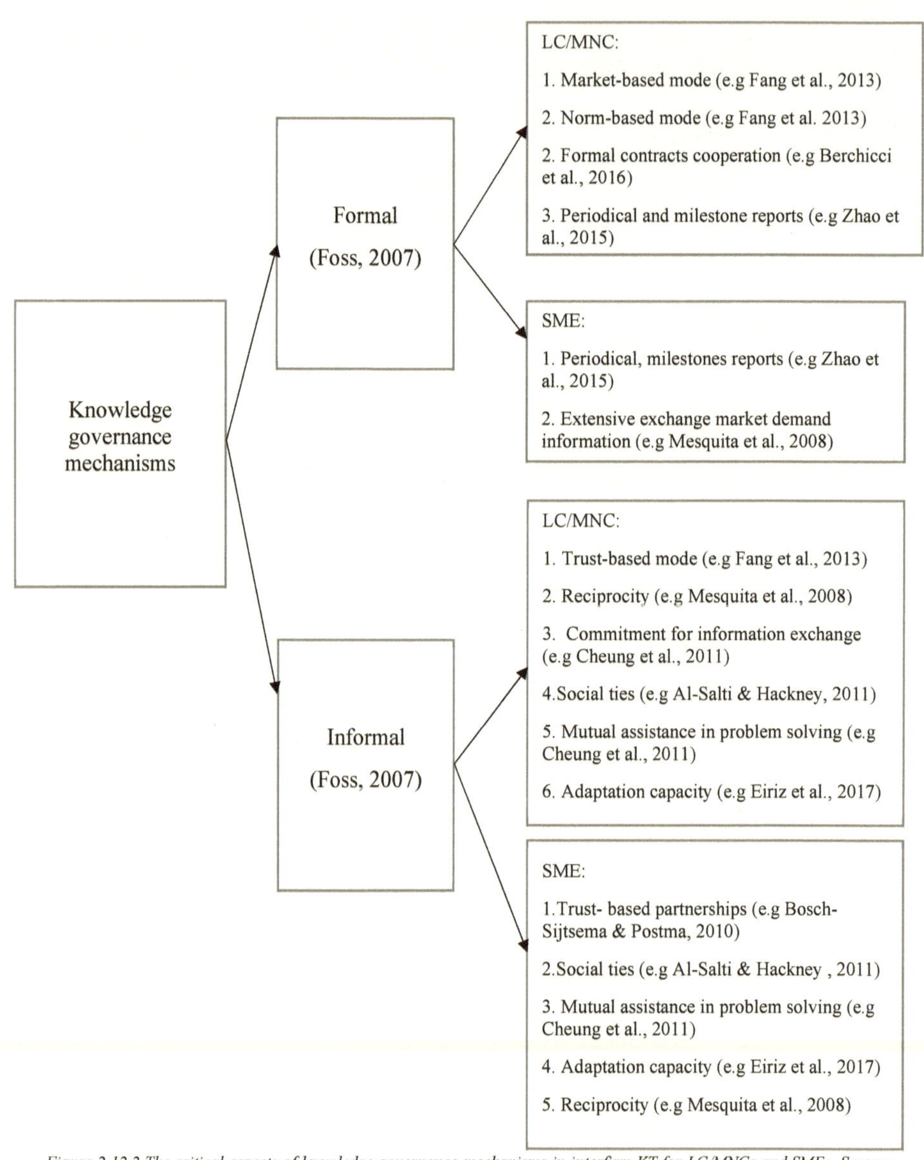

Figure 2.12.2 The critical aspects of knowledge governance mechanisms in interfirm KT for LC/MNCs and SMEs. Source: Developed by the authors.

2.13 Cultural distance

The literature supports that due to an expansion of businesses across different countries cultural factors significantly affect knowledge transfer in terms of higher complexity for knowledge flow because diverse national cultures, cultural characteristics as well as organizational culture influence the employee's ultimate behaviour in the interfirm knowledge transfer process (Milagres & Burcharth, 2019; Cheung et al., 2011; Easterby-Smith et al., 2008).

Hofstede (1994), who did a major contribution to studying national culture aspects for the international business management field, defines the national culture as *"the collective programming of the human mind that distinguishes the members of one human group from those of another"*. Similarly, House et al. (2004) under the "Globe study" (Global Leadership and Organizational Behaviour Effectiveness Research Programme) describes culture as "shared motives, values, beliefs, identities, and interpretations or meanings of significant events that result from common experiences of members of collectives and are transmitted across age generations". Hence, the national culture in terms of the shared vision, beliefs, and mutual understanding provide a fundamental base for transferring and integration the different partners' knowledge (Battistella, 2016). Essentially, the national cultural distance arises different barriers for the KT among international partners regardless of the firm size such as (1) higher cost of entry and hinders the transmitting the core competencies to the international markets; (2) facing operational challenges in terms of lacking common understanding of norms, values, and motivation; (3) misunderstanding between foreign partners can limit sharing of critical organizational knowledge (Wijk et al., 2008). This is supported by the study of Ahammad et al. (2016) that argue culture distance has a negative effect on KT between the acquirer and cross-border M&A target as a result, to mitigate this adverse effect, managers should conduct different cultural awareness workshops among employees.

Equally important, the majority of scholars examine national culture characteristics that affect KT through applying Hofstede's (1994) cultural framework in terms of the five dimensions that describe the culture in different countries such as power distance (PD), Individualism/collectivism (IC), Uncertainty avoidance (UA), Masculinity vs Femininity (MF), and Long-term Orientation (LO). As a result, Li et al. (2014) suggested in their study to apply power distance (PD), Individualism/collectivism (IC) dimensions due to better applicability to examine knowledge transfer context. Similarly, Wilkesmann et al. (2009) proposed to use "Globe study" (House et al., 2004) approach that is in line with Hofstede's (1994) cultural framework but additionally includes the performance orientation (PO) dimension. In table 2.13.1 below, a more detailed description is provided.

Author (Year)	Dimension	Description	From KT perspective
Li et al. (2014)	Power distance (PD)	"The extent to which the members of a society accept inequality in an organization. It reflects the non-symmetrical nature of relationships that may exist between knowledge provider and recipient"	In low power distance societies, people are treating each other as equal in a democratic way democratic among employees (e.g., France, Sweden, Norway, US); whereas societies with high PD (China, Japan) is associated with a strong hierarchical subordination and less likely to ask question or critique those of higher hierarchical positions and interactions between hierarchical levels tend to be more authoritative and paternalistic.

		"The extent to which a person sees himself or herself as an individual rather than part of a group. In individualistic cultures, ties among individuals are very loose. Everyone is expected to look after himself or herself. Collectivist societies reinforce the notion of group. Such cultures are generally driven by group interest rather than by self-interest"	In an individualistic society (e.g., France, Sweden, Germany, US), workers view knowledge as their own property, and they treat knowledge creation as the intervention of individual effort. As a result, they might not see the need to transfer the knowledge with others but can do if they see clear benefits from KT activities for them. In contrast, employees under collectivism (China, Japan) regard knowledge as the organization's property, and they regard integration, modification and creation of the existing knowledge as the group's efforts.
Wilkesmann et al. (2009)	Uncertainty avoidance (UA)	The extent to which the member of a society feels uncomfortable with uncertainty and ambiguity	People in high uncertainty avoidance cultures (France, Germany, Sweden, Norway) actively seek to decrease the probability of unpredictable future events that could adversely affect the operation of an organization or society and remedy the success of such adverse effects. The opposite is true for lower uncertainty avoidance societies such as China.
	Performance orientation (PO)	"Performance orientation is the degree to which an organization or society encourages and rewards group members for performance improvement and excellence. Status is not inherited but can be earned by success and achievement".	A society with a high-performance orientation like China will try to reward or punish transferring or not transferring knowledge. In cultures with lower performance orientation (Germany), monetary rewards play merely a secondary role for motivation KT while recognition and appreciation of other members of society are also important to enhance willingness for KT.

Table 2.13.1 Cultural dimensions description for KT according to Li et al. (2014) and Wilkesmann et al. (2009).

By applying Hofstede's and GLOBE study cultural dimensions, the literature found mixed results on how different national cultures affect knowledge movement (see table 2.13.2).

Author (Year)	Perspective	Key Findings
Li et al. (2014)	Meta-analytic comparison on the influencing factors of knowledge transfer in various cultural contexts that applicable for	(1) *"transfer capacity is not significantly correlated with knowledge transfer performance in individualism and low power distance. Moreover, the results show that as long as people in this cultural context have strong transfer willingness, there will be decent transfer performance"* (2) *"the role of transfer willingness is more important and there is a positive correlation between knowledge ambiguity and knowledge transfer performance in collectivism and high-power distance"*

	LC/MNCs and SMEs.	(3) *"credibility, accepting willingness and absorptive capacity play more important roles in individualism and low power distance than they do in collectivism and high-power distance"* (4) *"tie strength, trust and common cognition have greater effect in collectivism and high-power distance when compared with that in individualism and low power distance"*
Wilkesm ann et al. (2009)	The paper examines the cultural characteristics of knowledge transfer in German and Hong Kong small and medium sized companies.	(1) In high PD (Hong Kong company), *"a culture like that implies that subordinates transfer their pieces of knowledge only if they are asked for"* in other words, subordinates mostly transmit knowledge on orders."* (2) *"In high power distance culture employees will prefer to give their superiors face instead of providing knowledge freely. The clear hierarchical gap between superior and subordinate hinders bottom-up knowledge transfer"* (3) *"In Hong Kong subordinates (high PD) are afraid to transfer knowledge in a bottom-up direction because they fear that somebody else (mostly the superior) could take advantage of the information they have provided"* (4) *"The collectivistic attitude provides a chance for teams and departments to benefit from individual knowledge but also represents the risk of enclosed fractions within a company where members are not willing to transfer knowledge with outsiders"* Employees from collectivism is motivated with the selective incentives, which seem to have a stronger impact there than for employees in Germany with individualism culture. (5) In high PO society salary growth is a main driver for transfer knowledge but career promotion and *"certificates of recognition"* also crucial factors for willingness for KT. On contrary, in lower PO society, employees showed a part of financial incentives recognition and appreciation from colleagues and top managers are essential for encouraging transfer core competencies. Hence, intrinsic and extrinsic motivation has to be balanced in lower PO society culture.
Lièvre & Tang (2015)	The study focused on knowledge transfer in a multicultural context (France and China) in the health sector thorough the socialization– externalization – combination – internalization (SECI) model, which concentrates on the interaction between tacit and explicit knowledge	(1) Employees with the individualistic culture (France), showed preference for an explicit knowledge, unlike Chinese workers which is centred on tacit knowledge. For example, French partners initiated long discussions and meeting with Chinese side for receiving explicit knowledge. (2) The lack of understanding due to a deficit of *"socialization"* between Chinese and French partners led to the failure in knowledge exchange.

Table 2.13.2 The main national cultural aspects affecting interorganizational KT in findings of scientific papers.

To summarize findings from table 2.13.2, the different national cultures distance influence KT between organizations regardless of the firm size, the critical factors are (1) willingness to transfer the knowledge (e.g., Li et al., 2014); (2) intrinsic and extrinsic incentives are also important for KT but differ due to the national culture context (e.g., Wilkesmann et al., 2009); (3) knowledge characteristics in different national culture aspect can be a barrier for effective inter-organizational KT (Lièvre & Tang, 2015); (4) cultural characteristics in terms of trust and credibility are crucial for effective KT between international partners (e.g., Li et al., 2014).

This section covered the main cultural distance as the relational aspect that influences interfirm KT, the next part will be focused on trust, which is another important relational factor that significantly impacts knowledge exchange between independent organizations.

2.14 Trust

The inter-organizational knowledge flow is significantly affected by the relational context between different partners (Mesquita et al., 2008) and one of the crucial aspects of the relational dimension is trust and trustworthy interaction between independent organizations (Martinkenaite, 2011). This is because trust between organizations affects the success of knowledge transfer, as trust correlates to the source's credibility and trustworthiness to deliver the adequate quality of both the knowledge and the KT (Al-Salti & Hackney, 2011; Chen et al., 2014). In line with this, Pérez-Nordtvedt et al., (2008) research of inter-organizational KT, states that trust in a relationship between organizations forms a common language, which facilitates KT. Thus, trust facilitates knowledge transfer by fostering openness in the inter-organizational relationships or the extent of bilateral disclosure (Squire et al., 2009).

According to Mayer et al. (1995), who proposed a model of trust between different parties by means of linking partners trustworthy behaviour and willingness to accept vulnerability of partnership, defined trust as *"the willingness of a party to be vulnerable to the actions of another party based on the expectation that the other will perform a particular action important to the trustor, irrespective of the ability to monitor or control that other party"*. Furthermore, these authors suggested that the motivation to accept vulnerability to the actions of another party results from the assessments that the *"trustor"* makes about the trustworthiness of the *"trustee"* in terms of the three critical dimensions such as (1) ability in terms of competence and expertise of the trustee; (2) integrity via sharing common principals and ethical behaviour between trustor and trustee; (3) benevolence through *"the extent to which a trustee is believed to want to do good to the trustor, aside from an egocentric profit motive."*.

Likewise, Rousseau et al. (1998) suggested the definition of trust as "a psychological state comprising the intention to accept vulnerability based upon positive expectations of the intentions or behaviour of another". In other words, a basement for trust is the social judgments in terms of evaluation of other partners' benevolence, expertise, competence coupled with assessment of the risk related to untrustworthy behaviour; hence, trust is a critical factor in the willingness of network actors to transfer knowledge (Inkpen & Tsang, 2005). In addition to this, the literature proposes to assess interfirm trust in terms of three dimensions of trustworthy behaviour such as (1) fulfillment of commitments agreed between partners; (2) honest negotiation; and (3) avoidance of receiving excessive benefits of ally organizations (Cummings & Bromiley, 1996).

Importantly, Becerra et al. (2008) applied Mayer's (1995) trust model (i.e. ability, benevolence, integrity) to explore the transfer of explicit and tacit knowledge found that indeed knowledge types have different influences on the trustworthy relationship between alliance partners because explicit knowledge requires greater willingness to bear higher risks than the transfer of tacit knowledge; as a result, former has insignificant impact on alliance performance, while tacit knowledge transfer has a greater positive influence partners' performance, which is aligned with a resource-based view, where tacit knowledge is a critical aspect for the competitive advantage of organizations. These findings are also supported by a more recent study by Qiu (2019) that applied Becerra et al. (2008) perspective, which stated that trustworthy relationships between alliance partners positively affect knowledge transfer and even more can prevent knowledge leakage among allies but only if they highly appreciate promotion, prevention and trustworthy approaches in the alliance relationships. Interestingly, this author determined in the study that firm size did not have any impact on knowledge leakage, but alliance age had a greater effect on tacit knowledge transfer between partners since successful tacit KT requires longer-term relationships (Qiu, 2019). However, Frishammar et al. (2015), which studied four large mineral and metal firms in Sweden, argue that trust can be a restricting mechanism for knowledge leakage at the interfirm level because trust is a common technique to manipulate willingness to prevent opportunistic behaviour among partners.

Equivalently, Al-Jabri & Al-Busaidi. (2018) suggests that SMEs organizations consider trust to be a very important aspect, since with trust comes risk- indicating "transferring too much knowledge may impose risk", which shows the delicate balance between trusting competitors with the knowledge an organization provides since it may be used against them and the organization may end up losing their competitive advantage (Maciejovsky & Budescu. 2013). Moreover, mutual trust for SMEs is also deemed to be an important aspect, since the exchange of favours or similar transactions tend to be met in return as "If we helped them today, they would help us later" was a response regarding trust in Al-Jabri & Al-Busaidi (2018) article. Besides, Corral de Zubielqui et al. (2015) argue that SMEs build university-industry pathways in terms of relying on published research outcomes rather than forming relationships, which is, in turn, requires less trust; whereas SMEs tend to form strong ties within their supply chain network (customers and suppliers) that entail a higher level of trust and eventually leads to more successful KT.

To conclude, trust is one of the critical foundations for effective KT between independent organizations for LC/MNCs and for SMEs.

2.15 Geographical distance/proximity

Geographical proximity or distance is another important relational factor that affects the inter-organizational knowledge flow because the transfer of knowledge slows down and is less effective as physical distance increases between source and recipient due to the difficulty, the time, and the cost of communication (Battistella. 2016; Cummings & Teng. 2003). Essentially, the literature applies two terms such as geographical proximity (Capaldo & Petruzzelli. 2014; Bell & Zaheer. 2007) and geographical distance (Garcia et al., 2018; Choi & Contractor. 2016); the former is assumed to foster knowledge spillovers, while later considered as a barrier for effective knowledge flow between different organizations (Phene & Tallman. 2014).

The term of geographical proximity referred to the spatial and physical distance between partners and is measured in terms of kilometric distance that separates two units (individuals, organizations, cities, countries) in geographic space (Capaldo & Petruzzelli. 2014). In other words, *"Geographical proximity refers to local co-presence"* (Schamp et al.. 2004).

First, according to Bell & Zaheer (2007) who explored knowledge flow in networks, which is affected by geographical proximity, found: (1) knowledge flow between different organizations enhances across geographically distant relationships if organizational ties are based on the friendship between various KT actors, which inherently trust-based rather than geographically proximate relationships; (2) firm size considerably predicted knowledge flow implying that larger firms engage more in KT rather than smaller counterparts, but organizational ties did not fortify knowledge flow by proximity, which possibly due to modern communication technology that makes irrelevant knowledge exchange via organizational ties.

Besides, Capaldo & Petruzzelli (2014) argue that geographical proximity between alliance partners provide better tacit KT due to richer network trust-based relationships within their business cluster in turn this reduces the obstacles for knowledge exchange; at the same time collaboration with geographical distant partners can provide access for a heterogeneous knowledge base that eventually increases innovation capabilities of allies.

Moreover, Korbi & Chouki (2017) who examined the knowledge transfer in asymmetric alliances (multinational companies and SMEs) stated that geographical proximity together with (1) the development of organizational proximity by means of nurturing cultural values, common knowledge base, and management style; and (2) implementing communication tools such as electronic data interchange for sharing information, common database, and enhancing interactions among partners, can significantly facilitate knowledge transfer between asymmetric partners. Similarly, Hughes et al. (2009) that studied the effectiveness of SME's knowledge network suggested *"closeness" of the parties within the network has an impact on the effectiveness of the way that knowledge is used as a resource in the SME sector... Proximity is not just about geography, it is also about mutual understanding and trust".*

Second, geographical distance negatively affects the transfer of tacit knowledge in inter-organizational KT since it contains a significant amount of complexity, by differences between the organizations in communication, coordination, routines, policies, management practices, language, cultural backgrounds, control and cohesion which leads to higher financial costs and time consumption in the transfer process of information outsourcing projects (Al-Salti & Hackney, 2011). As the geographical distance increase, the more dissimilar these processes will be, which decreases the learning process of the transfer of knowledge, which takes place (Bathelt et al., 2004). Therefore, *"the right choice of alliance governance mode"* in terms of more interactive, bilateral way of communication and organizationally integrated, can reduce geographical distance, which in turn enhances collaboration and knowledge exchange between distinct organizations (Choi & Contractor, 2016).

All in all, geographical distance negatively affects inter-organizational KT especially for tacit knowledge exchange but can be mediated in terms of (1) developing trust-based relations between distant partners (e.g.,Capaldo & Petruzzelli, 2014); (2) organizational proximity including common corporate culture and values (Korbi & Chouki, 2017); and (3) bilateral way of utilizing modern communication technologies (Choi & Contractor, 2016) in LC/MNCs as well as in SMEs.

2.16 Motivation

Motivation is an important aspect reoccurring in the literature of inter-organizational KT, the aspect occurs on an individual level, where motivation is key for example achieving effective networking between firms (Harris, 2008), in the collective context of the organization to promote acquisition of knowledge and distribution of it (Jantunen, 2005) and in alliances where learning intent as a motivational aspect facilitates higher efficiency of knowledge transfer (Mazloomi Khamseh & Jolly, 2008).

Taking a deeper dive into motivation itself, Martín Cruz et al. (2009) state that motivation consists of intrinsic and extrinsic motives, where the former takes place inside an individual, affected by per example pleasant work environment, atmosphere of mutual respect, accomplishment, self-respect, prestige, personal values and involvement. As for the latter, extrinsic motivation according to the authors is external motives which affects the individual, by per example monetary rewards that comes in direct or indirect ways, whereas the former relates to wages, incentives or bonuses and the latter relates to time not working, job education, health benefits or allowances.

The definition of motivation stretches back quite some time, to the early works of Sytrz & Porter (1372), their definition of motivation is that it's an intrinsic phenomenon which is affected by four aspects, the first is of the situation which takes the environment and external stimulus into consideration, secondly is of the temperament of the internal state of the individual, third is the purpose or goal which affects the individuals behaviour, lastly is which tools are utilized to reach the individual's motivation, combining these aspects results in the motivation of the individual. In more recent times, another perspective towards motivation according to Gollwitzer & Oettingen (2015) states that humans are *"construed as flexible strategists"* which originates from a study in where *"the focus is on the different tasks a person has to perform when transforming wishes into actions"* (Gollwitzer, 1990), shifting focus to which actions needs to be taken to achieve ones goal. As such, the following paragraphs will discuss the different levels of motivation.

Individual motivation of the employee to acquire new knowledge or skill-sets affect the transfer and efficiency of the KT process and its success, which relates to the individual's interest of what the organization is conducting their business in, as the individual need to have interest in the business field (Al-Salti & Hackney, 2011) while being motivated to share and collect knowledge (Lin, 2007). Moreover, when the needed knowledge has been acquired, there is no motivation from key employees according to Chen et al. (2006), to leverage knowledge from the other organization, which is increasing the difficulty of identifying the knowledge gap of choosing per example right channels or applying the knowledge efficiently in SMEs. Another reason could be because of how SMEs tend to focus less on development of their employees if you compare it to a much larger firm, affecting their motivation negatively (García-Morales et al., 2007). However, this is not exclusive to a specific firm size, as motivation may deteriorate when the recipient of knowledge realizes that the acquired knowledge no longer provides a learning scenario for the organization (Easterby-Smith et al., 2008), which state that the organization's motivation should not be ignored.

Organizational motivation relates to the firms motivation of acquiring knowledge and how to obtain it, the organization must invest time and effort to the process of knowledge transfer (Bengoa & Kaufmann, 2014), which correlates to the agenda of motivation, as the transaction must be seen worthwhile for the organization to invest in, along with the recipient to be motivated to gain knowledge (Easterby-Smith et al., 2008) and being motivated to learn (Pérez-Nordtvedt et al., 2008) while possessing learning intent (Mazloomi Khamseh & Jolly, 2008). Furthermore, these motivational aspects can be categorized in internal and external knowledge valuation, where market-based competition relates to external knowledge valuation and internal knowledge to the knowledge currently in the organization, whereas the relationship between these two categories, external market-based competition tend to devaluate internal knowledge because of flaws of the knowledge are revealed cause of the proximity to the knowledge, affecting the balance of how firms view their resources and motivation to acquire or dismiss the knowledge (Menon & Pfeffer, 2003).

In the cases of Cambra-Fierro et al., (2011), inter-industry relationships motives relates to exploration and co-discovery of new knowledge as a result of the KT process, as by exploration, one organization's motivation to seek out inter-relations was to find alternative means of communication, in this case, to explore the effectiveness of how channels such as the internet works, to induce prescriptions for their medicines. Moreover, another case from this author revealed that motivation in young firms comes from the need to absorbing new knowledge, since they possess limited access to knowledge and information. Thus, the motivation forms a learning process in inter-organizational KT where the firms absorb key knowledge from the transaction to enhance their own competences, forming a situation of co-discovery. On the other side of motivation, organizations may reject external knowledge as it was not made within the firm, since they might not understand the knowledge or possibly see it as a threat to the firm's own processes, this is known as the "not made here syndrome" (Lichtenthaler & Ernst, 2006).

2.17 Social ties

Strong social ties acts as a facilitator for knowledge transfer (Kang & Sauk Hau, 2014) and Liu et al. (2015) argues that there are three steps to establishing a high quality relationship, first step involves to locate a partner firm and identify its key contacts, secondly to develop and maintain the relationship established, last stage involves dispatching of knowledge from the source along with knowledge absorption of the recipient, which is how the knowledge is exchanged. Thus, strong social ties enhance the transfer of knowledge in inter-organizational relationships since it facilitates KT better than formal contracts by per example social events, interactions or team building (Chen & McQueen, 2010) where strong social ties promotes communication between organizations (Kraatz, 1998), which strengthens the sharing of knowledge in terms of interests and goals (De Luca & Cano Rubio, 2019) and aids the pursuit of gaining new knowledge from the relationship (Al-Salti & Hackney, 2011).

Social ties according to Tangaraja et al. (2016) is a relationship based concept, its either conducted directly or indirectly between two parties and differentiated by informal or formal ways. Direct contact refers to physical presence while indirect is communication through technological means (De Luca & Cano Rubio, 2019). These social ties exist on an individual, institutional or organization level (Bell & Zaheer, 2007).

Moreover, forming a relationship creates mutual understanding, whereas one of the steps is to emphasize to which application of learning along with what context its occurring in, meaning that the individuals involved should understand per example each other's environments as it eases the translation of theory into practice (Hughes et al., 2009). Additionally, as strong relationships forms a mutual understanding in communications, creating a common-language between the two parties, resulting in recipients of knowledge comprehends it easier and faster, this correlates to a more efficient KT process and leads to less financial costs in the process as well (Pérez-Nordtvedt et al., 2008). However maintaining strong social ties comes with other costs instead such as by meetings or visits, but strong social ties are still preferred since it decrease communication cost and increase trust between the parties (Hansen, 1999), whereas the quality of the relationship lays in identifying an appropriate connection of the external firm (Liu et al., 2015). Similarly, emphasized by the research of Cavusgil et al. (2003), as strong social ties facilitates more opportunities for individuals to share their knowledge, which leads to generation of new knowledge by obtaining tacit knowledge from external sources. Additionally, they also found that firm size does not affect tacit KT transfer between organizations, which could be because organizations tend to seek social ties with those who can strengthen their weaknesses (Bengoa & Kaufmann, 2014), while looking at the amount of social ties which correlates to more maintenance of the relationships, implies that more social ties could end up overloading firms instead, negatively impacting external knowledge acquisition, however the strength of a single social tie do facilitate KT better as stated previously (Wijk et al., 2008).

Large or multinational companies (LC/MNCs) tend to utilize social ties in networks of relations for means of knowledge acquisition, which further enhance their absorptive capacity (Jiménez-Jiménez et al., 2014). Where LC/MNCs use their social capital as an informal mechanism which coordinates the flow of tacit knowledge in their networks, by means of boundary spanners which is an employee role of socialization, they are responsible of building relationships and communication of such (Montazemi et al., 2012). Which in earlier research, Nelson (1989) found that when organizations lack a dominant representative of social ties (one who has strong connections with other groups), disruptive conflict arises, this could be because of how social interactions work between individuals, as they need a bridge to connect them with others. Thus, social capital facilitates external knowledge acquisition for LC/MNCs (Bapuji & Crossan, 2005), but only if the knowledge is deemed complementary to the firms existing knowledge, which then complements the firms absorptive capacity (Zahra & George, 2002).

In the case of small or medium enterprises (SMEs), informal relations between organizations is more occurring rather than formal ones (Al-Jabri & Al-Busaidi, 2018), since SMEs tend to have less formal managerial structures (Hutchinson & Quintas, 2008). Looking at international social ties, Ali et al., (2020) argues that SMEs should be selective in the process of choosing other firms to engage with, as forming strong social ties with new firms more often proves valuable when the firm in question has strong prior success experience, financially or internationally. As for the learning process to be beneficial, the knowledge of the chosen organizations need to be different to make the interaction worthwhile (Bathelt et al., 2004). Because the SME may benefit from their knowledge, prior success and learn from it themselves, this transfer of knowledge leads to enhancing the organizations absorptive capacity, as the organization learns from the relationship. As this relates to external knowledge acquisition, Chen et al. (2006) argues strong social ties with customers and suppliers for SMEs assist in gaining new knowledge which is utilized for developing their business practises and solving complaints, which stress the importance of the customer and supplier relationship and external knowledge acquisition for SMEs, as without proper knowledge from these relations may end up as costly mistakes (Wei Chong et al., 2011), since external knowledge acquisition relates to better performance for firms as they become better equipped to adjust to the market (De Luca & Cano Rubio, 2019).

2.18 Knowledge characteristics

The literature of knowledge characteristics has identified that tacitness, ambiguity and complexity exist for both tacit and explicit knowledge and has a significant impact in inter-organizational knowledge transfer and have multiple times proven that it affects the success and efficiency of knowledge transfer in LCs or MNCs (Simonin, 2004), SMEs (Corral de Zubielqui et al., 2015) and in alliances (Mazloomi Khamseh & Jolly, 2008).

Wijk et al. (2008) claims that knowledge ambiguity is defined by the uncertainty of the knowledge, what underlying components and sources are in play (content-dependency), along with how they interact with each other (context-dependency), this context makes the knowledge difficult to imitate, which is the combined effect of tacitness, complexity and specificity of the knowledge characteristics (Reed & Defillippi, 1990). Or by another definition, *"the existence of multiple and conflicting interpretations with respect to a given situation"* (Joia & Lemos, 2010).

As for tacitness, Kalling, (2003) found that tacitness exists for both tacit and explicit knowledge, whereas tacitness is affected by the individuals understanding of the knowledge they are handling *"While all respondents admitted that the knowledge required to run any particular machine efficiently was not completely explicit, there were some differences in whether there is a point in trying to articulate and make transferable such knowledge."*.

Complexity consider how the knowledge that is transferred is affected by different variations as it is comprehended by different skills or experiences of organizations (Stock & Tatikonda, 2000). Thus, complexity is increased by per example organizations different absorptive capacity (Mazloomi Khamseh & Jolly, 2008), social capital (Filieri & Alguezaui, 2014) and their technological capabilities or culture (Battistella, 2016). As when knowledge is of high complexity, it contains a sizable tacit component (Szulanski, 1996), which is not always negative since it also makes knowledge less inimitable (Pérez-Nordtvedt et al., 2008), which relates to knowledge acquisition or replication since the more complex it is the more difficult it is to obtain (Cohen & Levinthal, 1990).

Inkpen (2008) state these aspects arise as the complexity of a process increases, per example firms may not always comprehend the performance achieved from production, since lots of the knowledge used is intertwined within routines that involves several individuals. Where the author relates this situation to context-dependency, as the situation and result was achieved with how the individuals implemented these routines to reach the given performance, to overcome the barrier of ambiguity, organization can only transfer and replicate this context-dependency knowledge by implementing new routines, as it is by interactions with the routines that organizations learn from the context-dependency generated knowledge, which additionally increase the relative absorptive capacity of organizations (Schildt et al., 2012).

As previous paragraph describes the context-dependency of ambiguity, it relates to the content-ambiguity of the object at hand, as argued by Inkpen & Pien (2006), this is the vagueness in the transferred knowledge, which is triggered by tacitness since it is embedded knowledge. Reed & Defillippi (1990) further argue that content-ambiguity is affected by complexity since it affects the comprehension of the object as it relates to the routines and individuals. Moreover, the authors argue that this leads to uncertainty during the transfer of knowledge, as increasing tacitness that is involved regarding the object, it becomes subject to uncertainty (Battistella, 2015).

The typology of knowledge characteristics consists of two primary dimensions, tacit and explicit knowledge which were originally introduced by Polanyi (1966). Tacit knowledge is defined as experience, skills, intuition of individuals, also described as know-how and is communicated in an inform way, while explicit is defined as knowledge which is captured and store and has a universal character which is described as know-what (Brown & Duguid, 2000; Nonaka & von Krogh, 2009). Tacit knowledge transfer is affected by aspects as per example inter-firm relationship (Cavusgil et al., 2003), knowledge management strategies and organizational structure (Joia & Lemos, 2010), design capabilities (Narteh, 2008) and are made through the collaborative experience that an individual experiences from an event, which is difficult to formalize and articulate (Nonaka & Takeuchi, 1995). Whereas explicit knowledge transfer is affected by per example aspects such as cost of conversion and technological tools used for conversion (Schulz & Jobe, 2001) and transferred by codified materials (manuals, electronic systems), by formal communication means (Giannakis, 2008), and is made by a codifying process (making tacit knowledge explicit) which could be described as a "people-to-document" approach that involves storing knowledge from an individual into knowledge banks (Hansen, 1999). Aspects which are relatable to both knowledge types are capabilities or abilities of the sender and recipient (Garavelli et al., 2002), degree of technical knowledge (Tsang, 1999).

Moreover, knowledge characteristics in alliances between firms is characterized by tacitness, specificity, complexity and institutional embeddedness of knowledge, these characteristics are identified as key antecedents for knowledge transfer between the firms in the alliance (Kogut & Zander, 1992), as they affect knowledge accumulation, retention and diffusion externally outside of the firms boundaries (Argote et al., 2003). When these characteristics are deemed high in an alliance, the knowledge cause higher levels of ambiguity, which have benefits and downfalls, such as hindering the transfer of knowledge between the firms involved in the alliance but also acts as a barrier of imitation (Martinkenaite, 2011). In the research of Korbi & Chouki (2017) between different national organizations and firm sizes, they argue that characteristics in communication such as linguistic barriers can be mediated through artifacts, such as language translator, additionally, when organizations use the same standardized formats (documents, reports, information systems) it increase communication efficiency and the KT process's efficiency between organizations.

In the perspective of information outsourcing research, tacit knowledge is difficult to transport and be absorbed by the recipient, depending on how technical and embedded the knowledge is (Al-Salti & Hackney, 2011). Which depends on the absorptive capacity of the recipient, the more tacit and complex the knowledge is the more formalized the process needs to be for successful KT (Chen & McQueen, 2010). Similarly, Chen (2004) findings reveal that organization prefer other means over contract-based collaboration, ergo outsourcing, this indication is because organizations find explicit knowledge easier to transfer rather than tacit knowledge. Because of how explicit knowledge is codified, which influence the speed of the knowledge transfer positivity, especially when it regard manufacturing firms (Zander & Kogut, 1995) as explicit knowledge comes in a formalized and systematic language (Nonaka, 1991). According to Zander & Kogut's findings, they found clear evidence that aspects such as codifiability and teachability *"are tacit and difficult to communicate and have significant effect on the hazard of transfer; the more codifiable and teachable a capability is, the higher the "risk" of rapid transfer"*, which indicates that explicit knowledge is faster to transfer when it comes to manufacturers capabilities.

Pérez-Nordtvedt et al., (2008) research found that when knowledge is perceived to consists of characteristics such as value, rarity, inimitability or not substitutable by the recipients organization, they are more likely to want and attain the knowledge to enhance their own performance, which Lavie (2006) also agree on, as firms seeking external knowledge to capitalizing on the inbound spill-over rents from the KT process, which relates to the partners resources that is attained during the transfer and attain resources that aren't shared in the process, as both appear as spill-over rents as a bi-product of the knowledge transfer. Likewise, Menon & Pfeffer (2003) found that acquisition of external knowledge is more appealing since it could appear as scarce and unique, over the firm's internal knowledge, which states the importance of attractiveness of knowledge as it facilitates the process of KT and acquisition of knowledge.

2.19 Absorptive capacity

Absorptive capacity according to Cohen & Levinthal (1990) is defined as the firm's ability to recognize external information which is valuable for the firm, to assimilate the information and apply it to the firms commercial means, where the former is achieved with prior related knowledge in the area, as it allows the firm to recognize the value and then assimilate it. Similarly, Tripsas (1997)'s view of absorptive capacity, the capability to source and integrate external knowledge within the firm determines the firms absorptive capacity, which could be interpreted as the same definition as the previous one. Furthermore, Cohen & Levinthal's definition are based on the literature of external knowledge acquisition and how it impacts organization's innovation capabilities and performance which emerged in around the 70's, whereas innovation is seen as how to produce new knowledge from external knowledge. As such, the literature of absorptive capacity agrees that it consist of aspects such as acquisition of external knowledge and intra-firm knowledge transmission because after acquiring knowledge it needs to be diffused within the firm, which is how they are these aspects are interrelated (De Luca & Cano Rubio, 2019).

Further contributions to absorptive capacity was made by Zahra & George (2002) in the early 20th century, their approach to absorptive capacity differentiate from Cohen & Levinthal's as they claim that absorptive capacity is a dynamic capability and should be categorized by potential (acquisition and assimilation of knowledge) and realized absorptive capacity (transformation and exploitation of knowledge), where acquisition is defined as the ability to recognize external valuable knowledge, assimilation refers to the firms routines and processes to understand this knowledge, transformation is based on the ability of the firm to develop routines that is used for existing knowledge or the assimilated new knowledge, lastly, exploitation refers on the routines within the firm to leverage its competences or form new ones accordingly to the acquired knowledge and transform it into the firms processes. Moreover, Zahra & George approach to absorptive capacity focuses on internal processes, routines, systems and the firm's structure and how firms apply them to assimilate knowledge, then transform it and utilize the external knowledge acquired.

Additionally, further literature contributions to absorptive capacity by Lewin et al. (2011) attempts to separate the internal and external procedures (acquisition of external knowledge and transmission of the knowledge internally) of the absorptive capacity process into internal and external absorptive capabilities of the organization. What is true for all approaches to absorptive capacity is that it facilitates inter-organizational knowledge transfer (Mowery et al., 1996) and contributes to knowledge learning internally in firms (Szulanski, 1996). Therefore, no matter how one author develop different approaches to absorptive capacity, the KT process itself will always be important for absorptive capacity since it is used for exploitation and commercialization of the acquired knowledge (Easterby-Smith et al., 2008). Thus, for this book the importance of absorptive capacity is clear, organizations should understand how to manage and transfer knowledge from external sources and diffuse the knowledge internally to increase their innovative capabilities and organizational performance.

The concept of absorptive capacity exists on different levels of analysis, the individual level and organizational level, as the organization depends on the individual's absorptive capacity, making them interrelated, meaning that as the individuals absorptive increases, so does the organizations (Cohen & Levinthal, 1990).

As such, absorptive capacity of the source in the KT process is increased by inter-organizational KT, as in the research of Al-Salti & Hackney (2011), the authors found that individuals gain higher absorptive capacity in information outsourcing projects by acquiring external knowledge in the form of database knowledge or expertise during the project, also by continuous training employees to acquire new knowledge. Moreover, according to the author, seniority in the field tend to correlate to higher absorptive capacity due to their personal experiences and expertise which determines the organizations capabilities and credibility's, which affect the success of KT and improves the likelihood to choose the organization for the project. Similarly, the findings of Chen (2004) found that organizations learn knowledge more efficiently when they possess strong absorptive capabilities, as it leads to better understanding of context which leads to opportunities, resulting in organization acquiring new external knowledge through per example alliances as a channel of transfer or exchange of knowledge

(Khamseh & Jolly, 2014), where the process is further enhanced by technological diversity of the organizations in the early and late stage of knowledge transfer (Schildt et al., 2012). By looking at it from the recipients side, when absorptive capacity is deemed to be low, acquiring tacit and complex knowledge types is more difficult to obtain since they do not possess the capabilities to utilize the knowledge, which requires more formal (explicit) knowledge transfer process to achieve successful KT processes between the firms (Chen & McQueen, 2010).

Firm size and inter-organizational knowledge are positively related, benefitting large or multinational companies' (LC/MNCs) absorptive capacity over SMEs, since there are more options for LC/MNCs to acquire knowledge, either by their subsidiaries geographically dispersed knowledge resources (Kogut & Zander, 1992) and their external relations or other organizational partners (Jiménez-Jiménez et al., 2014) or by their influence on the market and production capabilities which brings more appeal in certain cases (Kaminski et al., 2008). Thus, the knowledge acquisition of subsidiaries specifically amplify LC/MNCs competitive advantage and absorptive capacity when the headquarters and subsidiary possess adequate abilities to transfer and manage intra-firm knowledge transfer, along with the employee's abilities and motivation was found to facilitate the firms internal knowledge transfer (Lee & Wu, 2006). With a scope on subsidiaries, the research of Minbaeva et al. (2003) focus on their employees abilities and motivation to learn from organization's internal network to utilize knowledge to improve their absorptive capacity, which is one example of how subsidiaries increase their absorptive capacity to enable knowledge flows from the LC/MNCs internal network.

Small or medium enterprises (SMEs) engaging in inter-organizational KT tend to see it as an opportunity to enhance their own absorptive capacity by indirectly gaining new knowledge in the transfer process, which in turn can be utilized to formalize new business opportunities (Al-Jabri & Al-Busaidi, 2018). As Fletcher & Prashantham (2011) research on SME's where they use knowledge assimilation to enhance their internationalization, this involves acquisition of knowledge which also enhance their absorptive capacity. Whereas they state that the assimilation consists of two aspects, sharing of tacit or explicit knowledge and converting tacit knowledge into explicit in a routine like manner, within these routines SMEs formalized processes to share the tacit knowledge in a systematic way, the second aspect conversion, convert the tacit knowledge to make it easier to transfer by per example training programs or information systems. Furthermore, a positive relationship between innovation and sharing knowledge in SMEs was emphasized by Harel et al. (2020), as it foster innovations in their processes, products and marketing and organizational. As such, absorptive capacity is further enhanced by individuals of SMEs who possess higher educations and focus on staff development (Gray, 2006).

2.20 Disseminative capacity

The literature agrees that disseminative capacity is an essential aspect for the interfirm knowledge flow because if the knowledge source fails to transfer the knowledge then the recipient cannot successfully assimilate and utilize this knowledge (Tang et al., 2010; Kuiken & van der Sijde, 2011).

Equally important, the literature supports that disseminative capacity is an organizational construct because knowledge donors' capabilities and motivation for further interfirm KT are initially demonstrated at the firm level and after can be transmitted to outside of the company's boundaries (Schulze et al., 2014). Furthermore, Yih-Tong & Scott (2005) that research KT barriers, found that various aspects at the organizational level can serve as barriers for inter-organizational knowledge transfer due to various organizational practices and behaviour that influence interfirm relationships. For instance, these authors argue that alliance partners can choose organizational behaviour: as knowledge donors, whether to be more or less "transparent" in terms of the degree of unique information disclosure to the partner that eventually leads to certain responsive behaviour from the recipient firm (Yih-Tong & Scott, 2005). Additionally, Milagres & Burcharth (2019), which conducted a systematic literature review of KT in inter-organizational partnerships, is also distinguished disseminative capacity as firm-level characteristics that affect interfirm KT. Hence, antecedents that affect disseminative capacity in this book will be examined not only from interfirm perspective but also from the organizational dimension that eventually affects inter-organizational KT.

Disseminative capacity defined as *"ability of knowledge holders to convey knowledge in a way that a recipient can comprehend it and put it into practice"* (Schulze et al., 2014). In other words, disseminative capacity is *"a combination of the sender's ability to codify and articulate knowledge, the sender's willingness to share knowledge, and the sender's propensity to create and use opportunities for knowledge acquisition by the receiver"*(Minbaeva et al., 2018). Hence, disseminative capacity is the ability of network actors (knowledge source) to efficiently and effectively codify, articulate, communicate and teach knowledge to other network actors (Mu et al., 2010). All things considered; aspects related to disseminative capacity are important for examination that consequently influence knowledge flow between different sized organizations.

First of all, Tang et al. (2010) suggested a powerful knowledge holder, which owns social knowledge and deep technical competency can transfer knowledge at a faster speed at the organizational network due to the ability to minimize misunderstanding and duplication of the KT. These findings are also supported from the interfirm perspective by Schulze et al. (2014) who argued the following aspects positively affect KT in alliances: (1) the degree of partner's expertise and competence; (2) the extent of the source partner's evaluation of the recipient learning capacity; and (3) the degree of the source firm's ability to codify knowledge by developing common glossaries that can facilitate mutual understanding. Furthermore, Easterby-Smith et al. (2008) state that knowledge donor's capacity to teach the recipient company is an important factor for the inter-organizational KT.

Besides, Minbaeva et al. (2018), which focused on the research of disseminative capacity in the international joint ventures (IJVs), stated that not only the knowledge sender's motivation (via mutual trust) can enhance interfirm KT but also the application of various communication channels between partners (videoconference calls, face-to-face meetings, and mentorship by employees of the partner) are vital mediators for interfirm KT, as these authors argue that *"larger IJVs may have less incentive to seek knowledge from foreign partners because they have greater self-sufficiency"*.

In addition, joint cooperation between a supplier-buyer relationship in terms of direct interaction and mutual experience sharing can be a driver for the motivation of the source firm to share its internally generated knowledge because this can ensure more successful KT (Squire et al., 2009). Furthermore, Pérez-Nordtvedt et al. (2008) argue that the attractiveness of source companies in terms of competence and knowledge holding value (rareness, inimitability, and non-substitutability) are also critical aspects for better knowledge absorption that consequently affect overall interfirm KT success.

As opposed to larger firms, which have a higher ability for knowledge codification and teaching ability and capacity, SMEs transfer tacit knowledge that is inherently complex and difficult to share with other partners, as a result the disseminative capacity of SMEs is amplified through strengthening their social ties in terms of the trust, reciprocity, and collaboration with their network actors (Noblet & Simon. 2012). Particularly, according to Al-Jabri & Al-Busaidi (2018), which explored factors influencing inter-organizational KT for SMEs by applying a research model of Easterby-Smith et al. (2008), found that donor firms should possess the following characteristics as (1) absorption capacity; (2) intra-organizational capability to diffuse external knowledge within its boundaries (3) motivation to teach and (4) leadership support during the knowledge donation to other firms. Essentially, these findings support the interconnection between absorptive capacity and disseminative capacity because in spite of the fact that absorption abilities are an essential condition for successful knowledge transfer (Apriliyanti & Alon. 2017), but not sufficient, due to the disseminative capacity of the knowledge source in terms of ability, motivation and further distribution of the knowledge across the organization (Golzadeh Kermani. 2019).

Besides, Zapata Cantú et al. (2009), which examined the generation and transfer of knowledge in information technology-related SMEs, found that resistance of knowledge source that can be the outcome of losing control over knowledge ownership or centrality in network position can affect greatly KT within SMEs: (1) if the knowledge sender is low resistant then this facilitates KT; (2) if the knowledge holder has good reliability in terms of willingness, feeling comfortable and possessing enough resources to share knowledge, these factors also positively affect KT in SMEs. Moreover, according to Joshi et al. (2007), the knowledge source, who shows credibility in terms of trustworthy behaviour and high performance, has a strong ability to transfer knowledge, these authors suggest that if the knowledge holder frequently interacts and communicates with the recipients, he/she is perceived to transmit a substantial portion of the knowledge to other team members.

Despite that some studies are done at the organizational level; in this book, we will explore the use of these concepts in an inter-organizational knowledge transfer context.

To conclude, aspects that are related to the disseminative capacity in LC/MNC, as well as SMEs, are summarized in figure 2.20.1.

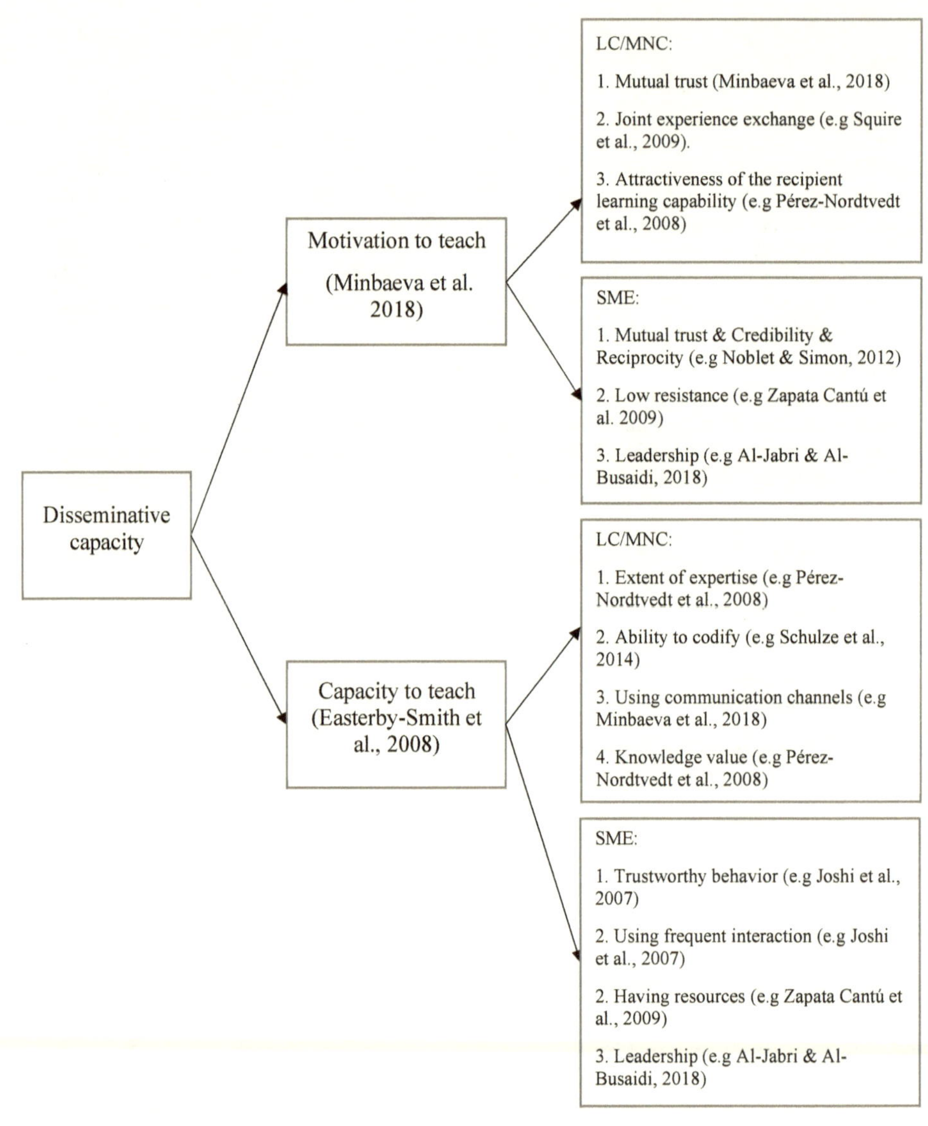

Figure 2.20.1 The critical aspects of disseminative capacity in interfirm KT for LC/MNCs and SMEs. Source: Developed by the authors.

Chapter III. Theoretical framework

Looking back at the purpose and research question of the book (section 1), whereas the purpose was to examine the various aspect that affects knowledge transfer (KT) in the perspective of different firm sizes, and the following research question was proposed:

What aspects affect inter-organizational knowledge transfer in different company's sizes?

To answer our research question, the first chapter provided how the literature of KT identified that it is paramount for firms. Which lead to the second chapter, where we identified the most reoccurring critical aspects in the literature affecting interfirm KT in large or multinational companies and small or medium companies, such as cultural distance, geographical distance/proximity, knowledge governance mechanisms, , trust, motivation, social ties, knowledge characteristics, absorptive and disseminative capabilities and capacity.

These aspects are categorized as follows: (1) characteristics of the relationship in terms of, cultural distance, geographical distance/proximity, knowledge governance mechanisms, trust, and social ties; (2) knowledge characteristics that include tacit and explicit knowledge, ambiguity, complexity, and codifiability; (3) organizational characteristics of the recipient and source in terms of absorptive capacity, disseminative capacity and motivation (see table III).

Dimensions/ Characteristics	Aspects that affect interfirm KT
Organizational characteristics of the recipient and the source	• Absorptive capacity (e.g., Cohen & Levinthal, 1990) • Disseminative capacity (e.g., Minbaeva et al., 2018) • Motivation (e.g., Martín Cruz et al., 2009)
Characteristics of the relationships	• Cultural distance (e.g., Wilkesmann et al., 2009) • Geographical proximity and distance (e.g., Choi & Contractor, 2016) • Knowledge governance mechanisms (e.g., Fang et al., 2013) • Trust (e.g., Becerra et al., 2008) • Social ties (e.g., Tangaraja et al., 2016)
Characteristics of the knowledge	• Tacit and explicit (e.g., Nonaka & von Krogh, 2009) • Ambiguity (eg., Wijk et al., 2008) • Complexity (e.g., Szulanski, 1996) • Tacitness (e.g., Kalling, 2003)

Table III. The critical aspects for inter-organizational KT in different firm sizes. Source: Developed by the authors of the book.

2.21 Characteristics of the relationship

The literature supports that transfer of the knowledge entails elements of the relational context because KT involves interaction between the source and the recipient (Battistella, 2016). Thus, aspects related to the cultural distance that includes knowledge transfer relationship across different national boundaries (e.g., Wilkesmann et al., 2009), geographical proximity/distance serves as facilitator or barrier for communication and interaction between knowledge transfer parties (e.g., Choi & Contractor, 2016), the knowledge governance mechanisms, which involve establishing rules for facilitating knowledge exchange and communication between partners (e.g., Fang et al., 2013; Foss, 2007), trust is considered as a foundation for any relationship (Chen et al., 2014; Becerra et al., 2008), and social ties contain social relationship for facilitating interfirm KT (e.g., Liu et al., 2015) are categorized into characteristics of the relationship of KT for this book. The following section describes in detail each of the mentioned aspects.

2.21.1 Cultural distance

House et al. (2004) describes culture as *"shared motives, values, beliefs, identities, and interpretations of meanings of significant events that result from common experiences of members of collectives and are transmitted across age generations"*.

The national cultural distance can serve as a barrier for the KT among international partners regardless of the firm size such as (1) higher cost of entry and hinders transmitting the core competencies to the international markets; (2) facing operational challenges in terms of lacking common understanding of norms, values, and motivation; (3) misunderstanding between foreign partners can limit sharing of critical organizational knowledge (Wijk et al., 2008).

2.21.2 Geographical distance/proximity

The literature applies two terms such as geographical proximity (Capaldo & Petruzzelli, 2014; Bell & Zaheer, 2007) and geographical distance (Garcia et al., 2018; Choi & Contractor, 2016), the former is assumed to foster knowledge spillovers, while later considered as a barrier for effective knowledge flow between different organizations (Phene & Tallman, 2014).

The term of geographical proximity referred to the spatial and physical distance between partners and is measured in terms of kilometric distance that separates two units (individuals, organizations, cities, countries) in geographic space (Capaldo & Petruzzelli, 2014). In other words, *"Geographical proximity refers to local co-presence"*(Schamp et al., 2004).

The geographical distance negatively influences inter-organizational KT especially for tacit knowledge exchange but can be mediated in terms of (1) developing trust-based relations between distant partners (e.g., Capaldo & Petruzzelli, 2014); (2) organizational proximity including common corporate culture and values (Korbi & Chouki, 2017); and (3) bilateral way of utilizing modern communication technologies (Choi & Contractor, 2016) in LC/MNCs as well as in SMEs.

2.21.3 Knowledge governance mechanisms

Knowledge governance mechanisms defined as *"the underlying and concrete management and control activities, which describe in detail how the required behaviour of the partner will become motivated, influenced, and established, or more generally, in which ways the desirable or predetermined gains are to be fulfilled"* (Hoetker & Mellewigt, 2009). Knowledge governance mechanisms can be formal and informal: (1) the formal governance mechanisms consist of management policies and procedures, organizational structure, incentive schemes, information systems, and other control and coordination systems, whereas (2) informal (also referred to as relational) includes corporate culture, trust, social ties and networks (Foss, 2007).

2.21.4 Trust

Trust is one of the critical foundations for effective KT between independent organizations for LC/MNCs as well as for SMEs, but different size companies exploit different modes of trust-based partnerships. According to Mayer et al. (1995), trust is defined as *"the willingness of a party to be vulnerable to the actions of another party based on the expectation that the other will perform a particular action important to the trustor, irrespective of the ability to monitor or control that other party"*. To put it differently, the trust is based on social judgments in terms of (1) evaluation of other partners' benevolence, (2) expertise and competence coupled with (3) assessment of the risk related to untrustworthy behaviour (Inkpen & Tsang, 2005). In addition, to this, the literature proposes to assess interfirm trust based on (1) fulfillment of commitments agreed between partners; (2) honest negotiation; and (3) avoidance of receiving excessive benefits of ally organizations (Cummings & Bromiley, 1996). Moreover, mutual trust is also deemed to be an important aspect, since the exchange of favours or similar transactions (e.g., reciprocity behaviour) tend to be met in return as "If we helped them today, they would help us later" was a response regarding trust in Al-Jabri & Al-Busaidi (2018) article.

2.21.5 Social ties

Social ties according to Tangaraja et al. (2016) is a relationship based concept, its either conducted directly or indirectly between two parties and differentiated by informal or formal ways. Direct contact refers to physical presence while indirect is communication through technological means (De Luca & Cano Rubio, 2019). These social ties exist on an individual, institutional or organization level (Bell & Zaheer, 2007).

Strong social ties acts as a facilitator for knowledge transfer (Kang & Sauk Hau, 2014) and Liu et al. (2015) argues that there are three steps to establishing a high quality relationship, (1) first step involves to locate a partner firm and identify its key contacts, (2) secondly to develop and maintain the relationship established, last stage involves dispatching of knowledge from the source along with knowledge absorption of the recipient, which is how the knowledge is exchanged.

As our thesis's scenario is based on two organizations already engaging in knowledge transfer, the book will not consider the first step of establishing a high-quality relationship, as the initial connection is already established from the scenario that is researched. Therefore, the first step is not the focus of this study, rather how the organizations develop/establish and maintain the social tie. To then measure social ties, we will take into consideration how they are initiated upon and how they are maintained by the different sized organizations.

2.22 Organizational characteristics of the recipient and the source

The literature agrees that organizational characteristics serve as antecedents of organizational knowledge transfer that contain (1) organizational abilities of the recipient in terms of the absorptive capacity and of the source in terms of disseminative capacity, and (2) organizational motivation for KT (Easterby-Smith et al., 2008; Wijk et al., 2008). As organizational motivation and abilities are greatly influenced by different firm sizes due to possessing different organizational capabilities and resources (Korbi & Chouki, 2017; Durst & Runar Edvardsson, 2012). Hence, absorptive capacity, disseminative capacity and motivation aspects are categorized as organizational characteristics of the recipient and the source (Battistella, 2016). The following section describes each aspect mentioned:

2.22.1 Absorptive capacity

Absorptive capacity according to Cohen & Levinthal (1990) is defined as the firm's ability to recognize external information which is valuable for the firm, to assimilate the information and apply it to the firms commercial means, where the former is achieved with prior related knowledge in the area, as it allows the firm to recognize the value and then assimilate it. Similarly, Tripsas (1997)'s view of absorptive capacity, the capability to source and integrate external knowledge within the firm determines the firms absorptive capacity, which could be interpreted as the same definition as the previous one. Furthermore, according to Cohen & Levinthal their definitions are based on the literature of external knowledge acquisition and how it impacts organization's innovation capabilities and performance which emerged in around the 70's, whereas innovation is seen as how to produce new knowledge from external knowledge. As such, the literature of absorptive capacity agrees that it consist of aspects such as acquisition of external knowledge and intra-firm knowledge transmission because after acquiring knowledge it needs to be diffused within the firm, which is how they are these aspects are interrelated (De Luca & Cano Rubio, 2019).

As our thesis's scenario is based on two organizations already engaging in knowledge transfer, the book will not consider acquisition of external knowledge for absorptive capacity as the acquisition of knowledge is already established from the scenario that is researched. Therefore, the acquisition of external knowledge is not the focus of this study, rather the implementation and dissemination of knowledge. Thus, as the research is done on an organizational level, to then measure absorptive capacity of SMEs and LC/MNCs this book will research how organizations go about implementing external knowledge to the organizational level, along with how the organization diffuse it in the organization from the individual.

2.22.2 Disseminative capacity

The concept disseminative capacity is defined as "a combination of the sender's ability to codify and articulate knowledge, the sender's willingness to share knowledge, and the sender's propensity to create and use opportunities for knowledge acquisition by the receiver" (Minbaeva et al., 2018).

Different firm sizes in terms of LC/MNC and SMEs have various aspects related to motivation and ability to teach as a source of knowledge due to having distinct resource and technology capabilities (Al-Jabri & Al-Busaidi, 2018; Schulze et al., 2014). These aspects related to disseminative capacity are summarized in figure 2.22.2.1.

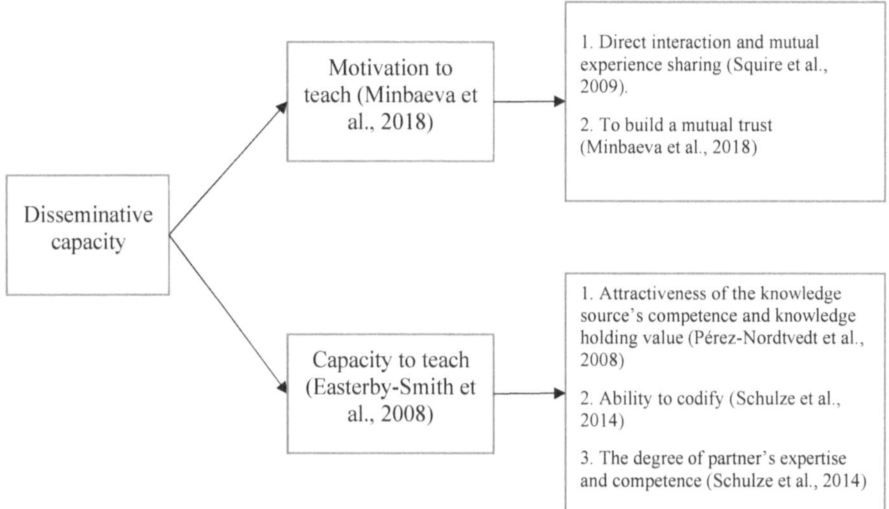

Figure 2.22.2.1. The critical aspects of disseminative capacity in interfirm KT for LC/MNCs and SMEs. Source: Developed by the authors.

2.22.3 Motivation

Sytrz & Porter (1372) defines motivation as an intrinsic phenomenon which is affected by four aspects: (1) the situation; (2) temperament; (3) purpose or goal and (4) tools used to reach the individual. More recently, Gollwitzer & Oettingen (2015) states that humans are *"construed as flexible strategists"* which originates from a study in where *"the focus is on the different tasks a person has to perform when transforming wishes into actions"* (Gollwitzer, 1990), shifting focus to which actions needs to be taken to achieve ones goal.

Motivation is an important aspect reoccurring in the literature of inter-organizational KT, the aspect occurs on an individual level, where motivation is key for example achieving effective networking between firms (Harris, 2008), in the collective context of the organization to promote acquisition of knowledge and distribution of it (Jantunen, 2005) and in alliances where learning intent as a motivational aspect facilitates higher efficiency of knowledge transfer (Mazloomi Khamseh & Jolly, 2008).

Martín Cruz et al. (2009) state that motivation consists of intrinsic and extrinsic motives, where the former takes place inside an individual, affected by per example pleasant work environment, atmosphere of mutual respect, accomplishment, self-respect, prestige, personal values and involvement. As for the latter, extrinsic motivation according to the authors is external motives which affects the individual, by per example monetary rewards that comes in direct or indirect ways, whereas the former relates to wages, incentives or bonuses and the latter relates to time not working, job education, health benefits or allowances.

Organizational motivation relates to the firms motivation of acquiring knowledge and how to obtain it, the organization must invest time and effort to the process of knowledge transfer (Bengoa & Kaufmann, 2014), which correlates to the agenda of motivation, as the transaction must be seen worthwhile for the organization to invest in, along with the recipient to be motivated to gain knowledge (Easterby-Smith et al., 2008) and being motivated to learn (Pérez-

Nordtvedt et al., 2008) while possessing learning intent (Mazloomi Khamseh & Jolly, 2008). Furthermore, these motivational aspects can be categorized in internal and external knowledge valuation, where market-based competition relates to external knowledge valuation and internal knowledge to the knowledge currently in the organization, whereas the relationship between these two categories, external market-based competition tend to devaluate internal knowledge because of flaws of the knowledge are revealed cause of the proximity to the knowledge, affecting the balance of how firms view their resources and motivation to acquire or dismiss the knowledge (Menon & Pfeffer, 2003).

As such, to measure organizational motivation of a firm, the book will research how the organization motivate their employees by intrinsic and extrinsic motivation to meet the needs of acquiring new knowledge from partners and sharing knowledge with partners.

2.23 Characteristics of knowledge
The literature has identifies various knowledge characteristics that distinguish the object of the transfer and that can influence the transfer process of knowledge (Battistella, 2016). Furthermore, the knowledge transfer process includes an exchange of the object between two actors, wherein the knowledge as a object is transferred with its nature characteristics (Ajith Kumar & Ganesh, 2009) and depends on context-specific properties (Cohen & Levinthal, 1990). The nature of the knowledge is characterized with two dimensions such as tacit and explicit (Nonaka, 1991), whereas the context of knowledge is described in terms of tacitness, ambiguity and complexity, which exist for both tacit and explicit knowledge because these context-specific properties can serve as main barriers for interfirm KT regardless of the knowledge type exchanged between different entities (Milagres & Burcharth, 2019), affecting the success and efficiency of knowledge transfer in LCs or MNCs (Simonin, 2004), SMEs (Corral de Zubielqui et al., 2015) and in alliance (Mazloomi Khamseh & Jolly, 2008). The following section describes each aspect mentioned.

Since every aspect of knowledge characteristic exists for both tacit and explicit knowledge, the theoretical framework will be established according to how ambiguity, complexity and tacitness is perceived and prioritized by SMEs and LC/MNCs for both tacit knowledge and explicit knowledge when they engage in a knowledge transfer to their partners.

2.23.1 Tacit and explicit knowledge
Tacit knowledge is defined as experience, skills, intuition of individuals, also described as know-how and is communicated in an inform way, while explicit is defined as knowledge which is captured and store and has a universal character which is described as know-what (Brown & Duguid, 2000; Nonaka & von Krogh, 2009).

2.23.2 Ambiguity
Wijk et al. (2008) claims that knowledge ambiguity is defined by the uncertainty of the knowledge, what underlying components and sources are in play (content-dependency), along with how they interact with each other (context-dependency), this context makes the knowledge difficult to imitate, which is the combined effect of tacitness, complexity and specificity of the knowledge characteristics (Reed & Defillippi, 1990). Or by another definition, *"the existence of multiple and conflicting interpretations with respect to a given situation"* (Joia & Lemos, 2010).

Inkpen (2008) state these aspects arise as the complexity of a process increases, per example firms may not always comprehend the performance achieved from production, since lots of the knowledge used is intertwined within routines that involves several individuals. Where the author relates this situation to context-dependency, as the situation and result was achieved with how the individuals implemented these routines to reach the given performance, to overcome the barrier of ambiguity, organization can only transfer and replicate this context-dependency knowledge by implementing new routines, as it is by interactions with the routines that organizations learn from the context-dependency generated knowledge, which additionally increase the relative absorptive capacity of organizations (Schildt et al., 2012).

As previous paragraph describes the context-dependency of ambiguity, it relates to the content-ambiguity of the object at hand, as argued by Inkpen & Pien (2006), this is the vagueness in the transferred knowledge, which is triggered by tacitness since it is embedded knowledge. Reed & Defillippi (1990) further argue that content-ambiguity is affected by complexity since it affects the comprehension of the object as it relates to the routines and individuals. Moreover, the authors argue that this leads to uncertainty during the transfer of knowledge, as increasing tacitness that is involved regarding the object, it becomes more difficult to replicate since it becomes subject to uncertainty (Battistella, 2015).

Given the statements above, ambiguity can be measured by content and context dependency (Fang et al., 2013).

2.23.3 Complexity

Complexity consider how the knowledge that is transferred is affected by different variations as it is comprehended by different skills or experiences of organizations (Stock & Tatikonda, 2000). Thus, complexity is increased by per example organizations different absorptive capacity (Mazloomi Khamseh & Jolly, 2008), social capital (Filieri & Alguezaui, 2014) and their technological capabilities or culture (Battistella, 2016). As when knowledge is of high complexity, it contains a sizable tacit component (Szulanski, 1996), which is not always negative since it also makes knowledge less inimitable (Pérez-Nordtvedt et al., 2008), which relates to knowledge acquisition since the more complex it is the more difficult it is to obtain (Cohen & Levinthal, 1990). Given the production example in the ambiguity section, complexity can be measured by how many different variables are involved (content and context dependency) in a certain situation (Milagres & Burcharth, 2019).

2.23.4 Tacitness

Tacitness according to the findings of Kalling, (2003), state that tacitness exists for both tacit and explicit knowledge, whereas tacitness is affected by the individuals understanding of the knowledge they are handling *"While all respondents admitted that the knowledge required to run any particular machine efficiently was not completely explicit, there were some differences in whether there is a point in trying to articulate and make transferable such knowledge."*. Therefore, to some measure tacitness in knowledge to some extent we will analyse the individuals understanding of the context at hand (context dependency).

2.24 Level of Analysis

The level of analysis includes how these aspects affects KT in two different scenarios. First, how aspects affect LC/MNCs engaging in inter-organizational KT with a partner (e.g., a SME), secondly, how aspects affect SMEs engaging in inter-organizational KT with a partner (e.g., a LC/MNC), whereas both are illustrated in figure 2.24.1.

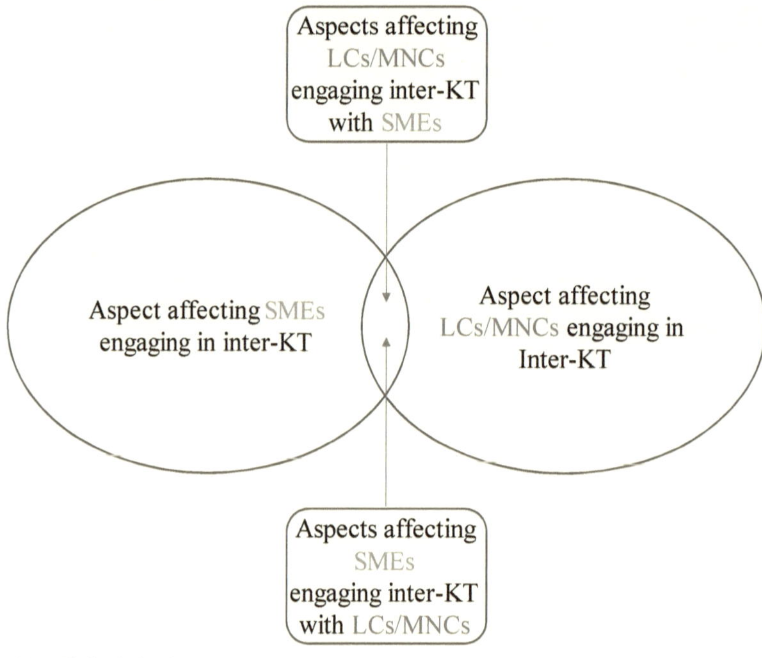

Figure 2.24.1 The level of analysis.

3. Methodology

This section will provide the reader with an ideation of how the authors conducted research for their book.

However, this study will conduct an exploration of aspects that affect inter-organizational knowledge transfer for different firm sizes such as LC/MNCs and SMEs. This is crucial because by overlooking knowledge transfer in the different firm size context, researchers and practitioners might lose a holistic picture of aspects that impact knowledge flow on inter and intra-organizational levels that eventually, might significantly influence firm's innovation performance and further success.

The aim of the research is to explore aspects that affect knowledge transfer (KT) in the context of different firm sizes, therefore authors chose to conduct a qualitative study. As such, the methodology section will discuss about the research design, which includes the following sections:

- Qualitative study
- Deductive approach
- Exploratory case study as a research strategy
- Data Collection: interviews, participant observation, visits, documents, focus group
- Data Analysis
- Quality of the Research: validate, transparency, reliability, triangulation

The authors of the book also divide the aspects in different levels and categories (see section 2, chapter 3) that might aid in structuring and apply a more systematic way for future research needs and managerial implementation.

3.1 Research Design

This book about knowledge transfer was developed to contribute to the literature by researching a phenomena in the field, whereas the identification of a problem related to different sized firms gave a setting to explore the perceptions of aspects that inference knowledge transfer.

In this sense, by taking a qualitative approach, a researcher must fulfill certain criteria's, according to Crescentini & Mainardi (2009) such as: first, is to propose a research question, whereas the research question *"evolves from the interaction between the researcher's goals (individual, ethical, . . .) and the researcher's theoretical frames"*, as the theoretical framework includes previous research in the field. As this study was in principle a journey to obtain deeper knowledge about how organizations interacts with others to gain new knowledge or learnings, in this sense the research field of knowledge transfer becomes the area in which the theoretical elements were selected. Which evolved from a prior literature review to the book and by interaction between literature research and the researcher's goal, this provided the research proposal to the thesis's problematization (see section 1.2). Furthermore, the interaction between the existing literature and the researcher's goals formulized the problematization. Based on the problematization the authors could draw the conclusion that (1) successful companies differentiates from others if they can effectively leverage knowledge transfer within and across the organization (Easterby-Smith et al., 2008), (2) Small and medium organizations does not manage knowledge similar to larger counter parts (Desouza & Awazu, 2006), (3) different sized organizations utilize and generate knowledge differently (Korbi & Chouki, 2017; Cerchione et al., 2020), (4) divergent dynamic capabilities, distinct corporate governance structure and culture, various extent of specialization (Korbi & Chouki, 2017; Wee & Chua, 2013), different complexities (Qian & Xu, 1998; Thompson, 1965) and flexibilities (Harel et al., 2020; Corral de Zubielqui et al., 2019; Durst & Runar Edvardsson, 2012). As all these variables include the different aspects that involves and affects knowledge transfer, the following research question was stated:

What aspects affect inter-organizational knowledge transfer in different company's sizes?

To adequately answer the research question a qualitative research approach (will be discussed in section 3.1.1) is a better fit since knowledge transfer is affected differently by the source organizational size and recipient organizational size as their aspects affect knowledge transfer differently, the research question suggest that there is "*a need to describe, verify or understand*" making it qualitative (Crescentini & Mainardi, 2009). As research questions that focus on discovery of experience, behaviour or fields where the phenomena is not well explored to provide an understand of why or how, are some qualitative research examples (Holstein & Gubrium, 1997), the research question proposed fit to this category since it describes the differences and similarities that emerge from engaging in a unidirectional knowledge transfer with a different sized organization.

Furthermore, by utilizing existing theory it is easier to organize the data as it illuminates the field of research which illustrates the phenomena that is being researched and it aids in (1) justification of the study, how it will propose and address the unanswered questions; (2) it informs the researcher about decisions of most adequate method to pursue; (3) the source of data can be used to test or make modifications to the theory (Maxwell, 2009).

Consequently, this book propose that (1) the aspects that were considered relevant to answer the research question was taken from the existing literature of knowledge transfer based on how often they were continuously reappearing, whereas the opportunities of the selected phenomena for this book can provide new knowledge in regard to knowledge transfer between different sized organizations. In this sense, the literature was collected from different data bases such as ScienceDirect, Sage Journals, Elsevier, Emerald Insight, Wiley Online Library, JSTOR and Springer Link, to gain an extended view and firm base of knowledge for the book, the collection of literature was spread wide to gain a vast spectrum of different publications. To then gain relevant publications, keywords was set in the process of data collection to find as relevant articles as possible, the keywords include "knowledge transfer in organizations", "knowledge transfer in firms", "knowledge transfer in small or medium enterprises", "knowledge transfer in SMEs", "knowledge transfer in large organizations", "knowledge transfer in large firms", "knowledge transfer in multinational companies", "knowledge transfer in MNCs". In turn, when the aspects were identified (see chapter 2), the authors selected specific keywords related to the aspects to identify in-depth knowledge regarding it, if the currently obtained literature felt lacking, that is.

As this relates to "*which case(s) are selected*" of a qualitative approach (Crescentini & Mainardi, 2009), the book then argue to study knowledge transfer in different organizational sizes (SME and LC/MNC), providing the book with two cases, then, the selected cases must show different types of firms and how the established aspects in the existing literature affect knowledge transfer (chapter 2) when engaging with a unidirectional knowledge transfer with one of their partner. This case selection provides a perspective that fits the field of knowledge transfer as it researches how the existing aspects which are established for small or medium enterprises and large or multinational companies in the literature are affected when they engage in a KT with their partners, as this interaction provides similarities and differences to the research question that was previously stated.

(2) As the mature literature of knowledge transfer deploys different levels of analysis as explained before on the individual or organizational level (see section 2.2), accordingly, the use of existing literature (chapter 2) was required in the selection of aspects that affect knowledge transfer in the level that the book is interested in exploring because they are already established by the existing literature. As this reveals which method towards the study is more relevant, it also directs to "*how the information is collected*" (Crescentini & Mainardi, 2009). As the information collection, this relates to the case selection for this book, as the two cases, one SME and one LC/MNC, are treated with independent roles in knowledge transfer to a partner organization on a inter organizational level, researching what aspects affect knowledge transfer for these organizations by a unidirectional knowledge transfer. Thus, the information is collected by interviews from the different sized interdependent organizations, additionally, as the book directs its attention to the organizational level, the information collected during the interviews will be directed toward this level of analysis.

As the information collection consists of a narrative approach of the interviewee's, which is of a qualitative approach whereas how the information is analysed (Easterby-Smith et al., 1991), this book will take the information that is narrated by the interviewee (coded primary data, see appendix 3) and transcript it by using computer aided/assisted transcription software called Kaltura and match it with existing reports (secondary data) of the organization along with the existing literature (see chapter 3) to establish a framework of how organizations aspects are affected similarly or differently when they transfer knowledge in a unidirectional way to their partners. Additionally, as the interviewee's narrative is highly dependent on their job description and responsibilities, the interviewee's have been selected from various departments of the case study, this is important to acknowledge since knowledge contains higher complexity, ambiguity and tacitness (Simonin, 2004) depending on the different job assignments that are delegated to the job position of the interviewee. This relates to how knowledge transfer is based on human behaviour and expression, as it contains non-numerical data and is based on a situation is subjectively perceives, which puts this research in a qualitative approach as it is based on social and behavioural science and is more suitable for qualitative research (Ghauri & Gronhaug, 2005). Notably, in this study the interviewees had different positions in the organization (such as manager, quality controller, etc) and to establish how they interpret the KT process according to their own involvement, as such, it is expected that the informants do not have the same experience and this makes it relevant to understand it in their context, as they might provide different results towards knowledge transfer. This means that the result obtained should be taken with a grain of salt, as it may appear differently in another situation.

(3) As this specific phenomenon is not researched to a well extent in the existing literature, the book believes to find something new by studying the specific phenomena previously described by using the existing literature and putting it under the scope of analysation in a different perspective. Whereas the book argues that it can extend the theory of aspects that affect knowledge transfer by adopting a deductive approach towards its research (discussed in 3.1.2), by analysing the similarities and differences that appear from the information collection and the developed theoretical framework and secondary data from the cases selected, which is "*how the kind of data analysis is chosen*" (Crescentini & Mainardi, 2009).

This research design in turn reflects on how the study developed its purpose by discussing the problematization and proposing its research question, along with how the research question was developed and what was considered during the development. Additionally, it shows how this study follows a qualitative approach and taking a deductive approach towards it when collecting and analysing the existing and new data, which is further argued on in the upcoming sections.

3.1.1 Qualitative Study

To propose an answer to the research question in this book, a qualitative research strategy has been applied. Maxwell (2009) describes that qualitative research is aimed to explore *in what ways, why and how* a particular result is occurred rather *"determining whether a particular result is causally related to one or another variable, and to what extent these are related"*, in other words, a qualitative study utilizes a process orientation approach that refers to *"see the world in terms of people, situations, events, and the processes that connect these; explanation is based on an analysis of how some situations and events influence others"*.

This book focus on the phenomenon that occurs when organizations engage in a knowledge transfer with a partner from an inter organizational perspective, as the phenomena study consists of knowledge transfer where codified (explicit) and uncodified (tacit) knowledge is mobilized from one source to a recipient (Nonaka & von Krogh, 2009). In this sense, the theoretical assumption is that the aspects affecting the knowledge transfer process are considered mainly from the source to the recipient in a unidirectional way, whereas the source organization is only transferring knowledge to the recipient without receiving knowledge in return (Loebbecke et al., 2016), even though in practise it can be treated as bidirectional way, where the organization transfers knowledge to another organization and receives knowledge in return (Tangaraja et al., 2015).

Moreover, as knowledge transfer originates from the interaction between individuals but there are different paths it can be transferred to such as from individual to another individual, from the individual level knowledge can also be transferred to the organizational level implying knowledge may be transferred to another organization and this is where the book places itself in. Notably, knowledge transfer aspects exist both on the individual and organizational level, where differences occur to some extent (see chapter 3). Therefore, this book is aimed to examine how various aspects in terms of three critical dimensions such as organizational characteristics, relationship characteristics, and knowledge types that affect knowledge transfer at the inter-organizational level for large or multinational organizations and small or medium organizations that was discussed (see 2.21, 2.22, 2.23).

Furthermore, the existing bulk of literature focus on aspects that affect KT related to either LC/MNC or SME that are indistinctive and fragmented and there is further categories to this transfer, as there are intra and inter organizational perspectives (see section 2.7, 2.8, 2.9), where the book places itself in the inter organizational perspective which is due to the literature of intra organizational has had a great deal of focus and inter organizational has not, resulting in the literature regarding knowledge transfer from a specific sized company to their partners to still be fragmented. However, this study is aimed to recognize a relationship in terms of similarities and differences between specific aspects that influence knowledge transfer process from knowledge holder to a partner recipient, in our case from a SME and a LC/MNC.

Thus, a qualitative study can demonstrate *in what way* that small or medium and large or multinational organizations operate in the knowledge transfer process following these aspects discussed in the literature but they are not well-described in the case of comparison and how they apply is unclear. Hence, to contribute to the existing literature of knowledge transfer, this book selected a qualitative approach to make a framework by analysing the different aspects for large or multinational organizations and small or medium ones to draw they interact and transfer knowledge with partners of other size, to improve the understanding of how these aspects affect knowledge transfer.

3.1.2 Deductive Approach

Among the different approaches that represents the nature of relationship between theory and research the most adopted one is the deductive approach (Bryman & Bell, 2011). The aim of this research is related to the identification of aspects that affect knowledge transfer in different size firms. Considering that knowledge transfer is a relatively mature field, starting from Penrose (1959), with enough theoretical perspectives and approaches (e.g., Lucas & Ogilvie, 2006; Conner & Prahalad, 1996; Szulanski, 1996; Barney, 1991; Wernerfelt, 1984), most of the elements to answer the research question can be found in the literature. However, as we explained in the previous sections, some aspects regarding the perspective of the research is still unclear to some extent in the literature. The development of a theoretical framework is then more suitable for a deductive approach and to qualitatively test and analyse the specific phenomena selected in the book, as a deductive approach being with an analyzation of existing theories to reveal the relevancy of the research (Saunders, 2009).

This study uses the literature as the departure point and will analyse the theoretical framework created in two cases. As the deductive is an approach to testing a theory, initially the researcher makes theoretical considerations to test if the theory is applicable to a specific scenario (Hyde, 2000). As the knowledge transfer aspects (e.g., Minbaeva et al., 2018; Martín Cruz et al., 2009; Szulanski, 1996; Cohen & Levinthal, 1990) that explains the situation in the different sized organizations, it is however not connected in the way that this book suggests. Accordingly, as the phenomena is to study the level that considers the connection on how aspects can be understood in different sized organizations and test it. For this book, the empirical settings considers the specific scenario to be how aspects affects knowledge transfer by different sized organizations, which includes the two cases, providing two different scenarios, first how these aspects are affected when a SME transfers knowledge to a partner, secondly how the aspects are affected in the same sense but from a LC/MNC, both scenarios is analysed on an organizational level with an inter organizational perspective by a unidirectional transfer of knowledge. To then qualitatively test (Saunders, 2009) these phenomena's the research will include (1) if the literature of knowledge transfer is limited in regard to engaging with a knowledge transfer with partners (see chapter 1, 2); (2) if the identified aspects (see chapter 3) of similarities and differences can extend the literature by contribution to how these aspects are affected in the specific phenomena, which is revealed by triangulation of data (discussed in 3.8.5); (3) contribute to the existing literature by analysing and researching this phenomena by presenting how these aspects are affected when organizations of a different size, transfer knowledge to a partner organization.

3.2 Exploratory case study as a research strategy

In this book, an exploratory case study approach was used to fulfill the research purpose and to answer the research question.

First, it is important to recall that the case study method is defined as "a strategy for doing research which involves an empirical investigation of a particular contemporary phenomenon within its real-life context using multiple sources of evidence" (Robson & McCartan, 2017).

As it was shown in the problematization part of this study, the knowledge transfer process is complex (Tangaraja et al., 2016) and affected by different aspects and can be understood in various ways according to the empirical setting selected, for instance, the level of analysis (e.g., individual, organizational, inter-organizational), organizational characteristics (firm size, age, sector, absorptive capacity, etc), knowledge characteristics (explicit, tacit, ambiguity, etc), and so on (Wijk et al., 2008). The study of the knowledge transfer in the context of different firm sizes requires taking information and data from the real-world existing companies in order to offer additional information about what aspects are considered more relevant according to their processes that are impacted by their sizes. The current literature does not provide a full explanation and the source for the needed information in the firms, which are large or multinational corporations and small or medium organizations that are operating in interfirm knowledge transfer activities (Milagres & Burcharth, 2019; Cerchione et al., 2017).

Second, the exploratory case study method is aimed to collect data and explore different underlying themes, categories, and patterns (Yin, 2014) and it is a valuable technique of finding out "what is happening; to seek new insights; to ask questions and to assess phenomena in a new light" (Robson, 2002).

This book fulfills the exploratory case study criteria in the following ways (1) the literature fails to establish what are more relevant aspects for particular firm size and this book will strive to demonstrate through two cases, which operate in a different sector, various views of LC/MNCs and SMEs about those aspects that affect interfirm KT; (2) considering the exploratory nature of this study, the book will offer an enhanced theoretical framework that might fit better the situation in KT in different size firms by applying qualitative approach due to the needs to examine in what way different sized companies is influenced by aspects that impact interfirm knowledge transfer process; (3) case companies in this study belong to different sectors allowing to analyse the theoretical framework from various perspectives and identify new insights of inter-organizational KT in different firm sizes.

3.3 Case selection criteria and case description

This section of the study is focused on choosing samples and establishing case selection criteria for data collection to test the theory described in the previous sections of this book.

To start this book, we approached different companies located in Halmstad, Sweden, to establish potential dimensions for the research. As a result, one SME showed interest in the book project that has eventually become an SME case company for this study. This SME case company was interested in studying knowledge transfer between organizations to explore differences and similarities due to different firm sizes. Furthermore, the SME case company provided data access to another large-sized company partner to perform this type of research.

According to the literature, convenience sampling refers to *"a type of nonprobability or nonrandom sampling where members of the target population that meet certain practical criteria, such as easy accessibility, geographical proximity, availability at a given time, or the willingness to participate are included for the purpose of the study"* (Etikan, 2016). As was mentioned above, the authors selected a case company due to its geographical proximity and its interest in the knowledge transfer field, hence, the chosen sampling approach is in line with the convenience sampling method. Moreover, to receive data access to the large company size, SME case company suggested large company/MNC partner in order to participate in this research. Therefore, this study applied a similar approach to a snowball sampling that refers to the identification of a small number of cases that, in turn, suggests further new cases that can be suitable to achieve the research purpose (Saunders et al., 2009).

As a result, the decision of applying convenience and snowball samplings was based on the research objective of this book that is aimed to examine in depth the aspects, which affect inter-organizational knowledge transfer in different firm sizes, which requires to identify suitable cases that can contribute to the current literature in terms of better understanding of the aspects that more relevant for different types of firm's sizes, which engage in interfirm KT activities. Given the foregoing, the following criteria were defined to select company sample:

(1) The case companies should have different sizes such as LC/MNCs and SMEs because the research question states that the importance of different sized firms' context is needed to explore the aspects that affect knowledge transfer. As was discussed in section 2.10, the company size has a significant effect on knowledge exchange between firms due to various resources, capabilities, corporate structure, power, and so on (Korbi & Chouki. 2017; Wee & Chua. 2013; Durst & Runar Edvardsson. 2012; Desouza & Awazu. 2006). Therefore, the following criteria were established:

The SME is recognized if companies satisfy three criteria:

- The number of employees less than 250.

- Annual revenue does not exceed EUR 50 million.

The LC/MNC should meet the following criteria:

- The number of employees more than 250.

- Annual revenue exceeds EUR 50 million.

- Can be multinational company with one or more subsidiaries.

(2) In the theoretical framework of this study, aspects related to the organizational and knowledge characteristics (motivation, absorptive and disseminative capacities, tacit, explicit, complexity, knowledge governance, etc.) require from the case companies to have established processes with different mechanisms to store information that needed for keeping the memory of experiences in knowledge transfer with other partners, which can be informal and formal ways. To assure that the selected companies fulfill these criteria a meeting with members of the company was conducted, the main purpose was to establish in what way they are working in KT processes that show substantial information (in documents, platforms, or people) to describe the mentioned aspects. Considering the difficulties to do this type of meeting with many companies, the prioritization was made according to the interest that firms showed to participate in the study.

(3) Aspects related to relational characteristics such as trust and social ties and require the firms to have experiences in KT interactions with other organizations at least twice allowing to analyse how these aspects have changed after repetitive interaction in KT activities.

(4) Aspects such as cultural and geographical proximity/distance can be assessed in firms that have KT interactions at least twice in different locations including at the national but outside of one region and international level with other partners. When the potential companies were addressed, one characteristic explored was the diverse context in which they have participated in KT processes or projects, these contexts were defined in terms of different regions and

countries, implying that they have experience interacting with different cultures. Given the foregoing, in order to meet case selection criteria and to conduct this study, the two companies were chosen. The following information is provided on how the selected companies fulfill case selection criteria:

(1) Company A is a large/multinational (LC/MNC) with approximately 1000 employees and company B is a small and medium enterprise (SME) with about 100 workers. Company A (LC/MNC) is the world leader in access solutions for the hospitality industry, the total revenue made up about SEK 14 billion in 2020. Company B is a SME with a revenue of about SEK 190 million as of 2020. Both companies fulfill criteria related to the firm size.

(2) Company A was founded in 1974 and is the world leader in access solutions for the hospitality industry that have different projects and established processes to interact with wide-range customers and suppliers. Company B was founded in 2002 that supports 200 public municipalities and private care companies. Both companies have established processes with different KT mechanisms to store information needed for keeping the memory of experiences in knowledge transfer with other partners.

(3) Both companies confirmed that they permanently interact with different partners where social ties and trust are key aspects for partnerships.

(4) Both companies operate internationally and have foreign partnerships: more than 70 countries are covered by Company A operations and Company B works in Sweden, Norway, Finland, Denmark, the Netherlands, UK, and the US. Thus, both companies satisfy criteria related to geographical and cultural KT activities with their partners.

3.4 Data Collection

This section will illustrate how data collection has been achieved to adequately answer our proposed research question. As this research is an explorative case study of different sized organizations which includes different type of data collection, this section discusses how the data collection process will be achieved.

3.4.1 Primary data

To gather relative data, the more recent data received the more accurate it will be depending (Bryman & Bell, 2011). As such, for this book primary data collection will be obtained via interviews. Saunders et al. (2009) state that interviews comes in the form of structured, semi-structured or unstructured interviews, whereas the first involves a predefined questionnaire and the session is conducted in the same order and way as the other interviews, semi-structured interviews follows the same procedure as the former, however it is less strict in the sense of how the structure of the interviews in conducted, lastly unstructured interviews gives the most flexibility as it does not depend on a predefined questionnaire and is conducted in a informal way to enable a open discussion of said topic to be held.

Furthermore, as the different types of interviews gives various benefits depending on the topic researched, this book has selected semi-structured interviews since it allows the researchers to develop a predefined questionnaire and to be open for discussion during the interview, which might lead to another question in the questioner to capture the current thoughts of the interviewee, as "*it is the flexibility of the interview that makes it so attractive*" so "*the interviewee has a great deal of leeway in how to reply*" to capture the perspective of the interviewee (Bryman & Bell, 2011), as it may differ depending on their role in the company or their current thought process.

On a side note, due to the current pandemic we have selected to conduct these interviews via online meetings, on the platform Microsoft Teams instead of Face-to-Face interviews. The result of this may impact our data collection to yield differentiated results if we had selected a different approach (Crescentini & Mainardi, 2009) but it also provide less financial cost for the researchers and more potential to contact people in a wider perspective (Saunders et al., 2009).

The interviews were conducted according to the schedule illustrated in table 3.4.1 below.

Informant representing number	Roles and responsibility with respect to the company.	Number of years of experience (total and at the company)	Company type (LC/MNC or SME)	Interview type	Interview dates (y/m/d) and duration
Informant 1	Agile Coach within R&D for one team	Total experience is 19 years and 7 years with the company A	LC/MNC	Open video face-to-face interview via MS Teams	2021/04/12
Informant 2	Global Quality Systems Manager	25 years of overall experience and 4 years at the Company A	LC/MNC	Open video face-to-face interview via MS Teams	2021/04/13
Informant 3	Global Quality Director	14 years in total and 8 years at the Company A	LC/MNC	Open video face-to-face interview via MS Teams	2021/04/13
Informant 4	Quality Manager	25 years of overall experience and has been working at company B for one year.	Small and medium enterprise	Open video face-to-face interview via MS Teams	2021/04/14
Informant 5	Chief Product Officer	25 years of overall experience and has been working at company B since August 2020.	Small and medium enterprise	Open video face-to-face interview via MS Teams	2021/04/14
Informant 6	Head of Development	15 years of experience in total, whereas 7 years in case company B	Small and medium enterprise	Open video face-to-face interview via MS Teams	2021/04/14
Informant 7	Product Owner for two teams	13 years in total and 1 year at the Company A	LC/MNC	Open video face-to-face interview via MS Teams	2021/04/15
Informant 8	Operation Director	About 15 years overall experience and more than 2 years at the company B	Small and medium enterprise	Open video face-to-face interview via MS Teams	2021/04/20
Informant 9	Vice-president for research and development	More than 37 years' experience at the company A	LC/MNC	Open video face-to-face	2021/04/21

				interview via MS Teams	

Table 3.4.1 Interview schedule. Source: Authors.

3.4.2 Selection of the informants for the data

Since the research question propose plural amount of company's size, for this book the authors have selected a company with different sized divisions to collaborate with, which are independent units within the organization, where the authors will research on depending on their size, how knowledge transfer is affected by the identified aspects used in the theoretical framework. Accordingly, the informants that will be extracted of data from both a small or medium company and large or multinational company. The interviewee's profile that has been selected are managers and specialists who are responsible for coordination in knowledge management or practises knowledge management in either processes or projects.

3.4.3 Background and professional experience of interviewees from case company A

This section provides a short summary of the professional experience of the interviewees from the LC/MNC.

Informant 1

Informant 1 is currently working as an Agile Coach within R&D for one team. She started her career in 2002 and has a total experience of about 19 years, where 10 years she had been employed by Ericsson and held different positions. She has been working at company A for about 7 years. Her expertise falls into job roles such as a test engineer, developer, manual tester within the R&D field.

Informant 2

Informant 2 holds the job position as a Global Quality Systems Manager, she has been possessing for about 25 years of overall experience since graduation from engineering school. She has been working at *company A* since 2017. Her main expertise is operations including everything about processes and quality management in the businesses. She has an intensive knowledge and expertise in ISO standards (International Organization for Standardization).

Informant 3

Informant 3 is working as Global Quality Director in Global Solutions division; he has been working at case *company A* for about 8 years and overall experience made up approximately 14 years. Concerning his expertise, he graduated from school in robotics and automatization, he started his first job as a consultant for supporting customers in the automotive industry, then he became a project leader and a product manager. After he moved to case company A where he began to work as a technical engineer, then he became regional technical manager for the Middle East, Africa, and India, afterward he received the position of R&D quality manager where he had been working for 5 years and since 2020, he has been working as Global Quality Director.

Informant 7

Informant 4 holds the position of Product Owner for two teams and has been working at *company A* since March 2020. He also had been employed for two years as a business analyst consultant in a large organization such as EoN from the energy sector. Besides, he had been working for Sony Ericsson for 10 years by focusing on product management for apps and services.

Informant 9

Informant 9 holds the position of Vice-President for research and development within case *company A,* he has been serving for the company A since 1983 and worked in different position in product development units. He oversees 75 engineers who are directly reporting to him.

3.4.4 Background and professional experience of interviewees from case company B
This section provides a short summary of the professional experience of the interviewees from the SME.

Informant 4

Informant 4 is working as Quality Manager, her main job tasks related to quality management systems, reclaim, and different projects related to incident handling. Furthermore, she had been employed as a sustainability manager for 5 years. She has 25 years of overall experience and has been working at company B for one year.

Informant 5

Informant 5 is working as Chief Product Officer; she has been working at Company B since August 2020. Previously, she had been predominantly employed by large telecommunication organizations such as TerraCom, Sony Ericsson, Ericsson, Tele 2. Her background relates to civil engineering but mostly focused on project management, innovation and development including leading development projects and product management.

Informant 6

Informant 6 is Head of Development and oversees 30 employees who are directly reporting to him. His job duties mostly are focused on R&D management including product development, preparing requirements, and testing. He has approximately 15 years of experience in total, whereas 7 years in case company B. He possesses personal skills such as leadership, agile developments, and coaching teams as well as professional expertise in technology.

Informant 8

Informant 8 is Operations Director, and she has been working for case Company B about 3 years, overall experience is about 15 years. Her expertise mainly from supply chain field and focused on purchasing and logistics.

3.4.5 Framing the questions

To grasp for a better understanding of which perspective is taken towards the questions in the interview, the authors first ask the interviewee's some general information about themselves, such as their job position; years of experience in the field; their experience in years within the company; and about their expertise and skills.

To capture what affect and effect the different aspects, the interview questions have been formulized to be open for discussion, to avoid short answers. Additionally, depending on the interviewee's answer to a question, in some cases there was follow up questions on their answer since these aspects are interrelated as mentioned in the introduction, to capture valuable information. This is to make the interview discussion capture the thoughts of the interviewee's regarding the questions, as the interview is done by semi-structured method. The following part will provide the 17 questions used in the interview, briefly discuss how the question is formalized to be open for discussion and capture the essence towards the corresponding aspect.

As an example, Interviewee 1 works for a LC/MNC, the questions will ask how their interaction regarding KT was their partners. Accordingly, the theoretical definitions of aspects that affect knowledge transfer in the context of LC/MNC and SMEs and related questions were formulated (see table 3.4.5.1).

Dimension	Theoretical definition	LC/MNC	SME
Knowledge governance	Formal governance mechanisms consist of management policies and procedures, organizational structure (Foss, 2007).	In the project you have managed, how did the formal and informal procedures offer advantages and disadvantages? How often? Formal: reports, summaries. Informal: conversations to agree or follow the activities of the project (fikas).	In the project you have managed, how did the formal and informal procedures offer advantages and disadvantages? How often? Formal: reports, summaries Informal: conversations to agree or follow the activities of the project (fikas).
	Informal includes corporate culture, trust, social ties and networks (Foss, 2007).	If you have to recommend practices, activities, or aspects to improve the results of knowledge exchange between large companies and small or medium enterprises, what do you think is more important?	If you have to recommend practices, activities, or aspects to improve the results of knowledge exchange between large companies and small or medium enterprises, what do you think is more important?
Disseminative capacity	Source motivation to teach (Minbaeva et al., 2018)	In case you taught one of your business partners something, what motivates you to do so?	In case you taught one of your business partners something, what motivates you to do so?
	Source ability to teach (Easterby-Smith et al., 2008).	What did you need to know to be able to teach the other partner company especially, a small and medium sized firm?	What did you need to know to be able to teach the other partner company, especially large organization?
Cultural distance	The national cultural distance can serve as barrier for the KT among international partners regardless the firm size such as (1) higher cost of entry and hinders the transmitting the core competencies to the international markets; (2) facing operational challenges in terms of lacking common understanding of norms, values and motivation; (3) misunderstanding between foreign partners can limit sharing of critical organizational knowledge (Wijk et al., 2008).	In your experience, when you have interacted with people from different countries, what success (hits) and mistakes did you have? What did you learn?	In your experience, when you have interacted with people from different countries, what success (hits) and mistakes did you have? What did you learn?

Trust	Trust is based on a social judgements in terms of (1) evaluation of other partners' benevolence, (2) expertise and competence coupled with (3) assessment of the risk related to untrustworthy behaviour (Inkpen & Tsang, 2005). In addition, to this, the literature proposes to assess interfirm trust based on (1) fulfillment of commitments agreed between partners; (2) honest negotiation; and (3) avoidance of receiving excessive benefits of ally organizations (Cummings & Bromiley, 1996). Moreover, mutual trust is also deemed to be an important aspect, since exchange of favours or similar transactions (e.g., reciprocity behaviour) tend to be met in return as "If we helped them today, they would help us later" was a response regarding trust in Al-Jabri & Al-Busaidi (2018) article.	How do you know you trust other organization to exchange knowledge with them? How do you build trustworthy relationship during projects cooperation with small and medium sized partners?	How do you know you trust other organization to exchange knowledge with them? How do you build trustworthy relationship during projects cooperation with large organization partners?
Geographical distance/proximity	The geographical distance negatively influences inter-organizational KT especially for tacit knowledge exchange but can be mediated in terms of (1) developing trust-based relations between distant partners (e.g., Capaldo & Petruzzelli, 2014); (2) organizational proximity including common corporate culture and values (Korbi & Chouki, 2017); and (3) bilateral way of utilizing modern communication technologies (Choi & Contractor, 2016) in large or multinational companies as well as in SMEs.	In a globalized context of operation, does physical proximity or distance affect your collaboration with other companies, especially small and medium sized partners? In what way?	In a globalized context of operation, does physical proximity or distance affect your collaboration with other companies, especially large organization partners? In what way?
Motivation	Martín Cruz et al. (2009) state that motivation consists of intrinsic and extrinsic motives. Organizational motivation relates to the firms motivation of acquiring knowledge and how to obtain it, the organization must	What kind of improvements could your company do in order to motivate employees for knowledge exchange with partners?	What kind of improvements could your company do in order to motivate employees for knowledge exchange with partners?

63

	invest time and effort to the process of knowledge transfer (Bengoa & Kaufmann, 2014).	Does your organization offer incentives to employees to learn new knowledge from other companies?	Does your organization offer incentives to employees to learn new knowledge from other companies?
Absorptive capacity	The definition of absorptive capacity could be described as the firm's ability to recognize external information, which is valuable for the firm, to assimilate the information and apply it to the firm's commercial means, where the former is achieved with prior related knowledge in the area, as it allows the firm to recognize the value and then assimilate it (Cohen & Levinthal, 1990).	After you have engaged with a knowledge exchange especially with a small and medium partner, how do you implement and absorb the acquired knowledge so that the organization can utilize it?	After you have engaged with a knowledge exchange especially with a larger partner, how do you implement and absorb the acquired knowledge so that the organization can utilize it?
		Does your organization attempt to share the acquired knowledge amongst its employees? By what means?	Does your organization attempt to share the acquired knowledge amongst its employees? By what means?
Social ties	Social ties according to Tangaraja et al. (2016) is a relationship based concept, its either conducted directly or indirectly between two parties and differentiated by (1) informal or (2) formal ways. Liu et al. (2015) argues that there are three steps to establishing a high-quality relationship, first step involves locating a partner firm and identify its key contacts, secondly to develop and maintain the relationship established, last stage involves dispatching of knowledge from the source along with knowledge absorption of the recipient, which is how the knowledge is exchanged.	How does your company establish communication with potential and known small and medium-sized business partners?	How does your company establish communication with potential and known large-sized business partners?
		What do you think is effective communication between your organization and potential or known small and medium-sized partners?	What do you think is effective communication between your organization and potential or known large-sized partners?

Characteristics of Knowledge	**Tacit and explicit knowledge:** Tacit knowledge is defined as experience, skills, intuition of individuals, also described as know-how and is communicated in an inform way, while explicit is defined as knowledge which is captured and store and has a universal character which is described as know-what (Brown & Duguid, 2000; Nonaka & von Krogh, 2009). **Ambiguity:** Wijk et al. (2008) claims that knowledge ambiguity is defined by the uncertainty of the knowledge, what underlying components and sources are in play (content-dependency), along with how they interact with each other (context-dependency). **Complexity:** Complexity consider how the knowledge that is transferred is affected by different variations as it is comprehended by different skills or experiences of organizations (Stock & Tatikonda, 2000). Complexity can be measured by how many different variables are involved (content and context dependency) in a certain situation (Milagres & Burcharth, 2019). **Tacitness:** Tacitness according to the findings of Kalling, (2003), state that tacitness exists for both tacit and explicit knowledge, whereas tacitness is affected by the individuals understanding of the knowledge they are handling.	In your experience, when you shared knowledge what was the most difficult to transfer to the partner? Did you notice any difference when knowledge was transferred to small and medium enterprises?	In your experience, when you shared knowledge what was the most difficult to transfer to the partner? Did you notice any difference when knowledge was transferred to large enterprises?
		By what criteria do you customize the knowledge you have learned when you apply to another firm? Why and how?	By what criteria do you customize the knowledge you have learned when you apply to another firm? Why and how?
		What was the lesson learned (e.g., difficulties or opportunities) to successfully implement acquired knowledge?	What was the lesson learned (e.g., difficulties or opportunities) to successfully implement acquired knowledge?

Table 3.4.5.1 Interview questions related to corresponding aspect. Source: Authors.

3.4.6 Preparation of the interviews

To prepare for the interviews the authors developed a manuscript and PowerPoint presentation (see appendix 1 and 2), with the purpose to follow the same procedure of introducing our research topic, to provide the logic systematically in order to give the same understanding to each interviewee. The authors also welcomed the interviewee to the interview and gave them a brief guideline to the interview itself which includes the approximate length of each questions, confidentiality issues and recording permissions. As for the questions provided in the interview, the authors prepared a PowerPoint presentation which includes the previous parts mentioned as well as questions regarding the interviewee which then leads to the questions that relates to the aspects that this book is researching. If the interviewee did not understand a question or was unable to answer, the authors also prepared other ways of formulating the questions to make it easier or more applicable to the individuals job description or so forth. To summarize, the interview consists of four parts, (1) presenting the team and welcoming the interviewee; (2) introducing our research topic along with the purpose of it and what it is used for and the guidelines to the interview; (3) job related questions to the interviewee; (4) questions related to the aspects in section 3.6.3.

3.4.7 Secondary data

As a part of the data collection, secondary data can be combined with primary data which is one of the elements that validates the findings further, as it provides a longitudinal analysis opportunity, it also takes less time and cost to collect (Bryman & Bell, 2011). The secondary data of the company is contained in (1) documentary data such as recordings, writings; (2) survey-based data such as organizational, employee or customer surveys; (3) complied data which is formed by industry reports and publications (Saunders et al., 2009).

3.5 Data Analysis

As the aim of the book is to explore how aspects of knowledge transfer affect a unidirectional knowledge transfer from either a small or medium enterprise (SME) or a large/multinational organization (LC/MNC) to their partners, on an organization level, the data collected must be analysed. As such, this section will proceed to explain how the structuring of the data was done regarding both primary and secondary data.

To then structure and classify the code collection of primary data and analyse it with the literature and secondary data collection, this section will illustrate how these two separate processes was conducted. First process consists of two aspects, the primary data collection (in the form of interviews) and relating it with the existing literature (established in section 2, chapter 3) to draw conclusions or analysis of the information. Second process, which has two main aspects, the secondary data collection (in the form of documents from the case selection companies), relating it with existing literature (established in section 2, chapter 3) to draw conclusions or analysis of it. In summary, the data analysis will be based on the following:

1. Each main aspect that affects interfirm knowledge transfer, which was identified in theoretical framework of this study, was divided into sub-aspects (to measure) that were collected from interviewees and deploy for the two different companies.
2. The identified sub-aspects was made into short summaries from the interviews conducted and are combined from those informants that are collectively agree that they are related to the main aspect of the theoretical framework, while the ones that do not support the same statement are also mentioned.

3. Collected sub-aspects from primary data was also in some cases were supported by secondary data.

3.5.1 Primary data collection analysis
This section will illustrate how the structuring of the first process with primary data collection analysis was done with an example from the LC/MNC (see table 3.5.1.1).

The first column of the table to the left provides the categorization of the aspects (for further details see chapter 3, section 2). The second column contains the aspect within the category of the first column. Looking at the third column of the table, the subcategory of the aspect, this column provides what aspects are affecting the aspect in the second column that is used in the research and found in the literature. The fourth column states simply which informant it is and within brackets if the informant works for a LC/MNC or a SME. Looking at the fifth column, quote of the informant, which was originally left blank, provides the question that corresponds to the researched aspect within brackets and possible follow-up questions also in brackets (see 3.4.5 for further details on the questions). As for the sixth column, relating literature, originally filled with the corresponding literature, to match it with the theoretical framework. The seventh column, conclusion or analysis, originally left blank, puts the fifth and sixth column under a scope for analysing to draw a conclusions or analysis between them to state if it is consistent or inconsistent with the literature, either explicitly or implicitly (more details on how the coding was done in 3.6). The last column simply states the overall summary of the conclusion.

1	2	3	4	5	6	7	8
Dimensions/ Characteristics	Aspect that affect KT	Subcategory of the aspect	Primary data source	Quote from the informant	Relating literature	Conclusion/ Analysis	Summary
Characteristics of the relationship	Social ties	**(SOCT1)** How to establish new relationships and **(SOCT2)** How to maintain relationships	Informant 1 (LC/MNC)	(How does your company establish communication with potential and known business partners of a small or medium size?) *"Yeah, this is kind of outside of my role since I'm more embedded within the development department. It's unfortunately outside my role to interact with new partners or even existing ones... unfortunately a few layers of different roles, which is outside my role. I'm the Product Owner for a product and a team. And then (**SOCT1+2**) I have like a technical contact person and a key account manager. And they in turn have people. Who then interact with the customers."*	Strong social ties acts as a facilitator for knowledge transfer (Kang & Sauk Hau, 2014) and Liu et al. (2015) argues that there are three steps to establishing a high quality relationship, **(SOCT1)** first step involves to locate a partner firm and identify its key contacts, secondly to develop and **(SOCT2)** maintain the relationship established, last stage involves dispatching of knowledge from the source along with knowledge absorption of the recipient, which is how the knowledge is exchanged.	The informant directs **(SOCT1+2)** to be handled by a role of socialization within the organization. This is consistent with the literature.	Social ties are established and maintained by a role of socialization in the LC/MNC **(SOCT1+2)**

Table 3.5.1.1 Primary data collection structuring. Source: Authors.

3.5.2 Secondary data collection analysis
This section will illustrate how the structuring of the second process with secondary data collection analysis was done with an example from the LC/MNC (see table 3.5.2.1).

The first column of the table to the left provides the categorization of the aspects (for further details see chapter 3). The second column contains the aspect within the category of the first column. Looking at the third column of the table, the subcategory of the aspect, this column provides what aspects are affecting the aspect in the second column that is used in the research and found in the literature. The fourth column states which organization the data was taken from and the name of the found document. Looking at the fifth column, paragraph of the secondary data, which was originally left blank, provides the quote from the secondary data source that was used for the analysis. As for the sixth column, relating literature, originally filled with the corresponding literature, to match it with the theoretical framework. The seventh column, conclusion or analysis, originally left blank, puts the fifth and sixth column under a scope for analysing to draw a conclusions or analysis between them to state if it is consistent or inconsistent with the literature, either explicitly or implicitly (more details on how the coding was done in 3.6). The last column simply states the overall summary of the conclusion.

1	2	3	4	5	6	7	8
Dimensions/ Characteristics	Aspect that affect KT	Subcategory of the aspect	Secondary data source	Paragraph from the secondary data	Relating literature	Conclusion/ Analysis	Summary
Characteristics of the relationship	Social ties	**(SOCT1)** How to establish new relationships and **(SOCT2)** How to maintain relationships	(LC/MNC) Document: External communications guideline 2020	*"The Disclosure Policy: (SOCT1+2) Who may act as a spokesperson for ihe (LC/MNC)"*	Strong social ties acts as a facilitator for knowledge transfer (Kang & Sauk Hau, 2014) and Liu et al. (2015) argues that there are three steps to establishing a high quality relationship, **(SOCT1)** first step involves to locate a partner firm and identify its key contacts, secondly to develop and **(SOCT2)** maintain the relationship established, last stage involves dispatching of knowledge from the source along with knowledge absorption of the recipient, which is how the knowledge is exchanged.	The secondary data state that who may act as **(SOCT1+2)** a spokesperson for the LC/MNC, this indicates that the organization utilize a role of socialization to establish and maintain social ties.	Social ties are established and maintained by a role of socialization in the LC/MNC **(SOCT1+2)**

Table 3.5.2.1 Secondary data collection structuring. Source: Authors.

3.6 The Coding procedure

This section will demonstrate how the coding procedure of the data collection (primary and secondary data) was done. For the complete coding information see appendix 3 and 4.

3.6.1 Primary data coding

The primary data coding was done by generalizing the process for all the data received from the interviews. To demonstrate how, we use the example from table 3.5.1.1, whereas the interviewee from a LC/MNC was asked a question regarding the aspect of social ties in knowledge transfer:

"How does your company establish communication with potential and known business partners of a small or medium size?"

Note, to clarify for the readers, brackets have been filled in to clarify what the theoretical framework attempts to identify from the informant's quote. The answer received from first informant was:

*"Yeah, this is kind of outside of my role since I'm more embedded within the development department. It's unfortunately outside my role to interact with new partners or even existing ones... unfortunately a few layers of different roles, which is outside my role. I'm the Product Owner for a product and a team. And then (**SOCT1+2**) I have like a technical contact person and a key account manager. And they in turn have people. Who then interact with the customers."*

The authors then proceeded to draw a conclusion of the "**SOCT1+2**" (in this case, **SOCT1** means establishment of social ties and **SOCT2** means maintenance of it) and analyse the relating literature with the informants quote, to draw a conclusion in terms of if it is consistent or inconsistent with the literature and if it was draw in an explicit way or implicit.

As this example on social ties has the relating literature:

*"(**SOCT1**) first step involves locating a partner firm and identify its key contacts, secondly to develop and (**SOCT2**) maintain the relationship established"*

Whereas the following conclusion or analysis was drawn:

*"The informant directs (**SOCT1+2**) to be handled by a role of socialization within the organization. This is explicitly consistent with the literature."*

To then make the finding section easier to organize, the last step involves drawing a summary of the conclusion or analysis, which was established to be:

*"Social ties are established and maintained by a role of socialization in the LC/MNC (**SOCT1+2**)"*

By following the same procedure for each question asked during the interview (see questions at 3.4.5) that was developed for the identified aspects (see section 2, chapter 3), the coding process was generalized in this sense and follows the same logical procedure to achieve the findings of the research.

The secondary data coding was done by generalizing the process for all the data received from the case studies involving the study. To demonstrate how, we use the example from table 3.5.2.1, whereas the secondary data from the LC/MNC was taken regarding the aspect of social ties in knowledge transfer:

*"The Disclosure Policy: (**SOCT1+2**) Who may act as a spokesperson for the (LC/MNC)"*

Note, to clarify for the readers, brackets have been filled in to clarify what the theoretical framework attempts to identify from the secondary data.

The authors then proceeded to draw a conclusion of the "**SOCT1+2**" (in this case, **SOCT1** means establishment of social ties and **SOCT2** means maintenance of it) and analyse the relating literature with the secondary data at hand, to draw a conclusion in terms of if it is consistent or inconsistent with the literature and if it was draw in an explicit way or implicit.

As this example on social ties has the relating literature:

*"(**SOCT1**) first step involves locating a partner firm and identify its key contacts, secondly to develop and (**SOCT2**) maintain the relationship established"*

Whereas the following conclusion could be drawn:

*"The secondary data state that who may act as (**SOCT1+2**) a spokesperson for the LC/MNC, this indicates that the LC/MNC utilize a role of socialization to establish and maintain social ties."*

To then make the finding section easier to organize, the last step involves drawing a summary of the conclusion or analysis, which was established to be:

*"Social ties are established and maintained by a role of socialization in the LC/MNC (**SOCT1+2**)"*

By following the same procedure for each question asked during the interview (see questions at 3.4.5) that was developed for the identified aspects (see section 2, chapter 3), the coding process was generalized in this sense and follows the same logical procedure to achieve the findings of the research.

3.7 Quality of the research findings
The quality of this study is a central issue that should be addressed to ensure that the book meets standard criteria of quality and rigor across the research dimension. According to Guba & Lincoln (1981) the qualitative research can be assessed in terms of four criteria that help to enhance trustworthiness of the findings. Furthermore, the quality of this book will be strengthened through conduction of the triangulation technique that refers to using multiple sources for data collection that will amplify credibility (Kuzel & Like, 1991). Therefore, the next sections are focused on describing each approach in -depth for this book.

3.7.1 Credibility of research findings
The credibility of the research refers to the criterion for establishing trustworthiness from the perspective of members or study participants (Farquhar, 2012). The credibility can be achieved via *"iterative questioning in data collection dialogues"* (Shenton, 2004). Moreover, credibility

should be evaluated in terms of two dimensions such as reliability and validity (Saunders et al., 2009). Particularly, reliability refers to *"the absence of random error so if the research was repeated, researchers would arrive at the same insights"* (Farquhar, 2012), whereas validity *"is concerned with whether the findings are really about what they appear to be about"* (Saunders et al., 2009).

In this research, semi-structured individual interviews allow to collection of dependable data from the participants. Likewise, semi-structured interviews were conducted to help interviewees providing more reliable and unbiased information. Besides, informants' profile and their job positions with various roles including R&D, quality, business development, supply chain, project management, which belong to different organizations such as LC/MNC and SME (for more info, see section 3.4.3 and 3.4.4) provide broader perspectives for the study as a result more independent views are generated to strengthen the credibility of this study. Moreover, the primary data collected from the informants is backed by the secondary data sources (for more info, see section 3.6) such as the internal websites, knowledge databases, reliable documents, the annual reports.

The credibility of the study is also enhanced by means of applying data collection techniques. Thus, in order to ensure that informants provide perplexed and optimistic answers, the questions were designed such that interviewees were asked from the lens of their experience and projects involved including hypothesized scenarios. Moreover, there was specificity in the framed questions, which were designed based on the particular topic area that can be answered by a specific interviewee (for more info, see section 3.4.5). Besides, the interview process was designed based on iterative questioning dialog that assured to get feedback on data collected. The language used was English to facilitate mutual understanding. All in all, this ensured effectiveness and credible results in the interview process.

The validity of this book was enhanced by applying the relevant literature wherein selected concepts were related to knowledge transfer at the inter-organizational level that were used as foundation for theoretical framework of this study (for more info, see chapter 2 and chapter 3 in section 2).

3.7.2 Transferability

Transferability is associated with the extent of findings that can be transferred or generalized to other settings, contexts, or populations (Guba & Lincoln, 1981). To put it differently, transferability is one of the variations to generalizability as it is also referred to as external validity, which relates to the belief that the research findings can be equally applicable to other research contexts, e.g., other organizations (Saunders et al., 2009).

This book was not aimed to reach generalizability, but the study applied two different and independent firms (e.g., LC/MNC and SME) were selected to improve problems with transferability. Furthermore, to enhance transferability larger number of informants with various job roles were chosen from LC/MNC and SME that can supply different perspectives for the research findings (for more info, see section 3.4.3 and 3.4.4).

However, even though this study was not designed to be transferred or generalized (see section 3.1) but the findings can be useful for companies that share the similar characteristics and have similar criteria that was established to perform the case sampling for this book (for more info, refer section 3.3).

3.7.3 Dependability

Dependability is also referred to as reliability that is concerned about the consistency of observing the same finding under similar circumstances (Guba & Lincoln, 1981). The dependability of this book is confirmed by virtue of:

(1) To build the theoretical framework in this book, the literature was selected by applying relevance to the research question of this study that states: *"aspects that affect inter-organizational knowledge transfer in different firm sizes"*. Particularly, literature related to the inter-organizational level of knowledge transfer and different units of analysis such as LC/MNC and SME were chosen to achieve consistency and reliability (for more info, see chapter 2 in section 2).

(2) Formulating interview questions through connecting and replication to the theoretical framework of this study (see section 3.4.5).

(3) Transparency by carefully documenting all findings and keeping all records of the primary data and secondary data (see section 3.6).

3.7.4 Confirmability

Confirmability relates to the degree of the study findings that can be confirmed or corroborated by others (Guba & Lincoln, 1981). According to Bryman & Bell (2011) the research findings are threatened by the researcher's personal beliefs and values on the trustworthiness or confirmability of the study findings that can distort the research results. Moreover, this implies that the research findings must reflect the opinion and the information provided by interviewees.

In this book, confirmability was achieved by connecting the theoretical part and empirical part of this study by means of applying a logical approach including framing questions that are consistent with the theoretical framework of this study (see section 3.4.5). Moreover, data coding and transcribing processes were done iteratively, and each quote of the informant was compared with the theoretical framework (for more info, refer section 3.6) as a result the higher confirmability was achieved. Furthermore, the interviews were recorded with a video that ensured the higher extent of the confirmability, and authors of the book could observe the expressions of the informants.

3.7.5 Triangulation

Triangulation is a critical principle of case study research and refers to applying multiple sources (e.g., archival, interview, video), involving multiple informants (e.g., various key informants), and multiple methods (e.g., participant observation, focus groups) for data collection, which strengthens and validates the research findings (Farquhar, 2012).

In this study, triangulation is achieved by means of using primary data such as interviewing multiple informants from two different-sized organizations. Furthermore, the raw data supplied by informants are backed by secondary data such as knowledge bases from the company's websites, annual reports, policies, press releases, internal blogs, which provide various perspectives for the research findings of this study.

3.8 Ethical considerations

Ethics in any research is important as it directs the results of the study and the participants who provided data should be introduced properly to the research project (Bryman & Bell, 2011). Accordingly, when the interviews were conducted, the interviewees were informed thoroughly what the topic of the study is about, what information will be published (e.g., their professional background), if they agreed to be recorded in the sense that it will make the coding procedure easier and that they will be anonymous in the study. Furthermore, when this study was finished, a final presentation of the findings was presented to all who participated in this study, including the interviewees, academic supervisor, company supervisors and the presentation was further opened to all members who wanted to participate from our case study A and B.

4. Findings

This section will provide the findings which was collected and analysed from the primary and secondary data, whereas the former is based on the interviews which were conducted from case study A and B and the latter is based on documented data which were obtained from case study A and B. The structure of the findings will be presented in the following way, (1) the title of the finding, (2) primary data from the interviewee's and which quotes were used and not used as base for the finding and (3) secondary data from case study A and B if there was available data to support the finding.

4.1 Interview results of primary and secondary data

This section is aimed to present findings from both primary and secondary data of the different sized companies (LC/MNC and SME). However, some aspects that affect interfirm knowledge transfer were not possible to find for the secondary data due to unavailability and confidential information restrictions at the companies.

Furthermore, this section is organized as follows:

1. Based on the theoretical framework, the analysis of the data was deployed using the same logic, which means aspects and sub-aspects were considered.
2. For each sub-aspect, we have presented findings.
3. We have shown quotes of informants and secondary data that support each of the findings.

4.1.1 Primary and secondary data about Aspect: Knowledge governance

According to the literature, the knowledge governance mechanisms that affect KT are defined as (1) Formal knowledge governance mechanisms and (2) informal knowledge governance mechanisms (Foss. 2007), for more information see section 2.21.3. The following section describes findings from both type of knowledge governance mechanisms.

Sub-aspect: (1) Formal knowledge governance mechanisms

We have found in this study based on collectively agreed informants that:

Finding 1: Formal knowledge governance mechanisms are used for KT coordination between partners.

Finding 2: Formal is more appropriate for KT with external organizations whereas informal within the company for LC/MNC, whereas for SME it can help to show that company is larger than it is to its partners.

Finding 3: As firm sizes increase the more formal knowledge governance mechanisms should be applied.

Finding 4: Formal knowledge governance procedures are needed for LC/MNCs to manage their cost structure, while SMEs are focused on growth and customer base expansion that needs more informal procedures.

Finding 1: Formal knowledge governance mechanisms are used for KT coordination between partners.

This was concluded based on the following information:

Primary data:

Informants (1 and 9) from LC/MNC and Informant 8 from SMEs collectively support that formal knowledge governance mechanisms come first in KT to structure and coordinate KT process at the projects, then informal knowledge governance mechanisms are in place to improve the progress of projects via informal knowledge transfer activities.

As an illustration of the type of quotes we analysed and their connection with the theory, we show table 4.1.1.1 that explains quotes from informants (1 and 9) from the LC/MNC and (8) from SME. The rest of the Informants including (2, 3, and 7) from the LC/MNC and Informants (4, 5, and 6) from SME mentioned that informal knowledge governance mechanisms are more important for KT activities that are discussed in more detail at the next section of this study.

LC/MNC	SME
"So, I like formal procedure on a very high level. I object to a formal procedure on a very low level. I want processes to be Lean, be Agile. I still wanted to be a process that everyone can understand how it works and so on. But a lot of reports or paper writing or anything like that, I try to avoid I want to avoid. But still simplistic systems where we are able to decide, collect facts and decide upon those facts. So, let's say on a high level, I would like something formal, visible to everyone. So, to understand how we work and how we do stuff. But in their daily work, they shouldn't really notice that much of it. But of course, there are documents, there are things that you have to do. There are the deliverables. There are plans to our contracts. There is confidential information... So, you are in control"- Informant 9	*"From a general perspective. I think both a formal and informal are really important to create a good environment and create good prerequisites for a successful project. In general, I think the formal part is more important in the beginning. Because that kind of sets the expectations throughout the project or from the time plan to agreed activities, etc...it decreases in the overtime. Ask whereas the informal... when you create the good cooperation within the team. And if you have to manage, then you have to create a good networking between the different team members because then they kind of make the informal communication as all things throughout the way. And then the formal points more become of a follow-up to ensure that everybody still on the same page in what to do and when" – Informant 8*
"...I think that the formal meetings, the discussions maybe come first, to be honest, I think when the development team is involved, I think it's more informal for me because then they just focus on the activities of the project... I mean, the informal meetings are more when the R&D teams are involved. The formal are when maybe the people outside of the development team are discussing things	

like contracts and stuff like that" – Informant 1	

Table 4.1.1.1 Quotes from informants in the LC/MNC and SME. Source: developed by book' authors.

Primary data illustrates that formal knowledge governance mechanisms are starting point for establishing KT activities in terms of defining goals, roles, responsibilities, and expectations for collaboration projects while informal knowledge governance mechanisms help to enhance the progress of the projects.

Secondary data:

Secondary information for this finding 1 from the LC/MNC shows that the company uses project planning tools as formal procedures that help to acquire knowledge and information from the customer and share within teams to transform it into clear goals and having common understanding before starting any projects.

As an illustration of the type of secondary data we analysed and its connection with the theory, we show table 4.2.1.2 that describes information obtained from the internal website of the case companies.

LC/MNC	SME
A workshop to clarify what the project is supposed to deliver. It is done in order to transform voice of the customer data into actionable, prioritized product design requirements that maximize customer value, market acceptance and profit. It also ensures that all participants and stakeholders agree on what is part of the delivery of the project, and what is not. And it needs to be clarified and agreed before project planning can begin, otherwise the team doesn't know what to plan for." Source: News/ The internal website of the LC/MNC, "Avenue"	None.

Table 4.1.1.2 Secondary data from LC/MNC and SME. Source: developed by book' authors.

In conclusion, primary and secondary data support for the LC/MNC, the formal knowledge governance procedures are important for projects that facilitate knowledge transfer coordination that helps to build a common understanding between partners. However, for SMEs, only primary data was obtained without secondary data confirmation for this finding.

Finding 2: Formal is more appropriate for KT with external organizations whereas informal within the company for LC/MNC, whereas for SME it can help to show that company is larger than it is to its partners.

This was concluded based on the following information:

Primary data:

Informant 7 (from LC/MNC) support that informal communication is applied knowledge transfer within the organization, while then moves to the formal way when knowledge is needed to transfer to the external organizations such as customer wherein knowledge should be clear which can help reduce misunderstanding, for example, due to cultural and geographical distance.

According to Informant 4 (from SME), formal knowledge governance mechanisms are used by SMEs to show that the firm has the competence and commitment like the LC/MNC, in other words, to show the firm is bigger in size than it is. This could imply that having formal procedures are considered by SME as tools to attract solid reputation for KT process with external companies that is inter-organizational KT, while the informal knowledge governance mechanisms are used by SMEs to interact within the company or with the same size.

As an illustration of the type of quotes we analysed and their connection with the theory, we show table 4.1.1.3 that explains quotes from an informant (7) from LC/MNC and (4) from SME. The rest of the Informants including (1, 2, 3, and 9) from the LC/MNC and Informants (5, 6, and 8) from SME did not provide an opinion on this finding.

LC/MNC	SME
"...we have a lot of very informal conversational style communication within the development teams. When we create our source code that we compile into a product with them deliver to the customer, right? So for that part, a lot of chatting and talking and if we were at the office, we would have, as you said, fikas and so on... So that's internally in the project. I think it's really good. Also, I think shows that people are feeling comfortable. They don't have to type in legalize when they talk to each other. But then when we are finished and we should deliver it and send it out to the rest of the world. Then it becomes, much more formal... Because then it needs to be understood in different cultural contexts, like hotels in Brazil, US, Spain, and China. They all have the same instructions set. So then it has to be very concrete and clear" – Informant 7	*"...Informal is more important in small companies; you have to cooperate very tight, and you have to share information all the time. You cannot be too formal, but you also have to think about that some structure is necessary if small companies are going to succeeding there in their job and if they want to grow, they have to have some kind of structure... If you interact with a larger company, you have to show that you are very committed, that you have the knowledge necessary. Maybe you have to show that you are bigger than you really are. To get the business... I think you need to show that you have structure in the documentation. You keep version handling. And so, you have some kind of orderliness in the company and show that you have employees that are very committed and so on. Informally, I think, show that you are a good partner. Fun to cooperate with. Are willing to make extra things to make it possible" – Informant 4.*

Table 4.1.1.3 Quotes from informants in the LC/MNC and SME. Source: developed by book' authors.

Secondary data:

Secondary information for this finding 2 from the LC/MNC and SME shows that formal knowledge governance mechanisms in terms of manuals and instructions are important guidelines to transfer knowledge to external organizations such as customers.

As an illustration of the type of secondary data we analysed and its connection with the theory, we show table 4.1.1.4 that describes information obtained from the internal website of the case companies.

LC/MNC	SME
"Instructions and manuals that are aimed to transfer knowledge about different products to the customers and suppliers of the large organization" Source: The webpage of the LC/MNC has special section aimed to transfer the knowledge about services and products to the partners.	*"Welcome to the Knowledge base, a site where you will find lots of inspiration and knowledge in digitalization and e-health"* Source: The webpage of the SME company has special section as Knowledge base with different documents that is aimed to transfer the knowledge to the partners.

Table 4.1.1.4 Secondary data from the LC/MNC and SME. Source: developed by book' authors.

To sum up, according to informants' opinions formal knowledge governance mechanisms are applied for KT to the external organizations, whereas informal knowledge governance mechanisms are utilized at the company level. Secondary data in form of service and product instructions and manuals are formal knowledge governance mechanisms that facilitate KT activities to the partners of the LC/MNC as well as SME.

Finding 3: As firm sizes increase the more formal knowledge governance mechanisms should be applied in SME.

This was concluded based on the following information:

Primary data:

Informant 5 from SME gives an opinion that SME should apply more trust-based informal knowledge governance mechanisms to facilitate KT, but also the informant suggests that as the company grows, they need formal knowledge governance mechanisms rather than informal ones.

As an illustration of the type of quotes we analysed and their connection with the theory, we show table 4.1.1.5 that explains quotes from informant 5 from SME. This viewpoint was not supported by informants (1, 2, 3, 7, and 9) from the LC/MNC probably because the size of the firm is naturally large and as a result, it is not applicable. Informants (4, 6, and 8) from SME did not provide any opinion on this perspective.

LC/MNC	SME
None.	*"We need to be more formal since we are growing, we should establish dedicated roles on who tells what during meetings"* – *Informant 5*

Table 4.1.1.5 Quotes from informants in the LC/MNC and SME. Source: developed by book' authors.

Secondary data is also not available to support the finding 3.

Finding 4: Formal knowledge governance procedures are needed for LC/MNCs to manage their cost structure, while SMEs are focused on growth and customer base expansion that needs more informal procedures.

This was concluded based on the following information:

Primary data:

Informant 5 from SME gives an opinion that SMEs should apply more trust-based informal knowledge governance mechanisms to facilitate KT, but also the informant suggests that as the company grows, they need formal knowledge governance mechanisms rather than informal ones.

As an illustration of the type of quotes we analysed and their connection with the theory, we show table 4.1.1.6 that explains quotes from informant 5 from SME. This viewpoint was not supported by informants (1, 2, 3, 7, and 9) from the LC/MNC. Informants (4, 5, and 6) from SME did not provide any opinion on this perspective.

LC/MNC	SME
None.	*"And I would say is both say smaller customer usually pays more attention to. This is my way of thinking as well. As a smaller customer pays more attention to customers because they are more relying on them. They need them to be able to grow. And so suppliers are not so important in the beginning. And then I think that it shifts over time. Where when you are in a larger organization, there are probably fewer customers who have that large impact on the total result. And instead, you grow. You try to consolidate your supplier base so that you have*

	fewer and fewer suppliers that you're relying on and who is a bigger part of your delivery... I would say so because the journey is usually when you're a small company, you do need to find your incomes and what, what pays your growth. And when you are a larger company, you need to work more with being cost efficient in formalizing processes and consolidating volumes to suppliers, etc" – Informant 8

Table 4.1.1.6 Quotes from informants in the LC/MNC and SME. Source: developed by book' authors.

Secondary data is not available to support the finding 4.

Sub-aspect: (2) Informal knowledge governance mechanisms

We have found in this study based on collectively agreed informants that:

Finding 5: Informal knowledge governance mechanisms are essential for LC/MNC for easier navigation, while for SMEs they help to enhance adaptability and flexibility capabilities.

This was concluded based on the following information:

Primary data:

Informants (1, 2, 3, 7, and 9) from the LC/MNC and Informants (4, 5, 6, and 8) from SMEs supported that informal knowledge governance mechanisms are essential mechanisms for KT activities.

Particularly, Informant 2 (from the LC/MNC had the experience to work in both firm sized firms) supports that nurturing social ties in terms of networking is more relevant for KT in LC/MNCs due to difficulties with reaching out to people since the large number of people involved at the KT process as compared to SMEs because SMEs have fewer employees therefore informal knowledge governance naturally exists there. Moreover, Informant 3 (from the LC/MNC) provided an opinion that informal knowledge exchange can bring hidden (e.g., tacit knowledge) to the surface and can affect the project performance.

Informant 4 from SME gives an opinion that informal knowledge governance allows SMEs to possess high adaptability in terms of flexibility capacity as compared to LC/MNCs that have more formal knowledge governance procedures in place. As an illustration of the type of quotes we analysed and their connection with the theory, we show table 4.1.1.7 that explains quotes from informants (2,3) from LC/MNC and (4) from SME.

LC/MNC	SME

"One thing that's really important is networking. Whether if it's very small company, it's not hard to network, if there's only like 20 people in the company, but in large companies... you need to network and you need to know people... It's very informal. And from this informal networking, you can put together formal trainings. We do that as well but informal can lead to formal" – Informant 2	*"...Informal is more important in small companies; you have to cooperate very tight, and you have to share information all the time. You cannot be too formal, but you also have to think about that some structure is necessary if small companies are going to succeeding there in their job and if they want to grow, they have to have some kind of structure... I think about one more thing if you work in a small company, you can change direct directions the next day if you would like, it's no problem if you have maybe five or 10 employees just change direct direction and start to run. But in large company, they are often very structured. They have a lot of rules. They have to ask the boss, the big boss, and so on... If you're a small company, you should work with a larger company that you really point out that you are very flexible" – Informant 4*
"... But I would say the informal part is very important as well. And because the different people act differently when they are sitting in a larger group, if it's a formal with a formal agenda, not everyone feels comfortable in speaking out freely. And this is where the informal part comes, which is very important and essential. So if you're heading project, you want to have that small talk with different people. If it's development, designer, if it's the testers. Because you know, you sit in meeting, test results, show perfectly. And this is actually my hands-on experience. Everything looks perfect. It's like a dream scenario. And then because I know that the test manager well, I go to him and then we'll just talk about anything and everything and then we go into the specific project. And then, yeah, laughing, said Yeah, but we had that tested, captured this or that test will not fly this part of the region because of humidity or heat. Or, you know, we will never be able to sell in Brazil because the compliant that we will achieve does not cover Brazil and Japan. That's a completely different one. So that informal is very important, I would say" – Informant 3	

Table 4.1.1.7 Quotes from informants in the LC/MNC and SME. Source: developed by book' authors.

There is no secondary data found that can support the finding 5.

Overall, informal knowledge governance mechanisms are important for both LC/MNC and SMEs, but LC/MNCs need them to manage tacit knowledge flow via networking, while SMEs apply to increase its adaptability and flexibility that implicitly can serve as a competitive advantage as compared to the LC/MNC.

According to the literature trust is defined and measured in terms of 7 sub-aspects that were illustrated in theoretical framework of this study (see section 2.21.4)

1) Fulfilment of commitments agreed between partners (Cummings & Bromiley, 1996).
2) Honest negotiation (Cummings & Bromiley, 1996).
3) Avoidance of receiving excessive benefits of ally organizations (Cummings & Bromiley, 1996).
4) Mutual trust (Al-Jabri & Al-Busaidi, 2018).
5) Evaluation of other partners' benevolence (Inkpen & Tsang, 2005).
6) Expertise and competence (Inkpen & Tsang, 2005).
7) Assessment of the risk related to untrustworthy behaviour (Inkpen & Tsang, 2005).

Sub-aspects: (1) Fulfillment of commitments agreed and (2) honest negotiation between partners

We have found in this study based on collectively agreed informants that:

Finding 6: Trust for knowledge transfer activities between partners is based on formal documents.

Finding 7: LC/MNCs have the power to enforce the fulfillment of formal trust agreements whereas SMEs have less power and considers enforcement as costly and ineffective.

Finding 6: Trust for knowledge transfer activities between partners is based on formal documents.

This was concluded based on the following information:

Primary data:

Informants (1, 2, 7, and 9) from the LC/MNC agree that to build trust between organizations it is important to sign a non-disclosure agreement (NDA), which can be considered as establishing a formal document to have the trust for beginning the process of the knowledge transfer between partners.

Moreover, Informant 7 mentioned that apart from signing a formal document for KT, there is a need to get approval from a higher level of manager. This can be considered as a limited trust for KT due to compromising the interests of the knowledge source.

Informant 5 mentioned NDA for KT in SME who joined recently to SME with a solid background from LC/MNC. Informant 6 from SME also confirmed that the company signs NDA to KT activities.

As an illustration of the type of quotes we analysed and their connection with the theory, we show table 4.1.2.1 that explains quotes from informants from LC/MNC and SME. The rest of the Informants including (3) from the LC/MNC and Informants (4 and 8) from SME provided opinions that trust is built on mutual benefits and assessed in terms of previous experience and depending on the firm size that is discussed in more detail at the next sections of this study.

LC/MNC	SME
"...I mean, both our company and other partnering company sign NDAs" – Informant 1	*"First, you need to have NDA or legal document. So that you can talk to each other. Sometimes our partners are our competitors, they have their development and their side, their roadmaps and so on" – Informant 5*
"So, we do exchange knowledge with them (external recipient organizations) and we do trust because we have a legal recourse in case if they do not follow through. So, the trust with outsiders is a legal binding trust" – Informant 2	*"...from my experience, paperwork is not very effective. Because to enforce any paperwork, I mean, to enforce any paperwork and hire lawyers, to sue them or whatever. That's a very costly process. So, I mean, we signed NDA with you. It's not very powerful. I need to trust you anyhow, because I don't think if you leak any information, I don't think I will file a lawsuit against you" – Informant 6*
"...we always sign NDAs, and we always make contracts. Not always contracts, but at least NDAs that sometimes contracts or two" – Informant 9	
"...Are we allowed to share that kind of knowledge within the corporation? I mean, I think you should be able to do without an NDAs, but we still had to check and maybe even double-check that. So, there isn't like a 100 percent trust directly even though you are part of the same company. In order to trust the other organizations, you need to have a higher level of approval..." – Informant 7	

Table 4.1.2.1 Quotes from informants in the LC/MNC and SME. Source: developed by book' authors.

There is no secondary data found that can support the finding 6.

Finding 7: LC/MNCs have the power to enforce the fulfillment of formal trust agreements whereas SMEs have less power and considers enforcement as costly and ineffective.

This was concluded based on the following information:

Primary data:

Informant 7 (from LC/MNC) mentioned that LC/MNCs have the power to push the fulfillment of commitments by smaller organizations in case if they breach trust agreements. In contrast, Informant 6 from SME suggests that they sign NDA, but is considered as ineffective way because enforcing the fulfillment of commitments is costly that implies SME does not possess enough power and resource to punish its partners for untrustworthy behaviour.

As an illustration of the type of quotes we analysed and their connection with the theory, we show table 4.1.2.2 that explains quotes from informants from LC/MNC and SME. The rest of the informants including (1, 2, 3, and 9) from the LC/MNC and Informants (4, 5, and 8) from SME did not provide any viewpoint on this perspective.

LC/MNC	SME
"...So that was very like a formal way of doing it right? Then if they breach the contract, we are a giant corporation with a lot of legal power. I would assume that we could take powerful action on a smaller" – Informant 7	*"...from my experience, paperwork is not very effective. Because to enforce any paperwork, I mean, to enforce any paperwork and hire lawyers, to sue them or whatever. That's a very costly process. So, I mean, we signed NDA with you. It's not very powerful. I need to trust you anyhow, because I don't think if you leak any information, I don't think I will file a lawsuit against you. I don't think so, I think something that you could bring with you is that you should establish trust and engaged in intensive knowledge sharing. You will need to have the dual value of doing this"* – Informant 6

Table 4.1.2.2 Quotes from informants in the LC/MNC and SME. Source: developed by book' authors.

Secondary data:

The secondary data for findings 6 and 7 shows that both LC/MNC and SME have templates for non-disclosure agreements. This implies that both sized companies are using NDAs for KT activities that help them to manage risks related to knowledge leakage.

As an illustration of the type of secondary data we analysed and its connection with the theory, we show table 4.1.2.3 that describes information obtained from the internal website of the case companies.

LC/MNC	SME
Group Legal has developed two general NDA templates - one mutual and one unilateral. The mutual agreement is intended to be used if both the large company and the counterparty will provide confidential information. The unilateral agreement is intended to be used if only the large organization will provide confidential information (to the partner). *Each of the NDA template is available in a version governed by Swedish laws and a version governed by the laws of the State of New York. An NDA guideline is available.* Source: The internal website of the LC/MNC, "Avenue"	*From NDA template (translated from Swedish): The undersigned is aware that during my collaboration with Company. I will receive technical, scientific, financial and strategically important information regarding the company. I am also aware that Company considers such information to be valuable trade secrets, the value of which depends on it not being disclosed or disclosed. I undertake that both during the time the cooperation / assignment relationship is ongoing and for 5 years thereafter not for any unauthorized disclosure or disclosure of such information. The above does not apply to such information as:* *• on receipt was demonstrably publicly known or who subsequently became publicly known without breach of this Agreement.*

	• *at the reception I was demonstrably already in possession of* • *I have legally demonstrably obtained from a third party without a duty of confidentiality* • *developed independently of company's data.* *If and to the extent that I need to discuss confidential information in order to fulfill my tasks in collaboration with the company, I undertake to ensure that the persons concerned are authorized to take part in such information and bound to keep it secret to the same extent as myself according to this confidentiality link. I am obliged to store and handle all documentation concerning such information in a secure manner, inaccessible to unauthorized access. I undertake, at the request of the company, to return it and not to keep any copies thereof. Should I violate this confidentiality agreement, I am aware that I may be liable for the damage that the company may suffer as a result.* Source: The company provided by email.

Table 4.1.2.3 Secondary data from the LC/MNC and SME. Source: developed by book' authors.

Overall, trust for knowledge transfer activities between partners is based on formal documents that imply there is a limited trust between external partners for KT activities.

Sub-aspect: (3) avoidance of receiving excessive benefits of ally organizations and (4) mutual trust

We have found in this study that:

Finding 8: Mutual trust for KT between partners is built in a form of mutual benefit and reciprocity.

This was concluded based on the following information:

Primary data:

Informant 3 from the LC/MNC suggests mutual trust between partners comes in form of mutual benefit and reciprocity. So, this implies that trust in LC/MNC is also based on achieving mutual goals for SMEs. Implicitly, this implies that one of the components of trust to share knowledge is the reciprocity behaviour of both partners. Similarly, informant 6 from SME provides the opinion that SME needs trust to build relationships with the larger firms that are based on an "iterative trust-building" process through providing small deliverables to larger partners or customers that brings win-win scenario for engaged KT parties.

As an illustration of the type of quotes we analysed and their connection with the theory, we show table 4.1.2.4 that explains quotes from informants from LC/MNC and SME. The rest of the informants including (1, 2, 7, and 9) from the LC/MNC and Informants (4, 5, and 8) from SME did not provide opinions on this perspective.

LC/MNC	SME
"…You do a risk assessment based on the knowledge that you already have. So, this is, again, the partnership here is important. If we look into our company historically. If a supplier did not deliver or the cost was high, you would just swap the supplier. And when you do that too much, you lose that partnership with the supplier. So, what we do now, we tried to look ahead. The lowest cost should not be the main driver because you want your partner to earn money as well. It's a partnership. And when you establish that also the trust comes. And then you can trust your partners more. And then we have actually such a good relationship with some of them when we have issues, we ask them to come back with a solution for us to support our R&D department in regard to design" – Informant 3	*"…I think something that you could bring with you is that you should establish trust and engaged in intensive knowledge sharing. You will need to have the dual value of doing this" – Informant 6*

Table 4.1.2.4 Quotes from informants in the LC/MNC and SME. Source: developed by book' authors.

There is no secondary data found that can support finding 8.

In summary, mutual trust for KT between partners can be built in a form of mutual benefit and reciprocity.

Sub-aspect: (5) Assessment of benevolence of other partners for KT

We have found in this study that:

Finding 9: Mutual business benefits evaluation is more prevalent for KT rather than an assessment of the partner's benevolence.

This was concluded based on the following information:

Primary data:

Concerning aspects related to the evaluation of other partners' benevolence for KT, there was not provided direct information from informants (1, 2, 3, 7, and 9) from LC/MNC and informants (4, 5, 6, and 8) from the SME. However, Informants (3 and 9) from the LC/MNC and Informants (4 and 6) from SME collectively support that motivation for KT is driven by improving the business performance of their organization as well as partners. Informant 3 suggests that sharing knowledge with partners will enhance the partner's (supplier) competence that is needed for mutual business performance. Informant 9 provided the viewpoint that the main motivation for KT is a business purpose that eventually can be turned in a form of product to the partnering organization (e.g., customer). Implicitly, this might be interpreted that motivation for interfirm KT is to create business value through mutual sharing knowledge that can lead to mutual trust. This can imply that assessment of benevolence of partners for KT evaluated in terms of mutual business goals achievement that can bring successful knowledge transfer.

As an illustration of the type of quotes we analysed and their connection with the theory, we show table 4.1.2.5 that explains quotes from informants from LC/MNC and SME. The rest of the informants including (1, 2, and 7) from the LC/MNC and informants (5 and 8) from SME did not provide opinions on this perspective.

LC/MNC	SME
"...If you're looking into a business partner, then you're thinking about a long-term relationship that will benefit both. So, achieving that in addition, absolutely to see some results out of what you are teaching away. And see that bringing them either successful prosperity absolutely motivates me" *"...if suppliers are not performing and possess higher risks, you want to go out there, build again the relationship and you start to teach them what we can teach them to bring them up to a level where we want them" – Informant 3*	*"...It would be to help the situation or improve our relationship or make them grow as our, if it is a supplier, and by that way, also improve the incoming quality or product deliveries" – Informant 4*
"...So, there are different type of partners and there's different motivations for me why I partner up with them. But it's always a cause, a business reason or technology reason. We are very seldom to select the path and just to learn something, we always try to learn doing actual projects or products or components even. So, that is actually our main motivation is always behind some kind of a business case of business relations that we want to turn into a product" – Informant 9	*"...My answer to that you could be more efficient and make more money by learning people within my business, things that I do, or we do in a better way* *... And I will be able to recruit more talented people if we work efficiently and make things at a lower cost or better velocity" – Informant 6*

Table 4.1.2.5 Quotes from informants in the LC/MNC and SME. Source: developed by book' authors.

There is no secondary data found that can support the finding 9.

Sub-aspect: (6) Competence and expertise of partners for KT

We have found in this study that:

Finding 10: Trust in partnerships for knowledge transfer is based on an assessment of the competence and expertise of partners.

This was concluded based on the following information:

Primary data:

Informant 4 and Informant 8 from the SME provided the viewpoint that to trust the other organization for knowledge transfer activities is important to evaluate the previous experience and reputation of the partner. Informant 9 from the LC/MNC suggests that when the LC/MNC engage in KT activities, they assess partners for KT in terms of capabilities and competence.

As an illustration of the type of quotes we analysed and their connection with the theory, we show table 4.1.2.6 that explains quotes from informants in the LC/MNC and SME. The rest of the informants including (1, 2, and 7) from the LC/MNC and Informants (5 and 6) from SME did not provide opinions on this perspective.

LC/MNC	SME
"...the different type of trust I would say, a huge, huge company, I think you can trust in many ways, but they can also have a different agenda and they can hurt you like that. When it comes to a smaller company, they may try to steal some of your knowledge or some of your IP I should say. Or they're just want to get into the business and they oversell what they have, some different kind of trust, different kind of levels you need to be aware of. When you go into. Smaller is more like, you know, are they capable? Do they have the competence will it benefit us? The larger is more about what can they do with us? Depending on how they act? So there are different kinds of approaches to trust, I think" – Informant 9	*"I think if you have had a long cooperation, you then know that you can trust them and you can share information and get good information back. It can also be that you have to ask other organizations about their experience. You can also ask around in your network. If it is a quite big organization with a lot of the well-known customers, I think that's a good credit. To know that you can trust them" – Informant 4*
	"I would say that it's by experience that you might have worked with them. I mean, it's not complete common that you continue to work with suppliers even when you switch jobs that you bring relationships. So, I think that experience over time is one important factor. And then I would say the other one is reputation" – Informant 8

Table 4.1.2.6 Quotes from informants in the LC/MNC and SME. Source: developed by book' authors.

There is no secondary data found that can support the finding 10.

Sub-aspect (7) Assessment of the risk related to untrustworthy behaviour

We have found in this study that:

Finding 11: The degree of trust for KT depends on a company's size:

11.1. For LC/MNC, the greater the size of the partner the higher risks are associated with the KT activities.

11.2. For SMEs, the larger organization the higher credibility and trust for KT activities.

This was concluded based on the following information:

Primary data:

Informant 9 from the LC/MNC argues that there can be different types of trust depending on the partner's size: the greater the size of the partner the higher risks are associated with the KT activities. If the company is partnering with a larger organization as compared to their company, then these partners are considered in terms of what threats they can bring due to different goals that these allies have for knowledge transfer. In contrast, SME's partners are evaluated in terms of capabilities and competence for performing KT and how this process can benefit the LC/MNC. Furthermore, informant 9 mentioned that SMEs are assessed for KT in terms of in terms untrustworthy behaviour related to the knowledge leakage and further utilization of transferred knowledge.

Concerning SME perspective, Informant 4 and Informant 8 give an opinion that the LC/MNC s have higher credibility and trust for KT because they have a broader range of customer base. Furthermore, Informant 8 suggests that it is important to establish common understanding and expectations in advance to build trustworthy relationships that can facilitate interfirm KT.

Informant 6 argues that SMEs need trust to build a relationship with the larger firms that are based on an "iterative trust-building" process through providing small deliverables to a larger partner or customer that will bring a win-win scenario for both parties. This implies that SMEs are "gaining the trust" to establish a reputation by providing small deliveries that reduce risks related to untrustworthy relationships, especially with larger organizations.

Informant 5 mentioned that the KT process is easier to perform with SMEs rather than large organizations since building trust for the knowledge transfer with larger organizations requires SMEs to find several contacts as compared to small and medium-sized companies, where few contacts are needed to establish trustworthy relationships.

As an illustration of the type of quotes we analysed and their connection with the theory, we show table 4.1.2.7 that explains quotes from informants from LC/MNC and SME. The rest of the Informants including (1, 2, 3, and 7) from the LC/MNC did not provide opinions on this perspective.

LC/MNC	SME

"...the different type of trust I would say, a huge, huge company, I think you can trust in many ways, but they can also have a different agenda and they can hurt you like that. When it comes to a smaller company, they may try to steal some of your knowledge or some of your IP I should say. Or they're just want to get into the business and they oversell what they have, some different kind of trust, different kind of levels you need to be aware of. When you go into. Smaller is more like, you know, are they capable? Do they have the competence will it benefit us? The larger is more about what can they do with us? Depending on how they act? So there are different kinds of approaches to trust, I think" – Informant 9	*"If it is a quite big organization with a lot of the well-known customers, I think that's a good credit. To know that you can trust them" – Informant 4*
	"I would say that it might be easier to trust a larger company because they usually have a broader customer base from the beginning" – Informant 8
	"...Concerning large customers, I have a few. I usually find the suitable role and I gain trust and build a relationship with that person or person. Then I worked with them. I think direct dialogue is best and gaining the trust of the other organization. Whereas the communication should be established on a very human level... I'm very fond of iterative trust building. So you need to start off with some small deliverables and saying, hey, now you gain something about this that I share with you. And then you share something with me I gain from this" – Informant 6
	"...I mean if you are in a smaller company, there are only five people of people while if I have a partnership with larger, you talk to one person in larger organization and he or she might not be the one deciding everything they are. They are the contact persons. I mean, that's the difference because you can get the decision right away in smaller company and they may be straightforward" – Informant 5

Table 4.1.2.7 Quotes from informants in the LC/MNC and SME. Source: developed by book' authors.

There is no secondary data found that can support finding 11.

In conclusion, the LC/MNC considers the larger partners as higher risks related to untrustworthy behaviour. In contrast, the larger the size of the firm's partner the higher degree of trust and less risk for SMEs.

4.1.3 Primary and secondary data about Aspect: Cultural distance

According to the Wijk et al. (2008), the cultural distance affects knowledge transfer between partners in terms of the following aspects:

1) Facing operational challenges in terms of lacking common understanding of norms, values and motivation.
2) Misunderstandings between foreign partners can limit sharing of critical organizational knowledge.

3) Increases the cost of entry and hinders the transmitting the core competencies to the international markets.

Sub-aspects: (1) Operational challenges and (2) misunderstandings in knowledge transfer between partners

We have found in this study based on collectively agreed informants that:

Finding 12: The language barrier can impede KT between partners.

Finding 13: A lack of understanding of the decision-making process in different cultures can negatively affect KT activities between partners.

Finding 12: The language barrier can impede knowledge transfer between partners.

This was concluded based on the following information:

Primary data:

Informants (1, 2, and 7) from the LC/MNC collectively support that the language barrier is a key aspect that can hinder knowledge transfer and can lead to misunderstanding between partners.

Particularly, Informant 1 suggests that meetings and interactions should be predefined and well organized to establish a common understanding that facilitates KT activities among partners

Informant 2 gives an opinion that the language barrier can be reduced if the organization transmits the knowledge in a simple form. Informants (2 and 7) also suggest that understanding the uniqueness of cultures through adjusting communication in accordance with cultural aspects can facilitate KT activities. There is no information provided by informants of SMEs regarding the language barrier probably because they do not often interact with partners at the international level due to smaller size business.

As an illustration of the type of quotes we analysed and their connection with the theory, we show table 4.1.3.1 that explains quotes from informants from LC/MNC and SME. The rest of the informants including (3 and 9) from the LC/MNC and all informants from SME did not provide opinions on this perspective.

LC/MNC	SME
"I know that having like, online meetings... I mean, it's easy to misunderstand actually. I dont know if people are very good at English or not. But it's easy to misunderstand. But it's, it's really important that someone drives meetings and you keep track on what both of you are talking about" – Informant 1	None.

"So, when you communicate it's important that you remember who you are communicating with, and you adjust your communications with the people (from different culture). So when you communicate, for example, with colleagues in China, you need to make your communications very clear and not that you need to have. For example, short emails. Sentences, very simple. And like almost 1, 2, 3, 4, 5. If you have five questions, actually number them and they will answer. And you write your question and if it's a yes, no? And if it's a number you're looking for, you will write what number, you know, things like that, because you need to remember this. English is not their native language. And you need to make everything easy for them. And this will help you not be frustrated with their answer when, because sometimes people send a very long and complicated e-mail with like five" – Informant 2	
"I have interacted a lot with Japanese people when I worked, maybe I'll also a lot with German, where I worked with customers and operators. And I tried to joke a lot. And I've learned that doesn't fly at all. If you don't speak your native language to start with them in Japanese people kind of struggle with English. Maybe not so much for Swedes, Germans maybe as well. But then you also have the different cultural backgrounds. We don't really have the same like references and so on" – Informant 7	

Table 4.1.3.1 Quotes from informants in the LC/MNC and SME. Source: developed by book' authors.

Secondary data:

The secondary data for finding 12 from the LC/MNC supports that language skills can be a barrier for communication to perform their job responsibilities. Implicitly, improving communication via training means more effective knowledge exchange among employees and its partner's employees. There is no secondary data provided by SME regarding the language skills affected by the KT.

As an illustration of the type of secondary data we analysed and its connection with the theory, we show table 4.1.3.2 that describes information obtained from the internal website of the case companies.

LC/MNC	**SME**

| *"English opens doors for you in the company. The English language course is an interactive and web-based program designed to help improve your communication skills in English. English Opens Doors is targeted towards employees who need to improve their English skills to become more efficient in their jobs"*

Source: Internal website of the LC/MNC "Avenue". | None. |

Table 4.1.3.2 Secondary data from the LC/MNC and SME. Source: developed by book' authors.

Overall, cultural distance can be challenging for KT activities and one of the sub aspects that can lead to the misunderstanding between partners is the language barrier.

Finding 13: A lack of understanding of the decision-making process in different cultures can negatively affect KT activities between partners.

This was concluded based on the following information:

Primary data:

Informant 6 from SME suggests that understanding the decision-making process and motivations of partners can facilitate the KT process in different cultures.

As an illustration of the type of quotes we analysed and their connection with the theory, we show table 4.1.3.3 that explains quotes from informants from LC/MNC and SME. This viewpoint was not mentioned by Informant (1, 2, 3, 7, and 9) from the LC/MNC that is probably because their operations are inherently in the global context and as a result, LC/MNCs are already aware of the decision-making process of their international partners. Informants (4, 5, and 8) from SME did not provide any opinion on this perspective.

LC/MNC	SME
None.	*"So, I think when I work with international companies. Before starting to work with other people. I tried to understand their motivations and how they make decisions, how they get interested in things. So how they are motivated. So, this varies from different cultures. So, for instance, in Japan, when I work with them, I did this hierarchical structure to get attention from, from manager's managers. And then I found an individual that I worked with. And in American culture was quite similar, German*

| | *as well. But they have different requirements and how you should have a dialogue with them. So, Germans and Japanese people, they are very structured and like structured information. We're, whereas American. They love stories. I just loved the success stories"* *– Informant 6* |

Table 4.1.3.3 Quotes from informants in the LC/MNC and SME. Source: developed by book' authors.

There is no secondary data found that can support the finding 13.

To summarize, understanding the decision-making process and motivations of partnering organizations can enhance KT between firms.

Sub-aspect: (3) Cultural distance increases the cost of entry that hinders the transmitting the core competencies to the international markets

We have found in this study based on collectively agreed informants that:

Finding 14: High power distance in terms of hierarchical subordination can hinder KT.

This was concluded based on the following information:

Primary data:

Informant (2, 3, and 7) from the LC/MNC agree that a high-power distance in terms of strong hierarchical subordination among employees that exists in countries such as China and Japan can lead to misunderstanding due to less opportunity to ask questions or provide opposite opinion those managers who possess higher hierarchical position. As a result, interactions between hierarchical levels tend to be more authoritative and paternalistic, eventually, this can lead to impeding knowledge transfer between different organizations and increases the cost due to mistakes related to KT between different cultural societies. Informant 5 from SME confirmed that strong subordination societies with high power distance could hinder KT due to lesson learned process from mistakes is less acceptable and eventually, it can lead to hampering of core expertise transfer to international markets.

As an illustration of the type of quotes we analysed and their connection with the theory, we show table 4.1.3.4 that explains quotes from informants from LC/MNC and SME. The rest of the informants including (1 and 9) from the LC/MNC and informants (4, 6, and 8) from SME did not provide opinions on this perspective.

LC/MNC	SME
"Well, I think you need to consider it in the same way, I think the language and the culture as well. So, for example, in Japan, they always CC their boss. So, if somebody	*"...And, for example, I was working for a factory in China because we were producing mobile phone. And we are in Sweden, we talk to the people in the project manager and the*

is late in sending you stuff, I do report and things like that. I will always be careful not to hound the person too much because I don't want them to get in trouble. I just need my stuff, but I don't want them to be in trouble, you know. Yeah. So, I think it's, all about cooperating with people" – Informant 2	*team. And we had some opinions, want to change things. And then they are misunderstanding people, they had to leave because they thought they had done some mistake, and it wasn't a mistake. We want to correct things. And in Sweden you can talk to people, but there they fire people right away because we had said something. I had a bad experience, but then we corrected it, but we got to learn that way. You have to be careful how you pronounce thing, how you say things, we wanted some small correction and then they misunderstand us completely. I did a lot of experience there how to work with different culture in this way with China" – Informant 5*
"…So, I think actually this is one of the exciting parts of being a global company to be honest. Mistakes, so mistakes that I personally have encountered. So, for instance, we hired a consultancy company from India to develop a solution for us. So, they can provide you literally whatever expertise you want. I think they had like 5000 consultants in different areas. And when we started the project, what I'm used to combine an idea as a product manager, and you want to concept; you have a vision. And then the details are done by the teams. And this is where it was a bit different here because what my experiences is with people from India in general, they work a lot. They have really can take on a lot. But you need to be very detailed in what them to do. And if you want to get their ideas, you need to ask for them. They won't just say, well, you know, idea here what you say but you're completely wrong, that will never fly. You need to think about this. And that was absolutely a mistake from my side. I was throwing out ideas. They were just doing what I told them. I expected to get some feedback as well. So, certain part of the world. You need to be very specific and down to detail what to do and when. And you will get exactly what you asked for. And nothing more, nothing less. That's come to work ethics then the culture side. With some cultures you need to be direct. So, in Scandinavia you go around and round and round until we come to the point. But, for instance, if you took people from Germany, from US, you need to be direct and it's okay. It won't be frowned upon if you are to direct. So, that's something also that you need to consider, and you shouldn't get insulted if they aren't yet" - Informant 3	

Table 4.1.3.4 Quotes from informants in the LC/MNC and SME. Source: developed by book' authors.

There is no secondary data found that can support finding 14.

In conclusion, strong hierarchical subordination among employees can hinder KT between foreign partners.

4.1.4 Primary and secondary data about Aspect: Geographical proximity/distance
According to the literature and theoretical framework of this study, the Geographical distance negatively affects KT between partners but can be mediated in terms of:

1) Developing trust-based relations between distant partners (e.g., Capaldo & Petruzzelli, 2014).
2) Organizational proximity in terms of common corporate culture and values (Korbi & Chouki, 2017).
3) Bilateral way for KT by utilizing of communication technologies (Choi & Contractor, 2016).

Sub-aspect: (1) Development of trust-based relations between geographical distant partners

We have found in this study based on collectively agreed informants that:

Finding 15: Physical proximity is essential to build trust between partners.

Finding 16: LC/MNCs need to develop the adaptability capacity to transfer knowledge for geographically distant partners.

Finding 15: Physical proximity is essential to build trust between partners.

This was concluded based on the following information:

Primary data:

Informant 9 from the LC/MNC gives an opinion that physical proximity via face-to-face meetings can facilitate the transfer of knowledge related to the company's technology and products and solve different emergent situations. This implies that tacit knowledge and related complexity can be better transmitted to other partners only via real physical interaction. Implicitly, this means that trust-based relationships are an important aspect that can be built via physical proximity. In addition, Informant 1 from the LC/MNC states that they have dedicated personnel to easily reach the partners in a specific location of the world for learning and exchanging knowledge with those partners. Implicitly, this implies that physical proximity is still important to transfer tacit and complex knowledge.

Regarding SME, Informants (4, 5, and 6) from SME also support that geographical proximity and face-to-face interaction facilitate KT activities between partners, while distance can be a barrier for knowledge sharing. Informant 6 from SME provided the opinion that geographical proximity is important because it can facilitate establishing trust between partners for better learning capacity. This implies that to establish trust first it is needed to have physical proximity in terms of face-to-face interaction and then this collaboration can move to distant partnerships.

As an illustration of the type of quotes we analysed and their connection with the theory, we show table 4.1.4.1 that explains quotes from informants from LC/MNC and SME. The rest of the informants including (2, 3, and 7) from the LC/MNC and Informant 8 from SME did not provide opinions on this perspective.

LC/MNC	SME
"I think we have learned in the last year that we can solve most by actually teamwork and so on. And emails are good for nothing really. E-mails are just for exchanging the basic and then you meet either on a meeting like this (online). But occasionally you read, we at least we have to meet the customer, we have to sit face to face with our major customers or our partners to mostly why not is to avoid an emergent situation going on. It is required from a technology point of view, actually to be able to do some real life testing of what we have develop. We would need to have be physically present, but that's because we just cannot connect these two body parts over the Atlantic, we have to actually be physically present. But physical proximity can affect our collaboration both from a technical point of view, actually testing out, doing, finding out the product. This is the customer see the problems that I have physically understand them" – Informant 9	*"I think it's harder to, to cooperate with a large distance. Because if you really want to solve something, it's easier if you can go and meet in person. So that's the thing that I think is hard and the culture in some way but, but not necessary. So the main thing I think is that if you or when you need to meet the distance can be a problem" – Informant 4*
	"We work with a company that we bought in Netherland. I haven't met them before. I think, If you get to know people that you meet them face to face one or two times. I mean, it will always be easier to communicate. *You have you build a kind of relations. I always try to meet people for the first-time. That's why we went to China, for example, to meet people to be that they thought, then it was much easier than to have daily meetings where you think, you know, how they react. If they don't react, you know, how they pronounce thing, you understand more. I think that a good way to learn each other" – Informant 5*
"...I think so. I mean, we have a big part in the US, for instance, and I know that the people in our company, then we have like individuals that are very involved in different like the key accounts. I mean, I'm not talking about the key accounts' manager, So stuff like that. But all the bigger partners that we have are the big hotel groups, we have dedicated people to talk to them a lot and maybe try to see them and help them out with whatever it is. And I think that is very important to have at least a few people that know the customer and know what they want. And now they're using the systems. Because	*"Physical proximity has an effect. When I worked in an international company. I always tend to build the trust part by being present with the person, so the person could get a grasp of what I am...* *... But there are so much more in your communication that can be done over a physical meaning, the body language, the way we interact in social situations. That we become that come from the Savanna, right? We did not have video conferences on the Savanna. We worked as a group or as a group of people. And there were trust between these people. And that's due to many aspects of*

it's hard for us in Sweden to know exactly what America or China or India what needs they have, for instance. But I think we have to cope with it. ...I think, for instance, the last year, during the pandemic. I think our company had an ease transition from being located or working in the office to the online working when we're doing mutually sits where it is. It has been a very easy transition because we kind of use it. We have like partners and companies in China. We have US, Australia. Yeah. I mean, all over the place, we are used to it. But, of course, it's better to be in-person" – Informant 1	*what we did on the Savanna. What people knew what they had competence in and what if they were strong, if they were good then in crafting. If they were good speakers, they were good in finding water or whatever. That's something you gain from physical proximity" – Informant 6*

Table 4.1.4.1 Quotes from informants in the LC/MNC and SME. Source: developed by book' authors.

There is no secondary data found that can support finding 15.

Overall, physical proximity in terms of face-to-face interactions supports building trustworthy relationships between distant partners.

Finding 16: LC/MNCs need to develop the adaptability capacity to transfer knowledge for geographically distant partners.

This was concluded based on the following information:

Primary data:

Informant 1 from the LC/MNC argues that proximity to partners is important, but also, she mentioned that it is important to adapt to the new reality (implicitly, pandemic time) and find solutions how to help each other (between partners) for better collaboration for KT activities. An aspect related to trust-based relations was not explicitly described by the informant but rather it is more important to have adaptability capacity for partners for better KT activities.

As an illustration of the type of quotes we analysed and their connection with the theory, we show table 4.1.4.2 that explains quotes from informants from LC/MNC and SME. This viewpoint was not mentioned by informants (2, 3, 7, and 9) from the LC/MNC. Informants (4, 5, 6, and 8) from SME did not provide any opinions on this perspective probably because SME has already possessed adaptability capability as an inherent feature due to the smaller scale of the firm.

LC/MNC	**SME**
"... I mean, of course, I think it always matters actually because it's always simpler to be where the customers are, or it is depending on what you're going to like discuss. But yeah, of course it matters. But in	None.

the in the world, we live in, we have to kind of adapt to not being side-by-side all the time. So that's why I think it's important to kind of come up with what helped both the partners and our company in that collaboration" – Informant 1	

Table 4.1.4.2 Quotes from informants in the LC/MNC and SME. Source: developed by book' authors.

There is no secondary data found that can support this perspective.

In conclusion, adaptation ability is one of the important aspects that affect interfirm KT for distant partners.

Sub-aspect: (2) Organizational proximity in terms of common corporate culture and values to reduce geographical distance between partners

We have found in this study that:

Finding 17: Geographical distance/proximity does not influence interfirm KT.

This was concluded based on the following information:

Primary data:

Informant 2 from the LC/MNC suggests that proximity is not important for KT activities during the collaboration, but states that willingness for KT is more important to be motivated for collaboration rather geographical co-presence. Implicitly, this means willingness as organizational value is an important aspect for interfirm KT.

Informant 8 from SME suggests physical proximity is not essential for KT activities and confirms that LC/MNCs inherently have KT with distant partners due to widespread operations. However, informant 8 supports that physical distance can influence knowledge transfer in the case of manufacturing of physical products where tacit and complex knowledge inherently exists and by proximity can be easily transferred. However, for the service sector physical proximity does not have any importance.

As an illustration of the type of quotes we analysed and their connection with the theory, we show table 4.1.4.3 that explains quotes from informants from LC/MNC and SME. The rest of the informants including (1, 3, 7, and 9) from the LC/MNC and informants (4, 5, and 6) from SME did not provide opinions on this perspective.

LC/MNC	SME
"I would say physical proximity. It doesn't affect collaboration edges. That it's a bit tricky if you work with Australia, the time differences...when I was in Canada that we	*"I would say that it doesn't affect the collaboration with large organization so much because they're usually so widespread already from the beginning. I think my*

had exactly 12 hours with China. So, eight in the morning was eight in the evening for them. Somebody would not be in the right time. Not in the regular day time, right? It makes it a bit challenging, but you need to, and of course, if everybody's in one building, then it makes it easier. I don't think it affects collaboration because collaboration means people are willing to do things. So, I don't think people are either will or unwilling to do things if they're far apart. It's just that it makes it more difficult to do it at the same time, but it doesn't mean they're unwilling to do it" – Informant 2	*experience is that the proximity or physical distance isn't so important any longer. Even if it's nice to meet them from time to time, you still don't you usually don't have possibility to physical proximity to tool or parts of their organization. So you kind of learn how to deal with that on a distance... I think that depends a bit on what type of organization it is. If it's just a company where you work with on a more abstract level, with services or with general cooperation. And, you don't have so much physical on hand production, for example. Then I don't think that I would have visited them so much more, just even if they were very close. It's more of important once you actually, work with physical products for example. We have our largest manufacturing sites with the supplier just about an hour away by car. And that's really good for us because we can always go and help them. If they have issues in production, then it is easier for us to go there and help them to do the search and trying to figure out what's wrong and how to fix it and collaborate with them in setting up the testing equipment, etc. But, so that has a positive a side effect in having a shorter physical distance, then for example, looking and moving some of that production to an facility in China. And then of course it's a directly becomes more difficult to cooperate and figure out things from afar when you're actually not there and can see things. But if it's just general corporations, just meetings and calls and documentations and systems, then it doesn't matter. From my perspective, the physical distance doesn't have so much impact" – Informant 8*

Table 4.1.4.3 Quotes from informants in the LC/MNC and SME. Source: developed by book' authors.

Secondary data:

This secondary data provides information on finding 17 that to reduce the geographical distance it is important to establish training programs to reach a common knowledge base for different geographical distant parties.

As an illustration of the type of secondary data we analysed and its connection with the theory, we show table 4.1.4.4 that describes information obtained from the internal website of the case companies.

LC/MNC	SME
"The Global Product Management training program has been instrumental in building a common knowledge base and capabilities within product management and it has now been rolled out to over 200 participants globally. In the video, one of the participants suggests that this training helps to establish common knowledge for participants from different geographical regions." **Source:** The webpage of the LC/MNC.	None.

Table 4.1.4.4 Secondary data from the LC/MNC and SME. Source: developed by book' authors.

Overall, physical proximity is not essential rather the willingness of actors is a more important aspect for KT between different distant partners.

Sub-aspect: (3) Bilateral way for KT by utilizing of communication technologies

We have found in this study that:

Finding 18: Digital communication tools improve KT between physically distant organizations but as a complementary technique.

This was concluded based on the following information:

Primary data:

Informant 7 from the LC/MNC suggests that digital communication tools can enhance KT between distant organizations but as a complementary mechanism after establishing face-to-face meetings with partners. This is viewpoint was also supported by Informant 6 from SME by giving the opinion that digital communication tools cannot fully substitute face-to-face interaction for KT due to difficulties to establish trustworthy relations, but the informant supports that trust can be built if online meetings can be done longer period. Informant 4 from SME generally provided the opinion that digital communication tools can enhance KT between distant organizations.

As an illustration of the type of quotes we analysed and their connection with the theory, we show table 4.1.4.5 that explains quotes from informants from LC/MNC and SME. The rest of the informants including (1, 3, 7 and 9) from the LC/MNC and informants (4, 5 and 6) from SME did not provide opinions on this perspective.

LC/MNC	SME

"If you have asked me this a year and a half ago, I would probably give it a different answer than today. Because we are really practice this global and far away communication. And I think we have really improved and even become a bit tired of it as well. Using this kind of worn-out phrase, this is the new normal. So, I still see a great value in meeting people in person, definitely in the phase when you start to know them. And that's true for a mean, personal relationships, but also professional relationships... So, travel might be required in the future, but follow-up meeting. I think this kind of Teams meeting that we do now is very efficient and it's better for the environment as well" – Informant 7	*"I think this digital improvement or usage that we have learned now this year is very good because it's so easy to get to decide about the meeting and defined. And it's easier to plan since you can adjust according to their schedule if they have it online in other words, to find an available slot in the schedule of the other person (online)"* – Informant 4
	"I think I would have very hard to trust another people to full extent if I've only had one hour team's conference with that person at hand. So, I need to meet with the person to have full trust. So, I would say that yes, you need to meet at some point during your cooperation. Or you should have a longer session with some interactive social tools. Maybe you should show them around and your house, how you live. Maybe you should show your kids, or I mean to build deeper trust" – Informant 6

Table 4.1.4.5 Quotes from informants in the LC/MNC and SME. Source: developed by book' authors.

Secondary data:

Secondary data support this finding 18 because both sized companies use digital tools such as Microsoft Teams and other types of digital solutions for knowledge sharing with partners. As an illustration of the type of secondary data we analysed and its connection with the theory, we show table 4.1.4.6 that describes information obtained from the internal website of the case companies.

LC/MNC	**SME**
"Licenses for Microsoft apps such as Office 365 that includes MS Teams"	*"Licenses for Microsoft apps such as Office 365 that includes MS Teams"*
Source: The webpage of the LC/MNC.	Source: The external webpage of the SME company.

Table 4.1.4.6 Secondary data from the LC/MNC and SME. Source: developed by book' authors.

In summary, digital communication technologies such as MS Teams significantly facilitate KT between partners.

4.1.5 Primary and secondary data about Aspect: Disseminative capacity
According to the literature (Minbaeva et al., 2018; Easterby-Smith et al., 2008) and theoretical framework of this study, disseminative capacity is measured in terms of (1) motivation and (2)

ability to teach by the source of knowledge. More detailed information is provided in Chapter 3 of this study in section 2.22.3.

Sub-aspect: (1) Motivation to teach

We have found in this study based on collectively agreed informants that:

Finding 19: Mutual experience sharing, and learning are motivation drivers for KT between partners.

Finding 20: Mutual business performance enhancement motivates partners for KT activities.

Finding 21 SME as a source of knowledge is less motivated to exchange knowledge and it depends on size partner.

Finding 19: Mutual experience sharing, and learning are motivation drivers for KT between partners.

This was concluded based on the following information:

Primary data:

Informant (2 and 7) from the LC/MNC and Informant 5 from SME collectively agree that the motivation to educate their partners is mutual experience exchange and learning from each other that will benefit both organization itself and their partners. Moreover, Informant 7 from the LC/MNC also suggests that motivation to transfer knowledge is considered in terms of mutual experience sharing the process with partners rather than teaching which is regarded as the unidirectional way to share the knowledge. In addition, Informant 5 from SME noted that it is also essential to establish the learning culture values at the company that structured and it can drive the motivation of knowledge holder to share knowledge with its partners.

As an illustration of the type of quotes we analysed and their connection with the theory, we show table 4.1.5.1 that explains quotes from informants from LC/MNC and SME. The rest of the Informants including (1, 2, 3, and 9) from the LC/MNC and informants (4, 6, and 8) from SME did not provide opinions on this perspective.

LC/MNC	SME
"...What motivates me (to teach) is that we will go home, and we will both be able to do something better. Because when you teach someone or when you show somebody had to do something, you will do it better as well afterwards. Because you learn something when you teach... I'm going to learn something for sure. And they're going to learn and we're probably both going to elevate our level on that topic. So it's win-win" – Informant 2	*"...Now, but I think to grow together with your partners, it's a good thing. I mean, you should learn from each other and they tell their experience, we have experience from different people coming into the company... I think establishing culture of learning and experience is needed. In projects where you evaluate what went good, bad, how can we do differently next time And if you have a good structure of that and a good way of you think the experience for the next one, then you have*

"…from previous experience then so what motivates me to kind of teach or so, I mean, for me it's a way to make sure that we actually get what we ordered or what we wanted the other part to actually do for us. I have I think a few examples where we thought we ordered something; we get something else back. I think that's learning from that. Of course, the good motivation that to make it as clear as possible to explain what we need to be solved… I'm just bouncing a bit on the teach word because as a partner we are on equal levels, right? And teaching is I tell someone from an elevated position like you need to do that and this" – Informant 7	*a learning organization. I think the leadership team and the management need to work with this value at that learning from each other… For example, you should copy things you can learn from each other instead of just sitting there and just inventing everything" – Informant 5*

Table 4.1.5.1 Quotes from informants in the LC/MNC and SME. Source: developed by book' authors.

Secondary data

The secondary data for this finding 19 supports that mutual experience for knowledge sharing pushes the source LC/MNC to transfer knowledge to its external partners because of needs to avoid misunderstanding, eventually, it can lead to achieving better business performance due to creating value for customers. In contrast, SME provides the knowledge base to their customers to demonstrate the expertise and value of their innovative products to their partners via a knowledge-sharing base page at their website.

As an illustration of the type of secondary data we analysed and its connection with the theory, we show table 4.1.5.2 that describes information obtained from the internal website of the case companies.

LC/MNC	SME
"Growth through customer relevance" is one of the four strategic objectives defined by the company. That means Voice of the Customer (VOC) constitutes one of the foundations for the successful development of the Group as a global leader. *VOC is an in-depth process of capturing customers' expectations, preferences and aversions. In-depth means understanding why a customer has a specific preference, not just determining or knowing about it. VOC researchers need to understand drivers, barriers and pain points, keep an open mind and replace assumptions with understanding.*	*"Senior Housing Experts in Electronic Locking Solutions… Our locks are installed to over 100 000 senior's doors globally. Additionally, our innovative technology evolves with the organization's needs, increasing efficiencies, enhancing security, and providing seamless maintenance"* Source: Knowledge base section from the company website (SMEs).

| Source: Corporate library portal/ The internal website of the LC/MNC, "Avenue", posting date 17/01/2020. | |

Table 4.1.5.2 Secondary data from the LC/MNC and SME. Source: developed by book' authors.

Overall, knowledge source as LC/MNC is interested to transfer knowledge to external partners to avoid misunderstandings that could lead to higher firm performance due to lowering cost, while SME motivated to show its expertise and holding knowledge value to attract new customers.

Finding 20: Mutual business performance enhancement motivates partners for KT activities.

This was concluded based on the following information:

Primary data:

Informants (3 and 9) from the LC/MNC and Informants (4 and 6) from SME collectively support that motivation to teach is driven by improving the business performance of their organization as well as partners. Informant 3 suggests that sharing knowledge with partners will enhance the partner's (supplier) competence that is needed for mutual business performance. Informant 9 provided the viewpoint that the main motivation for KT is a business purpose that eventually can be turned in a form of product to the partnering organization (e.g., customer). Implicitly, this might be interpreted that motivation for interfirm KT is to create business value through mutual sharing knowledge that can lead to mutual trust.

As an illustration of the type of quotes we analysed and their connection with the theory, we show table 4.1.5.3 that explains quotes from informants from LC/MNC and SME. The rest of the informants including (1, 2, and 7) from the LC/MNC and informants (5 and 8) from SME did not provide opinions on this perspective.

LC/MNC	SME
"...If you're looking into a business partner, then you're thinking about a long-term relationship that will benefit both. So, achieving that in addition, absolutely to see some results out of what you are teaching away. And see that bringing them either successful prosperity absolutely motivates me" *"...if suppliers are not performing and possess higher risks, you want to go out there, build again the relationship and you start to teach them what we can teach them to bring them up to a level where we want them" – Informant 3*	*"...It would be to help the situation or improve our relationship or make them grow as our, if it is a supplier, and by that way, also improve the incoming quality or product deliveries" – Informant 4*

"...So, there are different type of partners and there's different motivations for me why I partner up with them. But it's always a cause, a business reason or technology reason. We are very seldom to select the path and just to learn something, we always try to learn doing actual projects or products or components even. So, that is actually our main motivation is always behind some kind of a business case of business relations that we want to turn into a product" – Informant 9	*"...My answer to that you could be more efficient and make more money by learning people within my business, things that I do, or we do in a better way* *... And I will be able to recruit more talented people if we work efficiently and make things at a lower cost or better velocity" – Informant 6*

Table 4.1.5.3 Quotes from informants in the LC/MNC and SME. Source: developed by book' authors.

There is no secondary data found that can support finding 20.

Finding 21: SME as a source of knowledge is less motivated to exchange knowledge and it depends on size partner.

This was concluded based on the following information:

Primary data:

Informant 8 from SME mentions that it is more difficult to share knowledge with partners that are larger-sized companies due to their rigid processes. Implicitly, this means that SMEs are less motivated to exchange knowledge with recipients as LC/MNC due to their inflexibility. This viewpoint is supported indirectly by Informant 9 from LC/MNC who mentioned that the LC/MNC is very selective in the learning process and mainly they want to acquire knowledge due to concrete business purposes.

As an illustration of the type of quotes we analysed and their connection with the theory, we show table 4.1.5.4 that explains quotes from informants from LC/MNC and SME. The rest of the informants including (1, 2, 3, and 7) from the LC/MNC and informants (4, 5, and 6) from SME did not provide opinions on this perspective.

LC/MNC	**SME**
"...So, there are different type of partners and there's different motivations for me why I partner up with them. But it's always a cause, a business reason or technology reason. We are very seldom to select the path and just to learn something, we always try to learn doing actual projects or products or components even. So, that is actually our main motivation is always behind some kind of a business case of business relations that	*"...It's mainly for the reason in improving our delivery from a qualitative perspective or making it more efficient so that it becomes more efficient in handling on our side* *... it's harder to work with larger companies due to the size of the company, their rigid processes make it difficult to teach new things because there are more set in their own ways" – Informant 8*

we want to turn into a product" – Informant 9	

Table 4.1.5.4 Quotes from informants in the LC/MNC and SME. Source: developed by book' authors.

There is no secondary data found that can support finding 21.

Sub-aspect: (2) Ability to teach

We have found in this study based on collectively agreed informants that:

Finding 22: The ability to transfer knowledge depends on the knowledge recipient's competence and maturity level.

Finding 23: The ability to teach depends on an assessment of risks related to knowledge leakage and protecting valuable knowledge.

Finding 24: The ability to codify the knowledge facilitates knowledge transfer between partners.

Finding 22: The ability to transfer knowledge to partners depends on the knowledge recipient's competence and maturity level.

This was concluded based on the following information:

Primary data:

Informants (1, 3, 7 and 9) from the LC/MNC and Informants (5, 6 and 8) from SME are collectively supported that it is important to assess the expertise of the recipient that measured by organizational structure and processes of the partner that help to the knowledge source to understand the needs and maturity level of the recipient, eventually, this can facilitate KT activities between different partners. Furthermore, Informant 6 from SME mentioned that: (1) time is the essential resource needed for KT and (2) coaching approach is a more important ability for KT rather teaching, which is considered as a bilateral way of knowledge sharing.

As an illustration of the type of quotes we analysed and their connection with the theory, we show table 4.1.5.5 that explains quotes from informants from LC/MNC and SME. The rest of the informants including (2) from the LC/MNC and informant (4) from SME did not provide opinions on this perspective.

LC/MNC	SME
"...First of all, you would need to know the current state of the company. If you are to fit that information and approach it and a way to teach and show. You need to know their	*"...There are two sides of that coins. I think one side is that yes, you need to get an understanding of what they what they do to give the correct advice. But on the other side*

| *ability to understand and the maturity level"*
– Informant 3 | *of that is that if you put the time and efforts into working with the people, working close with them that do the work. You will be able to coach them and learn and work with them so that it's more fruitful. So, I think time is it's an essential part because change needs time. And you will need to have to that very important dialogue, not just the PowerPoint slide to fulfill that change need" – Informant 6* |

Table 4.1.5.5 Quotes from informants in the LC/MNC and SME. Source: developed by book' authors.

There is no secondary data found that can support finding 22.

Finding 23: The ability to teach depends on an assessment of risks related to knowledge leakage and protecting valuable knowledge.

This was concluded based on the following information:

Primary data.

Informant 2 from the LC/MNC supports that the knowledge recipients should possess enough competence and valuable knowledge to share knowledge with other organizations. But informant did not specify what type of knowledge should be owned by the knowledge source.

However, Informant 9 specifies that the ability to teach is based on an assessment of risks related to knowledge leakage and protecting holding knowledge value by signing non-disclosure agreements. This perspective is also supported by Informant 4 from SME by suggesting that formal non-disclosure agreements (NDAs) are needed to transfer knowledge. Implicitly, this implies that the knowledge source due to possessing valuable knowledge wants to protect it by signing NDAs with the external partner.

As an illustration of the type of quotes we analysed and their connection with the theory, we show table 4.1.5.6 that explains quotes from informants from LC/MNC and SME. The rest of the informants including (1, 3, and 7) from the LC/MNC and informants (5, 6, and 8) from SME did not provide opinions on this perspective.

LC/MNC	SME
"...I think you just need to know a little bit more than the other person or you just need to know. You just need to have done it. And then you can show them what you know. And then, then that's a teaching moment" – Informant 2	*"... I think I will focus on what we are cooperating around. And I'm also thinking about NDA, like confidentiality and so on. So, if I should share something that is not known by everyone, I would like to do now that we have a secrecy agreement in place" – Informant 4*
"...I really don't share much unless we really have the basics in place as an NDA, which covers the area. And sometimes also a	

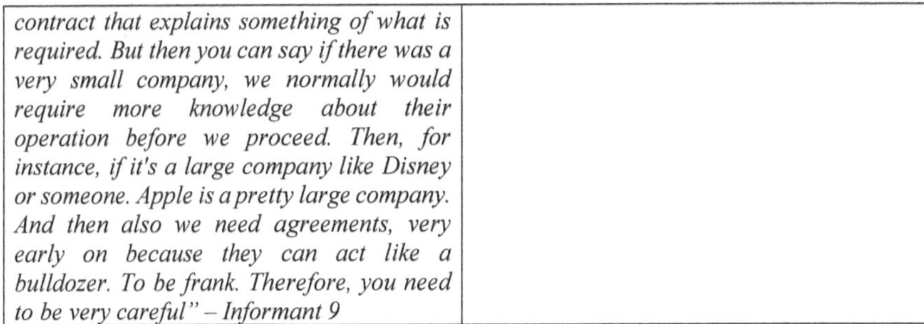

contract that explains something of what is required. But then you can say if there was a very small company, we normally would require more knowledge about their operation before we proceed. Then, for instance, if it's a large company like Disney or someone. Apple is a pretty large company. And then also we need agreements, very early on because they can act like a bulldozer. To be frank. Therefore, you need to be very careful" – Informant 9	

Table 4.1.5.6 Quotes from informants in the LC/MNC and SME. Source: developed by book' authors.

There is no secondary data found that can support the finding 23.

Finding 24: The ability to codify the knowledge facilitates knowledge transfer between partners.

This was concluded based on the following information:

Primary data:

Informant 1 suggests that the ability to codify the knowledge by the source helps to avoid misunderstanding and it is driven by the motivation to transfer knowledge. This might imply that ability to codify the knowledge facilitates KT due to the need for avoiding misunderstanding between partners. Furthermore, Informant 7 suggests that larger organizations have repeatable processes, which implies more formal procedures are in place that are related to the ability to codify the knowledge transferred, while SMEs possess adaptability and flexibility that might imply that they more rely on informal KT activities. This perspective is also indirectly supported by Informant 4 who suggests that KT exchange with SMEs is more difficult since "they are not so aware of the actual risk of sharing information" in contrast, KT with LC/MNC is easier because they know what they can share. This implies that formalized procedures of larger organizations facilitate KT that is related to the ability to codify the knowledge, while an informal approach that can imply less ability to the codification of SMEs impedes KT. In addition, Informant 6 from SME suggests that for better knowledge transfer, the coaching approach is applied by means of frequent interaction rather than an ability to codify knowledge. As an illustration of the type of quotes we analysed and their connection with the theory, we show table 4.1.5.7 that explains quotes from informants from LC/MNC and SME. The rest of the informants including (2, 3, and 9) from the LC/MNC and informants (5 and 8) from SME did not provide opinions on this perspective.

LC/MNC	SME
"…So, motivation for me to educate if it is a business partner or if it is like an internal stakeholder in some way, it would be for them to know (more clear understanding of the	*"…The large company knows what they can share and what not They used to have that kind of knowledge. And it can be harder to share information with smaller companies.*

product) the product that they're supposed to use" – Informant 1	*Also get information from smaller companies. That's my experience… I think it's harder because they (SMEs) often don't know the frames where they are able to share information and what would be effective without risk to share. So, it is simpler to work with larger companies in that way.* *… They (SMEs) are not so aware of the actual risk of sharing information" – Informant 4*
"…Needs to know "in structure and in my experience, larger companies, more structure, more process. And maybe to do things repeatable. Smaller company, maybe more adaptable and kinder of listening into our processes and needs" – Informant 7	*"…There are two sides of that coins. I think one side is that yes, you need to get an understanding of what they what they do to give the correct advice. But on the other side of that is that if you put the time and efforts into actually working with the people, working close with them that actually do the work. You will be able to coach them and learn and work with them so that it's more fruitful. So, I think time is it's an essential part because change needs time. And you will need to have to that very important dialogue, not just the PowerPoint slide to fulfill that change need" – Informant 6*

Table 4.1.5.7 Quotes from informants in the LC/MNC and SME. Source: developed by book' authors.

There is no secondary data found that can support the finding 24.

4.1.6 Primary and secondary data about Aspect: Social ties
Social ties according to Tangaraja et al. (2016) is a relationship based concept, its either conducted directly or indirectly between two parties and differentiated by (1) informal or formal ways. Additionally, Liu et al. (2015) argues that there are three steps to establishing a high quality relationship, (2) first step involves to locate a partner firm and identify its key contacts, secondly to develop and (3) maintain the relationship established last stage involves dispatching of knowledge from the source along with knowledge absorption of the recipient, which is how the knowledge is exchanged. The following section describes findings if the organization prefer informal or formal relations and how they go about establishing and maintain their relationships with other organizations.

Sub-aspect: (1) Formal or informal social ties

This study identified that informants collectively agreed upon:

Findings 25: LC/MNCs prefer both informal and formal knowledge transfers in social ties with partners but leaning more towards formal social ties, while SMEs prefer informal knowledge transfer in social ties with partners.

This was concluded based on the following information:

Primary data:

Informants (1, 2 and 3) from the LC/MNC collectively agree that they prefer both formal and informal knowledge transfer in social ties with partners, while Informant 9 from the LC/MNC leans more towards formal. Informant (4, 5, 6 and 8) from SMEs collectively agree that they prefer informal social ties with partners.

To illustrate the analysis of this finding, table 4.1.6.1 shows the quote of Informant (1, 2, 3 and 9) from the LC/MNC and Informant (4, 5, 6 and 8) from the SME. While the remaining Informant 7 from the LC/MNC did not mention formal or informal social ties with SMEs.

LC/MNC	SME
"Yeah. I mean, I think it's I think the difference is when it comes to like, for instance, new projects within the product. The ones that were involved in right now are quite big (the projects). And those are new sets of features for instance in both projects, which means that we don't have the what the customer (recipient) wants in the product as it is right now. Both of them are quite large. But I don't think it would matter in that case if the size of the partners actually, because it would be new things that have to be developed. Suddenly, I think that the difference is more when it comes to like, customers tend to buy the thing of the rack and are very satisfied with everything that you have in the product. And compare that to, yeah, we want your product, but we will all these kinds of things. You need to develop them for us to buy a product." – Informant 1	*"I think email is effective. Also, video meetings. Sometimes a real meeting or meeting in real life is more effective. It depends on if you know each other (sender and recipient). It can be, you can just send a question, I get an answer. If you don't know each other or if you discuss hard problems and so on. It can be more effective to meet, to have a real meeting. But it's also time-consuming. If I should meet you, I have to travel and so on." – Informant 4*
"Yeah, So, well, I think effective communication. The definition of effective communication is if at the end you get the result you were looking for. So, so that's, that's my answer. What do you think is effective communication? Effective communication regardless whether it's with smaller, medium size is getting the result you were looking for, not the answer, but the result. So, if you were looking to solve a problem or find people who can help you, things like that. I think that's effective communication. If you wanted to get a message through, if you were looking to	*"Schedule meetings. Same way I did my previous company. In the previous company I somehow know that these people that i needed to know could be good to know because they can lead me or helping to support me to get in contact with other people. So I try to establish a good network." – Informant 5* (Would you say your communication is more formal or informal when you tried to maintain those relationships?)

yeah, all these kinds of things. So effective communication is, is that." – Informant 2	*"I think informal. Yeah. Depends on who it is." – Informant 5*
"I mean, the last year has been teams. That's the only thing that we've had. And I will absolutely think that going forward, this the COVID year will absolutely change the future as well. Historically has been a lot of traveling, a lot of face-to-face. And we see now that there is not maybe the most effective way spending internal resources, but same time face-to-face is absolutely important and essential to maintain good relations with external companies especially. And also, I would mention is in regard to communication, what I often say, what happens in a meeting, it's very often that just to formalize what you often decide outside of that meeting. So that the formal meeting is, of course you can't avoid those, but also interacting in other forms is very important as well. If applicable that again, this then comes which part in the organizational we are talking about internally. Like, the interaction between R and D over a cup of coffee. During workshops, launch. Very important as well." – Informant 3 *"This is absolutely, I mean, this is also has to do with compliance, for instance. So I mean, if you have 200 plus suppliers and you have a small organization that cannot hand hold each supplier. You can't. Resources go out to visit them to do regular audits, you need to do a risk assessment. And then you do a risk assessment based on the knowledge that you already have. So this is, again, the partnership here is important. If we look into our company historically. If a supplier did not deliver or the cost was high, you would just swap the supplier. And when you do that too much, you lose that partnership with the supplier. So what we do now, we tried to look ahead. The lowest cost should not be the main driver because you want your partner to earn money as well. It's a partnership. And when you establish that, that also the trust comes. And then you can trust your partners more. And then we have actually such a good relationship with some of them when we have*	*"I think that this one is kind of tricky because with bigger organizations you work with KPIs (Performance indicators) (formal, standardized). That's an established way of getting an illusion of control. Illusional communication on how things go within different parts of the organizations. It's also not, from my perspective, not very strong. I think you should work lots more with storytelling (informal communication). So, if you have something that's smaller or bigger or whatever size company, if you, if you get a group or individuals to, to story tell about their achievements. And from this you gain commitment by other part, by other parts of the company and say, I want to do this as well. And then you get this interaction between these two individuals. So, these two teams, then you get the full-fledged effects. What you, what you need to do potentially." – Informant 6*

issues, we ask them to come back with a solution for us to support our R and D department in regard to design. Even though we designed it, they only the manufacturer. But based on their experience in other companies, they have a different type of experience. And then you have that type of relationship. And then you have suppliers that you consider as high risk. And those that is that those that we want to develop. So that's we had a team that works with their supplier development. We want to develop policy. So if they are not performing, you want to go out there, build again the relationship and you start to teach them what we can teach them to bring them up to a level where we want them. So the trust that's I think that's lacking personal life as well. There's something built over time." – Informant 3	
"… Start with a very informal discussion of what we need (the source), informal is good to get the impression (of the recipient), ends up with a very formal frame agreements and structure the way how to do that. And also of course we can assess their (recipients) capabilities or competencies and so on. But that's done more the informal way than actually formal ways like NDAs, contract and frame agreements." – Informant 9	*"I would say that it's probably easier for a larger company (source) going to a small company (recipient), than for a small company (source) going to the larger company (recipient) because it's from my perspective, it's usually harder being the person who runs different areas towards different persons, all in one person rather than on the other side when you only have one person to go to." – Informant 8* (Would you say that when you communicate with a LC/MNC is informal communication or formal communication more efficient?) *"I think I would say that in formal is more efficient. But to me, that might be more difficult to reach with a large company as well. Because when you then talk to, if you are, for example, in a small organization (source) and talk to someone in the larger organization (recipient), that person might not have the full view of the relationship and they might not have the full knowledge, on the complete setup. So, they might be more hesitant in maybe in taking action or promising things or changing things within the relation without coordinating it with the rest of the team evolved." – Informant 8*

Table 4.1.6.1 Quotes from informants in the LC/MNC and SME. Source: developed by book' authors.

There is no secondary data found that can support finding 25.

In summary, primary data supports finding 25 and it reveals the spectrum of what kind of social tie LC/MNCs wants with their partners and what SME wants, as they want different types of relationships, it illustrates that when they engage with each other they act differently in terms of per example, the LC/MNC puts a performance indicator for the SME to follow, whereas if they fail to achieve it they risk replacement. This indicates that when there is a social ties per example between the LC/MNC and SME, it is driven by the LC, leaving the SME to adapt to the LC/MNC needs, which could be why they end up taking the partners way of having formal social ties.

Sub-aspect: (2) Establishing social ties

The informants in this study collectively agree that:

Finding 26: LC/MNCs utilize a role of socialization to establish new social ties when engaging in knowledge transfers with partners. While the SME attempts to establish social ties with the role of socialization when engaging in knowledge transfer with the partner.

Finding 27: Both SMEs and LC/MNCs rely on the potential partner to engage in a knowledge transfer to establish a social tie with them.

Finding 28: LC/MNCs rely on informal knowledge transfer in meetings, digital or physical to establish new social ties with partners, while SMEs seeks to establish social ties with knowledge transfer via phone calls, emails or meetings to with partners.

Finding 26: LC/MNCs utilize a role of socialization to establish new social ties when engaging in knowledge transfers with partners. While the SME attempts to establish social ties with the role of socialization when engaging in knowledge transfer with the partner.

This was concluded based on the following information:

Primary data:

Informants (1 and 7) from the LC/MNC and Informant 8 from the SME supports that to establish social ties with the potential partner organization, the LC/MNC utilize a role of socialization, whereas the SME is aware of the role of socialization of the partner and attempt to establish a social tie with this individual.

An illustration of the quotes that were analysed to draw this connection is provided in table 4.1.6.2 Informants (2, 3 and 9) from the LC/MNC and Informants (4, 5 and 6) from the SME did not mention anything regarding this finding.

LC/MNC	SME

LC/MNC	SME
"I would assume. This is actually, I mean, since I'm a little bit deeper down in the organization, I really don't know what the first contact is. I assume that someone sells something. We have a sales organization. So, we have some kind of key account managers. And then it kind of drops down to the product management (source of KT) for the product. And then he has contact with the customers (recipient of KT). I think. I really don't know what the answer is otherwise, but that's my assumption." – Informant 1	*"I think that it's important to first, figure out who's the key contact (of recipient firm). And I would guess you could say, even if not all use that role that you're kind of key account manager who can help you and navigate through the organization (recipient) that you want to start a relationship with or that you might already happened, right?" – Informant 8*
"Yeah, this is kind of outside of my role since I'm more embedded within the development department. It's unfortunately outside my role to interact with new partners or even existing ones... unfortunately a few layers of different roles, which is outside my role. I'm the Product Owner for a product and a team. And then I have like a technical contact person and a key account manager. And they in turn have people. Who then interact with the customers." – Informant 7	

Table 4.1.6.2 Quotes from informants in the LC/MNC and SME. Source: developed by book' authors.

Secondary data:

The secondary data for finding 26 show that the LC/MNC utilize a role of socialization, while not specifically directed towards SMEs, it is assumed that this individual may speak for the organization in general. There is no secondary data from the SME to confirm this finding.

As an illustration on type of secondary data we analysed and its connection with the theory, we show table 4.1.6.3 that describes information obtained from the internal website of the case company.

LC/MNC	SME
"The Disclosure Policy: Who may act as a spokesperson for the (LC/MNC)" Source: The internal website of the LC/MNC: "Avenue"	None.

Table 4.1.6.3 Secondary data from the LC/MNC and SME. Source: developed by book' authors.

To conclude finding 26, primary and secondary data shows that a role of socialization assists the LC/MNC in terms of finding a connection point to establish social ties. Whereas SMEs are aware of the role of socialization, there was no secondary data to confirm this, however primary data shows that it assists them if they are the ones attempting to establish a social tie with the their potential partners, as the informants were aware of the role of socialization. Thus, the role of socialization serves as a connection point and bridge whereas a social tie can be established for both organizations.

Finding 27: Both SMEs and LC/MNCs rely on the potential partner to engage in a knowledge transfer to establish a social tie with them.

This was concluded based on the following information:

Primary data:

Informants 9 from the LC/MNC and Informant 5 from the SME supports that to establish social ties they are often the ones being approached by the partner organization.

An illustration of the quotes that were analysed to draw this connection is provided in table 4.1.6.4 Informants (1, 2, 3 and 7) from the LC/MNC and Informants (4, 6 and 8) from the SME did not mention anything regarding this finding.

LC/MNC	SME
"Yeah. Very often the first I mean, very often. It is them contacting us. Since we have this major position in the market. The big name, most other players wants to play with us. " – Informant 9	*"A lot of partners who wants to work with us. They contact me. I meet with them and see something we can work together on, do we, have any opportunities in the future, then try to learn what are they doing and If it's something we can learn from each other." – Informant 5*

Table 4.1.6.4 Quotes from informants in the LC/MNC and SME. Source: developed by book' authors.

There is no secondary data found that can support finding 27.

In summary, primary data supports finding 27, as it shows that a portion of the organization relies on the potential partner organization to contact them, this scenario can provide a conundrum for both organizations to establish the most suitable relationship, in the selection process of it, as they are not the one approaching the other, whereas one can not conduct prior research about the organization to evaluate if the social tie is most suitable.

Finding 28: LC/MNCs rely on informal knowledge transfer in meetings, digital or physical to establish new social ties with partners, while SMEs seeks to establish social ties with knowledge transfer via phone calls, emails or meetings to with partners.

This was concluded based on the following information:

Primary data:

Informants 2 from the LC/MNC and Informant 4 from the SME supports that to establish social ties with the partner, the LC/MNC rely on informal council meetings while the SME attempts to establish social ties via phone calls, email and meetings.

An illustration of the quotes that were analysed to draw this connection is provided in table 4.1.6.5 Informants (1, 7 and 9) from the LC/MNC and Informants (5, 6 and 8) from the SME did not mention anything regarding this finding.

LC/MNC	SME
"When we reach out to others, there's no established way (to contact recipient's firm). There's no one way that we establish communication. So, you end up knowing or meeting people through these council meetings. And often this is how we reach out to people (contact recipient's firm). We get to know people from other partners (other recipients) and we reach out through those. So, it's quite informal." – Informant 2	*"I think it's a phone call or email that start the communication or a Teams meeting. First contact (with the recipient) probably via email first." – Informant 4*
"I mean, the last year has been Teams. That's the only thing that we've had. And I will absolutely think that going forward, this the COVID year will absolutely change the future as well. Historically has been a lot of traveling, a lot of face-to-face. And we see now that there is not maybe the most effective way spending internal resources." – Informant 3	

Table 4.1.6.5 Quotes from informants in the LC/MNC and SME. Source: developed by book' authors.

Secondary data:

The secondary data for finding 28 show that the LC/MNC follows a strict external disclosure policy, which could imply that they cannot establish social ties by informal means. But regarding how they interact, it could be informal while the policy of continuation of the process of establishing social tie could be formal. Thus, the secondary data from the LC/MNC confirms that they could in fact establish social ties by informal means. There is no secondary data regarding technological means to establish social ties, nor is there any secondary data from the SME to validate this finding.

As an illustration on the type of secondary data we analysed and its connection with the theory, we show table 4.1.6.6 that describes information obtained from the internal website of the case companies.

LC/MNC	SME
"Communication to media and other stakeholders shall always be planned and	None.

<table>
<tr><td>carried out in line with the Group's External Disclosure Policy and Crisis Management Policy"

Source: The internal website of the LC/MNC: "Avenue"</td><td></td></tr>
</table>

Table 4.1.6.6 Secondary data from the LC/MNC and SME. Source: developed by book' authors.

Overall, primary data shows that informal meeting areas are essential for both the LC/MNC and SME as it allows for the organizations to broaden their networks and establish new social ties that are beneficial for the organizations. Secondary data shows that while the LC/MNC can engage in informal or formal means, they must follow formalized means of what they are allowed to share about the organization. Additionally, the means to establish social ties seems to be similar in both organizations, thus, findings 28 shows that both organizations are aware of how to establish new social ties.

Sub-aspect: (2) Maintaining social ties

The informants in this study collectively agree that:

Finding 29: LC/MNCs utilize a role of socialization to handle their knowledge transfers regarding maintain social ties with partners, while the SME attempts to maintain social ties by knowledge transfers to the role of socialization in the partner organisation.

Finding 30: LC/MNCs utilize more formal knowledge transfer regarding maintenance of social ties with partners, while SMEs prefers informal knowledge transfer in maintenance of the social tie but often ends up using formal means.

Finding 29: LC/MNCs utilize a role of socialization to handle their knowledge transfers regarding maintain social ties with partners, while the SME attempts to maintain social ties by knowledge transfers to the role of socialization in the partner organisation.

This was concluded based on the following information:

Primary data:

Informants (1 and 7) from the LC/MNC and Informant 8 from the SME supports that to maintain social ties with the partner organization, the LC/MNC utilize a role of socialization, whereas the SME is aware of the role of socialization in the partner organisation and attempt to maintain the social tie with this individual. As only one of the informants from the SME stated supports this finding, it shows that SMEs need to be more aware of the social capital of partners to maintain their social ties with them.

An illustration of the quotes that were analysed to draw this connection is provided in table 4.1.6.7 Informants (2, 3 and 9) from the LC/MNC and Informants (4, 5 and 6) from the SME did not mention anything regarding this finding.

LC/MNC	SME
"I would assume. This is actually, I mean, since I'm a little bit deeper down in the organization, I really don't know what the first contact is. I assume that someone sells something. We have a sales organization. So, we have some kind of key account managers. And then its kind of drops down to the product management (source of KT) for the product. And then he has contact with the customers (recipient of KT). I think. I really don't know what the answer is otherwise, but that's my assumption." – Informant 1	*"I think that it's important to first, figure out who's the key contact (of recipient firm). And I would guess you could say, even if not all use that role that you're kind of key account manager who can help you and navigate through the organization (recipient) that you want to start a relationship with or that you might already happened, right?" – Informant 8*
"Yeah, this is kind of outside of my role since I'm more embedded within the development department. It's unfortunately outside my role to interact with new partners or even existing ones... unfortunately a few layers of different roles, which is outside my role. I'm the Product Owner for a product and a team. And then I have like a technical contact person and a key account manager. And they in turn have people. Who then interact with the customers." – Informant 7	

Table 4.1.6.7 Quotes from informants in the LC/MNC and SME. Source: developed by book' authors.

There is no secondary data found that can support this finding 29.

Overall, finding 29 is supported by primary data and it shows that both the SME and LC/MNC are aware of the connection point of the social tie which they need to maintain it with. This implies that to maintain social ties, both organizations are aware of who to maintain it with inside of the partner organization.

Finding 30: LC/MNCs utilize more formal knowledge transfer regarding maintenance of social ties with partners, while SMEs prefers informal knowledge transfer in maintenance of the social tie but often ends up using formal means.

This was concluded based on the following information:

Primary data:

Informants (3 and 9) from the LC/MNC and Informant (4, 5, 6 and 8) from the SME supports that to maintain social ties with the partner organization, the LC/MNC utilize more formal means to maintain their social ties with partners, while the SME prefers informal maintenance they often end up using formal means to maintain the social ties with the partner organization. As Informant 6 state, that often the LC/MNC puts performance indicators for the partners to follow (to maintain the social tie) which according to the informant is not the optimal way to maintain social ties with per example an SME. However, Informant 8 state that the more frequent transactions the SME have with the partner, they need to maintain the social tie with a more formal governance structure regarding the cooperation. This indicates that SMEs need to be more adaptable to suit the maintenance structure of how the partner organization operates to have a strong social tie with them.

An illustration of the quotes that were analysed to draw this connection is provided in table 4.1.6.8 Informants (1, 2, and 7) from the LC/MNC and Informants 8 from the SME did not mention anything regarding this finding.

LC/MNC	SME
"…And also, I would mention is in regard to communication, what I often say, what happens in a meeting, it's very often that just to formalize what you often decide outside of that meeting. So that the formal meeting is, of course you can't avoid those, but also interacting in other forms is very important as well." – Informant 3	*"Yeah. In my situation, I think we often have meetings, minutes of meeting, and I've written actions and so on. Not so informal, more like documented meetings or minutes of meetings." – Informant 4*
"Regular basis, regular talks to follow up the contractors that we have or follow-up agreements (with the recipient). Agreements are signed for one year at a time (with the recipient). Frame Agreements are assigned for one year at a time. Assignments can be shorter. And every frame agreement and every assignment we meet up. So that's basically it, whenever it's needed." – Informant 9	*"Schedule meetings. Same way I did my previous company. In the previous company I somehow know that these people that I needed to know could be good to know because they can lead me or help to support me in getting contact with other people. So, I try to establish a good network." – Informant 5* (Would you say your communication is more formal or informal when you tried to maintain those relationships?) *"I think informal. Yeah. Depends on who it is." – Informant 5*
	"I think that this one is kind of tricky because in bigger organizations you work with KPIs (Performance indicator). That's an established way of getting an illusion of control. Illusional communication on how things go within different parts of the organizations. It's also not, from my perspective, not very strong. I think you should work lots more with storytelling. So, if you have something that's smaller or bigger or whatever size company, if you get a group or individuals to story tell about their achievements (to the recipient). And from this you gain commitment by other part (the

| | *recipient), by other parts of the company and say, I want to do this as well. And then you get this interaction between these two individuals. So, these two teams, then you get the full-fledged effects. What you, what you need to do potentially." – Informant 6* |
| | *"It depends on the frequency of the transactions with that company. Because if it's a company that you have many and even maybe even daily transactions with then I usually like to set up a more formal governance structure on how you cooperate and how you exchange information and knowledge. And the other one it's more of an ad hoc and on a need basis." – Informant 8* |

Table 4.1.6.8 Quotes from informants in the LC/MNC and SME. Source: developed by book' authors.

There is no secondary data found that can support this finding 30.

In summary for findings 30, primary data supports that the LC/MNC use formal means of maintenance of their social ties with partners, while SMEs want more informal means of maintenance, they end up following the partners formal means. This implies that the LC/MNC are the primary driver of the social ties and that the partner must adapt to the needs of the LC/MNC to maintain their social tie with them.

4.1.7 Primary and secondary data about Aspect: Motivation
According to Martín Cruz et al. (2009) motivation consists of intrinsic and extrinsic motives, where the former takes place inside an individual, affected by per example pleasant work environment, atmosphere of mutual respect, accomplishment, self-respect, prestige, personal values and involvement. As for the latter, extrinsic motivation according to the authors is external motives which affects the individual, by per example monetary rewards that comes in direct or indirect ways, whereas the former relates to wages, incentives or bonuses and the latter relates to time not working, job education, health benefits or allowances.

As such, the findings for motivation is (1) regarding whether intrinsic or extrinsic motivation facilitates knowledge transfer more for the SME or LC/MNC and (2) which type of incentives are used by the organization. Since organizational motivation relates to the firm's motivation of acquiring knowledge and how to obtain it, the organization must invest time and effort to the process of knowledge transfer (Bengoa & Kaufmann, 2014).

Sub-aspects: (1) Intrinsic or extrinsic motivation to facilitate knowledge transfer

We have found in this study based on collectively agreed informants that:

Finding 31: SMEs and LC/MNCs are more incentivises by intrinsic motivation in the form of pleasant work environment and accomplishment to acquire new knowledge from partners and sharing knowledge with them.

Finding 32: SMEs and LC/MNCs are not specifically motivated by incentives to acquire and share knowledge with partners, instead it is a part of the job description of the employees.

Finding 33: SMEs need extrinsic motivation in the form of job education (e.g. training or courses) which give them the understanding (e.g. framework) of how and why they should engage in acquisition and sharing of knowledge, while feel autonomous in KT activities from the partner.

Finding 34: LC/MNCs apply intrinsic or/and extrinsic motivation on partners to engage with acquisition and sharing of knowledge to fulfill the LC/MNC criterions of collaboration.

Finding 31: SMEs and LC/MNCs are more incentivises by intrinsic motivation in the form of pleasant work environment and accomplishment to acquire new knowledge from partners and sharing knowledge with them.

This was concluded based on the following information:

Primary data:

Informants (1 and 2) from the LC/MNC and Informant (5 and 6) agree that to acquire knowledge from partners and to share knowledge with them they are incentives by intrinsic motivation in the form of a pleasant work environment and a sense of accomplishment.

As an illustration on the type of quotes we analysed and their connection with the theory, we show table 4.1.7.1 that explains quotes from informants from the LC/MNC and SME. Informant (3, 7 and 9) from the LC/MNC and Informants (4 and 8) from the SME did not provide any opinion regarding this finding.

LC/MNC	SME
"Well, I don't really know, actually, to be honest with you. I think the employees know that the more information the partners have regarding our product, the less they come back to your R&D to ask questions." – Informant 1	*"With customers the more they know the less support they need so that results in less support issues since they know how to handle it." – Informant 5*
"So, we meet our quality standards, so with ISO or whatever and internally we do continual improvement." – Informant 2	*"I think I said that in the smaller organizations you need to motivate people. And maybe I didn't say this by them having a feeling that they, they provide value into the business. And I think more so in small than in bigger (organizations). But in smaller, the acknowledgement is the personal acknowledgment is very key. Because you are like I mentioned, the tribes (small*

	community). Like a tribe where the social interactions are very important." – Informant 6

Table 4.1.7.1 Quotes from informants in the LC/MNC and SME. Source: developed by book' authors.

Secondary data:

The secondary data confirms finding 31, both from the LC/MNC and SME, as it shows that they offer intrinsic motivation in the form of accomplishment, while more indirectly from the SME, it still goes within the boundaries of intrinsic motivation.

As an illustration on the type of secondary data we analysed and its connection with the theory, we show table 4.1.7.2 that describes information obtained from the internal website of the case companies.

LC/MNC	**SME**
"For Mike Ong, Opening Solutions Country Director for the Philippines, growth is a crucial factor in achieving your goals, and motivation is the key to that growth. "You must keep motivating, despite the challenges," he says." *"Last year was a year of limitations. Interactions with customers and each other, business conversations, and even fulfilling our overall objectives became more difficult than ever. For Mike's team, employee morale was low, and anxiety was high. Covid cases on the team meant providing strong emotional support to the sick and the healthy. These limitations had the potential to devastate Mike's team. "The Philippines has always been a growth engine for the SEA region, but 2020 was a bad year." Instead, Mike motivated his team through those challenges and is now looking forward to driving initiatives put in place to not only bounce forward, but to grow. "It's about the team working together. If each member grows, together we grow."* Source: The internal website of the LC/MNC: "Avenue"	*(Translated from Swedish) "The 1st June 2020, Swedens municipalities and regions organization (SKR) appointed ten municipalities that shall act as models for digitalization of elderly care. These role models will together with SKR support other municipalities with knowledge regarding digital services and welfare technology. Seven of these municipalities are customers to us (the SME) and we are really proud over this. Please, take inspiration and tips from these municipalities about how your municipality can strengthen the digitalization of elderly care. If you wish to contact any of these municipalities or if you have any questions, feel free to contact us (the SME) and we'll help you."* Source: The external website of the SME.

Table 4.1.7.2 Secondary data from the LC/MNC and SME. Source: developed by book' authors.

Overall, both organizations state the same regarding intrinsic motivation of a sense of accomplishment, this shows the importance of acknowledging employees for their accomplishments as it motivates them further to acquire and share knowledge with partners, which is based on both primary and secondary data. Regarding pleasant work environment, primary data illustrates the dilemma of sharing knowledge with partners, as both organizations state that the more the other party knows, the less they will ask, which formulates a pleasant work environment for them, this implies the importance of successfully transferring knowledge to one's partner, as if done correctly, it will lead to less time occupation on emphasizing details about said shared knowledge.

Finding 32: SMEs and LC/MNCs are not specifically motivated by incentives to acquire and share knowledge with partners, instead it is a part of the job description of the employees.

This was concluded based on the following information:

Primary data:

Informant (1, 2 and 9) from the LC/MNC and Informant 4 from the SME mention that to be incentives to engage in knowledge acquisition and sharing with their partners they are not incentivised by intrinsic or extrinsic motivation, instead it comes from the job description that is assigned to someone.

As an illustration on the type of quotes we analysed and their connection with the theory, we show table 4.1.7.3 that explains quotes from informants from the LC/MNC and SME. The remaining, Informants (3 and 7) from the LC/MNC and Informants (5, 6 and 8) from SME did not provide any viewpoint on this perspective.

LC/MNC	SME
"Yes... but I think I mean, it's like that's part of the job, to spread the knowledge about the system to stakeholder." – Informant 1	*"I don't think that's possible. Because you have your workday which is already full. So, I don't think it's possible. I think you need to make it available and discuss how to do it. I think so." – Informant 4*
"People have different kind of jobs. So, for example, if your job is coming to work and doing your work, you're not going to come to work and try to think, who can I reach out to today, right? So, whereas I come to work, and I have things to do, but then I'm trying to get people together, so that's my job. So, if your job is not trying to get together with people to do knowledge exchange?" – Informant 2	(Would you say different departments might be more incentivized to acquire knowledge from other company?) *"Yeah, definitely the sales department are already used to cooperate with external companies and external persons." – Informant 4*
"No, we have not incentives like this to learn knowledge from other companies. We have other types of incentives, know when it comes to patents and these kinds of things and IP	(Would you say that, for example, the sales personnel or more incentivize then to acquire knowledge for maybe personal gains or are there other motives?)

there are a lot of incentives, but not, not to learn new knowledge from other companies." – Informant 9	*"I think it's in their position (job role) that they have the availability to learn. I think it's my experience from my former company and also these companies that it would be very good if you had the developers to cooperate more with other companies. But it's, often hard to make that happen, but it would be really good. And also, product managers, product departments, to open up and meet more external organizations." – Informant 4*

Table 4.1.7.3 Quotes from informants in the LC/MNC and SME. Source: developed by book' authors.

There is no secondary data found that can support this finding 32.

Overall, the finding based on primary data shows the importance of the job descriptions for the employees, as in certain cases, they are not motivated by incentives to acquire and share knowledge with their partners, instead it is a part of the job description. This implies that if the individuals are expected to engage in these activities, they will not be needing specific incentives to do so, rather they will be expected to do so.

Finding 33: SMEs need extrinsic motivation in the form of job education (e.g. training or courses) which give them the understanding (e.g. framework) of how and why they should engage in acquisition and sharing of knowledge, while feel autonomous in KT activities from the partner.

This was concluded based on the following information:

Primary data:

This finding was based on primary data from the SME and none from the LC/MNC. Informant 4 from the SME mention that they need extrinsic motivation in the form of job education, which allows them to be aware of the framework of which they can acquire and share knowledge with their partners. Which relates to Informant 8's discussion about how when people feel involved (intrinsic motivation) they are more obliged to acquire and share knowledge, this relates to Informant 6's statement about how the SME also need to feel autonomous from the partner, which indirectly is intrinsic motivation.

As an illustration on the type of quotes we analysed and their connection with the theory, we show table 4.1.7.4 that explains quotes from informants from the SME and none from the LC/MNC, probably because they already offer job education. The remaining Informant 5 from SME did not provide any viewpoint on this perspective.

LC/MNC	**SME**
None.	*"Yeah, that can be in education, how can we possible share information? What are our rules? So, if you are sure about the frame*

	where you can share information then it's easier to do it. So, I think that's education." – *Informant 4*
	(Do you encourage sharing knowledge with partners then?) *"Yes, but not directly. That's really up to the individual or team to decide to do so. I mean, that's has traditionally been a good thing, with (the* LC/MNC*) acquiring smaller companies. You do whatever business you do. We (the large organization) will only be here to support you (the small organization). I think that's very good. But some trends are showing now that the organization says, you should do like this. You should use this system to mention is this process. And that's not very good for the smaller companies because they must choose themself. What tools, what processes, what knowledge to acquire from the bigger part."* – *Informant 6* (And this affects the motivation to be engaged with the larger organization?) *"Yes, definitely. So, if you feel autonomous, it makes your own decision what parts to use, then the adoption is significantly increased."* – *Informant 6*
	"I would say maybe two things. It's usually very common for an employee to have a very internal perspective. And maybe not always to reflect on what the customer sees or wants or experiences. So I think that we could maybe try to establish a culture (corporate) where more people are involved in talking to customers, for example, on a more informal basis that doesn't need to be so formerly talking to customers and figuring out and what they want and what they expect. So that you can use that knowledge within the organization and to build the processes and operations to best support that in a way And when it comes to business partner in the other end with suppliers, I think that we could do a better job in having more employees involved in the actual relationship with the supplier in talking to them so that you don't Channel everything through a few persons. And

	because my experience is also that this makes the understanding in smaller (organizations) of how things work and why they work they do. Because it's I think these ties I spoke about earlier that I think that to make a successful cooperation, you need to be clear on what to expect and how things are working and how you want them to work." – Informant 8

Table 4.1.7.4 Quotes from informants in the LC/MNC and SME. Source: developed by book' authors.

There is no secondary data found that can support this finding 33.

Overall, by combining the statements from the primary data from the SME that were mentioned, the finding can be drawn, that when the SME is extrinsically motivated by job education, they become aware of the framework where they can acquire and share knowledge, while intrinsically motivated by involvement in the knowledge transfer, they become more obliged to engage in knowledge transfer, additionally, while feeling autonomous from the partner, they feel that acquiring and sharing knowledge with the partner will not influence the remaining organization negatively since it wont force unnecessary changes.

Finding 34: LC/MNCs apply intrinsic or/and extrinsic motivation on partners to engage with acquisition and sharing of knowledge to fulfill the LC/MNC criterions of collaboration.

This was concluded based on the following information:

Primary data:

This finding on primary data from the LC/MNC and none from the SME. Informant 3 from the LC/MNC mention that they categorizer their suppliers in terms of performance and maturity level, whereas the recipient is intrinsically or extrinsically motivated to be put in a better categorization if they fulfill different criterions for the collaboration.

As an illustration on the type of quotes we analysed and their connection with the theory, we show table 4.1.7.5 that explains quotes from informants from the LC/MNC and none from the SME, as this viewpoint was not mentioned in the perspective from the SME, which could be because of the viewpoint of the relationship between the organization where the SME seeks approval. The remaining Informants (1, 2, 7 and 9) from LC/MNC did not provide any viewpoint on this perspective.

LC/MNC	SME
"So, a few years ago, we initiated this type of assessment of the supplier and then put them in different categories based on their performance and maturity level. And that is one way that we have pushed our suppliers to improve because that is presented to them as well. And if they are a certain level actually,	None.

they then have so and so much time to improve. Otherwise they will be exchanged. And also, we always set targets based on the assessment of their performance, both in regard to the development itself, but also how quickly they fix issues that appear, because we measured them at that as well. Both in how many cases and how quickly we always put KPIs (Performance indicators) on our suppliers as well and push for improvement. And then this is where supply development team comes in. We go in and we work with them. But the idea is that you collaborate with your partner and you improve your apartment to achieve the goal that we set." – Informant 3	

Table 4.1.7.5 Quotes from informants in the LC/MNC and SME. Source: developed by book' authors.

There is no secondary data found that can support this finding 34.

Overall, this finding based on primary data illustrates that the LC/MNC applies extrinsic and/or intrinsic motivation on partners to acquire and share knowledge to further advance their collaboration with the LC/MNC for this case study. This indicates that if the partner has higher performance and reaches a better maturity level, both the LC/MNC and partner is motivated to engage in knowledge acquisition and sharing with each other.

Sub-aspect: (2) Is intrinsic or extrinsic motivation applied to the employees from the organization

We have found in this study based on collectively agreed informants that:

Finding 35: Both SMEs and LC/MNCs attempt to indirectly motivate their employees via the job description to acquire and share knowledge with partners.

Finding 36: LC/MNCs offer direct extrinsic motivation in the form of job education and indirect intrinsic motivation in the form of accomplishment, while SMEs offer direct and indirect intrinsic motivation in the form of accomplishment to acquire and share knowledge with partners.

We have found in this study that:

Finding 35: Both SMEs and LC/MNCs attempt to indirectly motivate their employees via the job description to acquire and share knowledge with partners.

This was concluded based on the following information:

Informant (1 and 3) from the LC/MNC and Informant (4 and 8) from the SME state that they are not specifically incentivised by the organization in terms of intrinsic or extrinsic motivation to acquire or share knowledge with their partners, rather, it is a part of the job description.

As an illustration on the type of quotes we analysed and their connection with the theory, we show table 4.1.7.6 that explains quotes from informants from LC/MNC and SME. The remaining Informants (2, 7 and 9) from the LC/MNC and Informants (5 and 6) from SME did not provide opinions on this perspective.

LC/MNC	SME
"I don't know if we don't have any special incentives for it, but it's like a part of our job. Meaning, my direct colleagues tried to do this well within the teams to share knowledge both within teams and for them to learn more (colleagues motivates the team to acquire knowledge and share it). And the manager is very prone to, I mean, he wants people to know more to learn. I think its kind of more to try to get the teams to understand the benefits of knowing more. Tried to motivate them that way. It also depends on how much they must do to be honest." – Informant 1	*"I don't think that's possible. Because you have your workday which is already full. So, I don't think it's possible. I think you need to make it available and discuss how to do it. I think so." – Informant 4* (Would you say different departments might be more incentivized to acquire knowledge and share with or from other companies?) *"Yeah, definitely the sales department are already used to cooperate with external companies and external persons." – Informant 4*
"No, there are no incentive directly. No, I mean we, the collaboration is something that is very essential, we always strive for. So, we have all these different councils and forums where we bring people together that have the same functions across either business areas within one division, but also across divisions. So, there's a lot of a lot of focus on that. And then we learn from each other's, you know, so if one division have done something that is great, then you just copy directly. So we don't need to do the same thing, global solutions. So we have that type of collaboration a lot. But there are no incentives money-wise or something. It's just the way we work." – Informant 3	*"I don't think that we have any incentives as such. But I think that we tried to encourage people to actually to reach out to other companies and ask for knowledge sharing experiences. But I don't think that we have in a specific incentive, monetary or other related to that." – Informant 8* (It is just a part of your job?) *"Yeah." – Informant 8*

Table 4.1.7.6 Quotes from informants in the LC/MNC and SME. Source: developed by book' authors.

There is no secondary data found that can support this finding 35.

In summary, this finding shows that while not specifically mentioned if there are intrinsic or extrinsic motivation within the job description of the informants, their job role does not include the specific activities of knowledge acquisition and sharing with partners. Which could be one of the reasonings behind this finding, however, if the informant are involved with these activities, they state that they are not intrinsically or extrinsically motivated by either to acquire or share, rather they are motivated by the responsibilities of the job role they are assigned by the organization.

Finding 36: LC/MNCs offer direct extrinsic motivation in the form of job education and indirect intrinsic motivation in the form of accomplishment, while SMEs offer direct and indirect intrinsic motivation in the form of accomplishment to acquire and share knowledge with partners.

This was concluded based on the following information:

Primary data:

Informant (2 and 7) from the LC/MNC state that the organization offers direct extrinsic motivation in the form of job education and Informant (5 and 6) from the SME state that their organization offer both direct and indirect intrinsic motivation in the form of accomplishment to acquire and share knowledge with their partners.

As an illustration on the type of quotes we analysed and their connection with the theory, we show table 4.1.7.7 that explains quotes from informants from LC/MNC and SME. The remaining Informants (1, 3 and 9) from the LC/MNC and Informants (4 and 8) from SME did not provide opinions on this perspective.

LC/MNC	SME
"So, we do have access to something called LinkedIn Learning. And that's in our HR system. And you can sign up for as many as may well (training programs), probably not as many because we're kind of busy with regular work, but we can sign up to trainings and stuff like that. But it's up from well, I guess it's from another company, it's from LinkedIn. And we do also have some trainings that are obligatory, like we have to do these trainings. And everybody must do those every so often every two years, every three years. We must renew these trainings." – *Informant 2* *"So, we meet our quality standards, so with ISO or whatever and internally we do continual improvement."* – *Informant 2*	*"We don't offer incentives."* – *Informant 5* (Say the employee shows tremendous personal growth and are constantly motivated to learn new knowledge inside a company or outside the company and share it as well. Would you say that the organization would favour that person for maybe a promotion or other things?) *"Yeah. I think so. If you show that you are willing to learn and put in extra effort."* – *Informant 5* (You would say its more indirect motivation then?) *"Yes."* – *Informant 5*
"I mean, not sure. We have like a budget each year which you could use to buy courses or attend conferences. But, if you don't want to take any courses, that's also	*"No. If incentives are monetary that's a no. Today we only offer soft incentives, meaning the acknowledgements... So, in a smaller organization, I would say it's much more key.*

fine. No real like money motivation for that... I want to sometime in the future I would like to become a manager. And I know that there are like many managerial courses. So, for my personal motivation is to then be ready to take the next step, I would like to take those courses." – Informant 7	*So, it's like, I don't know where it is, but it's a way for success, is related to have good leaders that are good in motivating people and making them achieve more." – Informant 6*

Table 4.1.7.7 Quotes from informants in the LC/MNC and SME. Source: developed by book' authors.

Secondary data:

The secondary data confirms finding 36, from the LC/MNC shows that they offer direct extrinsic motivation in the form of job education, as one example states they offer formal media training. Additionally, by publishing an article regarding how a team overcame said challenges, the organization is also directly offering indirect intrinsic motivation in the form of accomplishment. The SME's secondary data shows how the organization is offering indirect intrinsic motivation in the form of accomplishment, while there is no secondary data for direct intrinsic motivation from the SME.

As an illustration on the type of secondary data we analysed and its connection with the theory, we show table 4.1.7.8 that describes information obtained from the internal website of the case companies.

LC/MNC	SME
"For Mike Ong, Opening Solutions Country Director for the Philippines, growth is a crucial factor in achieving your goals, and motivation is the key to that growth. "You must keep motivating, despite the challenges," he says." ... "Last year was a year of limitations. Interactions with customers and each other, business conversations, and even fulfilling our overall objectives became more difficult than ever. For Mike's team, employee morale was low, and anxiety was high. Covid cases on the team meant providing strong emotional support to the sick and the healthy. These limitations had the potential to devastate Mike's team. "The Philippines has always been a growth engine for the SEA region, but 2020 was a bad year." Instead, Mike motivated his team through those challenges and is now looking forward to driving initiatives put in place to not only bounce forward, but to grow. "It's about the team working together. If each member grows, together we grow."	*(Translated from Swedish) "The 1st June 2020, Swedens municipalities and regions organization (SKR) appointed ten municipalities that shall act as models for digitalization of elderly care. These role models will together with SKR support other municipalities with knowledge regarding digital services and welfare technology. Seven of these municipalities are customers to us (the SME) and we are really proud over this. Please, take inspiration and tips from these municipalities about how your municipality can strengthen the digitalization of elderly care. If you wish to contact any of these municipalities or if you have any questions, feel free to contact us (the SME) and we'll help you."* Source: The external website of the SME.

Source: The internal website of the LC/MNC: "Avenue"	
"All spokespersons must participate in a formal media training"	
Source: The internal website of the LC/MNC: "Avenue"	

Table 4.1.7.8 Secondary data from the LC/MNC and SME. Source: developed by book' authors.

In summary, finding 36 is shown from both primary and secondary data confirms that both organizations apply indirect intrinsic motivation in the form of accomplishment to engage in acquisition and sharing of knowledge with their partners and that the LC/MNC offer direct extrinsic motivation in the form of job education. Regarding direct intrinsic motivation in the form of accomplishments, it was only based on primary data, however it could implicitly be validated by the secondary data as well. What this finding shows is how the organizations go about motivating their employees and it illustrates the importance of both intrinsic and extrinsic motivation for the employees to engage in knowledge acquisition and sharing with their partners.

4.1.8 Primary and secondary data about Aspect: Absorptive capacity
According to the literature and theoretical framework of this study, absorptive capacity affects knowledge transfer in terms of how organizations (1) implement acquired external knowledge and (2) diffusion of the acquired knowledge in the organization (De Luca & Cano Rubio, 2019; Tripsas, 1997; Cohen & Levinthal, 1990).

Sub-aspects: (1) Implementation of acquired external knowledge

We have found in this study based on collectively agreed informants that:

Finding 37: SMEs and LCs/MNCs rely on implementing acquired knowledge from individuals to the organizational level in systematic data bases.

Finding 38: When the SMEs acts as a recipient of knowledge they cannot always fully implement the acquired knowledge from the source partner due to organizational limitations.

Finding 39: SMEs don't implement acquired knowledge to the organizational level, but rather to the individual level.

Finding 37: SMEs and LCs/MNCs rely on implementing acquired knowledge from individuals to the organizational level in systematic data bases.

This was concluded based on the following information:

Informants (1, 2, 3, 7 and 9) from the LC/MNC and Informants (4 and 8) from the SME collectively support the implementation of acquired knowledge is done to the organizational level by putting it in various systematic data bases. While not as present in the SME, it still confirms the literature regarding absorptive capacity.

As an illustration on the type of quotes we analysed and their connection with the theory, we show table 4.1.8.1 that explains quotes from informants from LC/MNC and SME. Remaining Informants (5 and 6) from the SME did not provided opinions on this perspective.

LC/MNC	SME
"I mean, either it comes from tech service and then it goes through a system called service now. And then into us in our R&D through Jira that is basically the tool we use. And then it really depends on if it's like, if it's a bug at the customer side, then the developers should fix it as soon as possible. But if it's like a new request for something, a new feature or something, then it goes into the backlog in Jira... and if something must be like more documented for the future, we use confluence." – Informant 1	*"Yes. We have a really good process-oriented management system where we visualize our processes. And we can connect activities to instructions and checklists and so on. So that's a really good tool that we are about to implement at the moment." – Informant 4*
"Yes, but not in a standardized way, unfortunately. So, I think once it's sent out by email, some people will probably either save it in their email, save it in a box folder somewhere. I kind of like to put it in folders. If we do a Teams meeting, I like to save the file in the Teams meeting. And in the global quality team, we do have folders. So, and for example, I worked with compliance, sustainability, and stuff like that. So, I have folders for all those topics. But there's no global way of saving that knowledge." – Informant 2	*"We have one process tool where we document all processes. And there we also create links to documents, for example, that can be used when you do a specific activity. Can be all from a presentation material to a template, to an instruction. And then we have two different documentation methods you can say so one is just a normal storing document in SharePoint library. And the other one, more of or less of a website that can b structured as a knowledge base, you can say, and you can make articles and connect documents and to some extent also create workflows." – Informant 8*
"Okay. So, tool wise, the changes are done in a system called Team Center. this is to let R and D use it for development where they store drawings. And that system also handles the revision handling of it as well." – Informant 3	
"I mean, like the database for me is my inbox. That's where I store most of information and then we have the, like. I'm not sure how familiar you are with software development, but we have these kind of user stories, virtual post-it notes that we store in a system and we	

move them between different statuses like to do, doing, done. These three. And our systems called JIRA in our case. And then we have a lot of information there. And that's also kind of very tied to our development process. So that one is, of course, I don't need to carry it in my head. It's on a computer somewhere." – *Informant 7*	
"... The one we get most often ends in a project or product. But sometimes it's the technology that goes across different types of products, different type of projects. And the way we do this is that whoever learned it. Internally, they start by actually documenting on it, on our documentation tool... And we also had workshops between key players within our departments where we try to explain the technology and where we try to establish a common knowledge on this. Everyone cannot learn everything. So, we must be careful, of course, not spending too many hours on it, to make sure that we preserve it within our organization. So, Confluence is a tool that we use a lot for conferences. It's a wiki for an internal form for us where we post the things that we know in different kind of areas and try to build up knowledge." – *Informant 9*	

Table 4.1.8.1 Quotes from informants in the LC/MNC and SME. Source: developed by book' authors.

Secondary data:

The secondary data for finding 37 shows that both organizations support implementation of acquired knowledge is stored in systematic data bases.

As an illustration on the type of secondary data we analysed and its connection with the theory, we show table 4.1.8.2 that describes information obtained from the internal website of the case companies.

LC/MNC	SME
"We are launching the Insights Database in an effort to increase the sharing of knowledge between divisions and business units," says *Klas Rydstrand from the Global Innovation Management team. "The Gateway process and generation planning process are just two examples where insights are critical to make fact-based decisions."*	The official website of the SME contains a Knowledge Bank, where various knowledge is accessible regarding services, manuals, recordings, products, customer articles and so on. Source: The official website of the SME.

The Insights Database's content is categorized in four groups:

Customer insights – Insights from Voice of Customer (VOC) research conducted by ASSA ABLOY and its entities. Discover what existing and potential customers say about us and our products and based on this, what business opportunities can be identified.

Market insights – Marketing intelligence including internal market studies, external market reports, and more.

Technology insights – Knowledge and research related to critical technologies such as sustainable energy solutions, wireless connectivity, software and identification.

News – Handy updates on competitors, partners, customers, tradeshows, standards and more.

Source: The internal website of the LC/MNC: "Avenue"

Table 4.1.8.2 Secondary data from the LC/MNC and SME. Source: developed by book' authors.

In summary, both primary and secondary data confirms finding 37 in terms of how the organizations implement acquired external knowledge into systematic data bases, this shows how the organizations applies knowledge transfer after acquiring said knowledge so that it may be utilized on an organizational level.

Finding 38: When the SMEs acts as a recipient of knowledge they cannot always fully implement the acquired knowledge from the source partner due to organizational limitations.

This was concluded based on the following information:

Primary data:

Finding 38 was based on primary data from the SME and none from the LC/MNC, probably because the LC/MNC does not share the same organizational limits as the SME, thus the informants from the LC/MNC do not necessarily think from this perspective when it comes to implementation of knowledge. Informant 5 from the SME discuss how they engaged in a knowledge transfer activity with a partner but could not fully implement the acquired knowledge as the organization does not operate in the same way and not to the same extent.

Thus, Informant 5 state that they took parts partially of the acquired knowledge to make it applicable for their organization by using the framework of the acquired knowledge as support and adapting it to their organization.

As an illustration on the type of quotes we analysed and their connection with the theory, we show in table 4.1.8.3 which shows the quote from the SME. Remaining Informants (4, 6 and 8) from SME did not provided opinions on this perspective and no informants from the LC/MNC.

LC/MNC	SME
None.	*"We tried to establish a new way of working in (SME source organization). And we learned from (LC/MNC recipient organization) and from Safe, a new agile way of working. And I think it's important to try to anchor it with a co-worker, with all the team members who have to make sure they have the they can get their opinion. Do you think this will work? We can't as I said, we can't copy everything and get to work in (the SME), because we need to find our own way... You look at the pain point, you see how we can utilize the method or the principle that's practical and use it in our organization. And take the framework of it. You take that framework and you have as support and how do we establish and implemented in our own organization." – Informant 5*

Table 4.1.8.3 Quotes from informants in the LC/MNC and SME. Source: developed by book' authors.

There is no secondary data found that can support the finding 38.

In summary, finding 38 is based on primary data from the SME, it state the importance of being aware of the extent and application of acquiring knowledge, as every organization is different and has various capabilities, which shows that to implement acquired external knowledge, the knowledge transfer must be partially accepted in this case to make it applicable for the organization.

Finding 39: SMEs implement acquired knowledge to the organizational level, but more towards to the individual level.

This was concluded based on the following information:

The primary data from the SME and none from the LC/MNC was used to base this finding. Informant 6 from the SME, discuss how after acquiring external knowledge, as being employed in a higher-level position, the informant does not implement the knowledge, but rather directs it to the employee that will be working with it, where it is up to them to decide.

As an illustration on the type of quotes we analysed and their connection with the theory, we show in table 4.1.8.4 that explains the quote from Informant 6 from the SME. Remaining Informants (4, 5 and 8) from the SME did not provided opinions on this perspective and none from the LC/MNC, as this viewpoint was not mentioned from the perspective of the LC/MNC, which could be because their organizational structure operates differently compared to a SME.

LC/MNC	SME
None.	*"I do not implement it. That's, that's the key difference in smaller organizations. So, the people that are working in the business, they are implementing it. So, in a smaller organization, I do not go about to implement the change. I ensure that the people that are to implement the change, they have the heart to do so. In small organizations you have like a shotgun of things to do every day. So, if you're not motivated to do this specific work task, you will not do it." – Informant 6* (Would you say that your organization tries to maybe store the information in any way? Say you learned about the method to conduct a process or so. Does your organization store that in a document or something?) *"Yeah, we do. But I mean, we have Google. So small organization, we don't work very much with storage and documentation of things we do, but we don't have time to keep it up to date and look at it. I mean, we when we had something at hand that we need to do, we Google it. And then we get inspired from that and we do some shots and then you say, hey, we do like this. This is much quicker decision process of implementing things." – Informant 6*

Table 4.1.8.4 Quotes from informants in the LC/MNC and SME. Source: developed by book' authors.

There is no secondary data found that can support the finding 39.

In summary, finding 39 was based on primary data which illustrates how SMEs implements acquired knowledge more towards an individual level rather than the organizational level. This implies that the knowledge transfer does not always become implemented to the organizational level which could make it more difficult to generate value from the knowledge long-term, however, as Informant 6 also mentions, employees in SMEs has more things to do (due to less niche job descriptions), which is the reason behind why the knowledge does not always become implemented to the organizational level.

Sub-aspects: (1) Dissemination of acquired external knowledge in the organization

We have found in this study based on collectively agreed informants that:

Finding 40: LCs/MNCs diffuse acquired knowledge amongst employees with meetings, emails, conferences, workshops and their technology network, while SMEs diffuse the knowledge by staff meetings, projects, recordings and emails.

Finding 41: LCs/MNCs rely on the individuals to access the knowledge from the data systems themselves rather than having it given directly to them, while SMEs diffuse knowledge more informally in the organization.

Finding 40: LCs/MNCs diffuse acquired knowledge amongst employees with meetings, emails, conferences, workshops and their technology network, while SMEs diffuse the knowledge by staff meetings, projects, recordings and emails.

This was concluded based on the following information:

Primary data:

Informants (1, 2, 7 and 9) from the LC/MNC and Informants (4, 5 and 8) from the SME collectively support the means of disseminating acquired external knowledge is done by meetings and emails. While the means are rather similar, the Informants from the LC/MNC further state it is disseminated by conferences, workshops and their existing technology network, while the SME focus more on projects and recordings.

As an illustration on the type of quotes we analysed and their connection with the theory, we show table 4.1.8.5 that explain the quotes from informants from LC/MNC and SME. Remaining Informants 6 from the SME and Informant 3 from the LC/MNC did not provided opinions on this perspective.

LC/MNC	SME
"I mean, we have in R&D, there's a system architect meeting. For instance, we have system architect meetings in the teams, and they gather, I don't remember how often but they have meetings where they share knowledge there. Within VOSTIO we have what we call tech-bord, which is the little bit smaller. But it is within the VOSTIO	*"Maybe by a project, defining a project and take advantage of this new knowledge. It can be a project, or it can also be some staff meetings, share the knowledge and exchange it."– Informant 4*

products. But there are several teams involved there as well, but they can discuss." – Informant 1	
"So usually if I find out something, I love to spread it out right away to everyone who might need it, even to those who might not need it (internally). So, two ways to do that. Send it out by email. Send it out by having a meeting with the people around the table. So, in global solutions for each area of interest. So, areas of interests are sustainability, quality, customer success. So, there's different areas that we highlight in our global quality team. So, we have people in each business area. We get them around a table, virtual table because of the pandemic. And then we just disseminate this information." – Informant 2	*"We have like different kind of meetings, where if someone, for example, have acquired knowledge, that was shared between the team. Though I think it's important that not everyone needs to learn everything you can share in-between. For example, if someone attending good seminar or good course, they can share the main topics. What's your takeaway from this occasion? I think its a good method to do."* – Informant 5
"If I take the conference example. Unfortunately, we don't really share that kind of knowledge. Sometimes, I guess we do, but there has not been that many conferences last year." – Informant 7 (But if we go back to the example of the conference. You said when you talk to your team, then that's your ways of sharing that knowledge?) *"Yeah. I think I mentioned that I was like this conference during our stand-up meeting in the morning. And at least previously there has been in other roles. Some of the companies I've had like maybe a team meeting where I have 20 minutes and talk about this was what the conference was about and it was pretty cool. Here's the link if you want to look at the keynote and so on."* – Informant 7	*"Yes, I would say so if the evaluation that to be worth to be known more persons than the ones who were involved in the exchange. So, we have a few different ways. We want to get good example now, for example, where we have had customer meetings with some, we call them model customers, not sure what the English term for this. But front runners as customers. And we have made in interviews with them to kind of figure out how they work and way they have made the choices they have made in choosing other solution. So maybe also choosing others. And then we recorded this and we should try to spread it within the organization. So that's more people get to know about our customers and how they work, for example. And then spread via the internet."* – Informant 8 (How do you spread it to the other employees? By e-mail or additional meetings?) *"I would say that it can either e-mail or via a post that is either written email or that you can watch a recorded content. I would say that in some cases to do additional meetings, but it's I would say that it's usually accesses to material or distribution by email."* – Informant 8
"Yes, for instance, there is an example now where we have a network, which technology network where we have been using ZigBee. We are going to upgraded to the latest standard of ZigBee. We buy competence to teach us how to do this from Reno CB consulting company, buy services from them." – Informant 9	

(You mentioned from the previous question where you said, after you have acquired knowledge you try to explain the technology with workshops and meetings, would you say these are the means you share the acquired knowledge to the employees?) *"And also, Confluence. I mean, everything that we do on this week, we will put on Confluence what the conference is like. You know, you have one engineer starting without something about this. Next thing is then another guy learned something, and they go in and edit it, updated and correct it. So, it becomes a common knowledge like in Wiki. But then yes, workshops. But actually, before we do the purchase, before we sign the contract with this company, we have a plan for how to share. Not like a written, very formal plan, but an agreement between the people and needs to know this and the management that buys this of how to actually spread it to the right people." – Informant 9*	

Table 4.1.8.5 Quotes from informants in the LC/MNC and SME. Source: developed by book' authors.

Secondary data:

The secondary data for finding 40 shows that both organizations support dissemination of knowledge via their systematic data banks (technology networks), additionally the LC shows that they do engage in dissemination of knowledge by projects and meetings, whereas there was no secondary data for the SME to support this.

As an illustration on the type of secondary data we analysed and its connection with the theory, we show table 4.1.8.6 that describes information obtained from the internal website of the case companies.

LC/MNC	SME
"We are launching the Insights Database in an effort to increase the sharing of knowledge between divisions and business units," says Klas Rydstrand *from the Global Innovation Management team. "The Gateway process and generation planning process are just two examples where insights are critical to make fact-based decisions."* The Insights Database's content is categorized in four groups:	The official website of the SME contains a Knowledge Bank, where various knowledge is accessible regarding services, manuals, products, customer articles and so on. Source: The external website of the SME.

Customer insights – Insights from Voice of Customer (VOC) research conducted by ASSA ABLOY and its entities. Discover what existing and potential customers say about us and our products and based on this, what business opportunities can be identified.

Market insights – Marketing intelligence including internal market studies, external market reports, and more.

Technology insights – Knowledge and research related to critical technologies such as sustainable energy solutions, wireless connectivity, software and identification.

News – Handy updates on competitors, partners, customers, tradeshows, standards and more.

Source: The internal website of the LC/MNC: "Avenue"

"If we work systematically to identify and close knowledge gaps our decision making is improved, resulting in solutions that delight customer, excellent product quality and efficent projects. A project kick-off meeting is an excellent opportunity to do some brainstorming and leverage the team members' experiences to build a good initial set of knowledge gaps. The team can immediately build the tracking, how and knowledge capture systems to manage the gaps."

Source: The internal website of the LC/MNC: "Avenue"

Table 4.1.8.6 Secondary data from the LC/MNC and SME. Source: developed by book' authors.

In summary, while primary and secondary data shows very similar means of dissemination of knowledge for both organizations, it is shown that SMEs focus more on small scale dissemination of knowledge, where the LC/MNC attempts to broaden the knowledge via larger scales of gatherings. This means for knowledge transfer when disseminating it in SMEs that the focus is narrower, while in the larger one, it is broader. Which is natural, based on the reasoning of the size of the organizations.

Finding 41: LCs/MNCs rely on the individuals to access the knowledge from the data systems themselves rather than having it given directly to them which indicates formal knowledge sharing, while SMEs diffuse knowledge more informally in the organization.

This was concluded based on the following information:

Primary data:

Informants 3 from the LC/MNC and Informants 6 from the SME collectively support the means of disseminating acquired external knowledge is approached differently in the organization. While Informant 6 discuss about how the knowledge is disseminated more informally in the organization, Informant 3 implicitly indicates a formal means of knowledge dissemination where employees can give their opinion on the knowledge before it is implemented.

As an illustration on the type of quotes we analysed and their connection with the theory, we show table 4.1.8.7 that explain the quotes from informants from LC/MNC and SME. Remaining Informants (4, 5 and 8) from the SME and Informant (1, 2, 7, 9) from the LC/MNC did not provided opinions on this perspective.

LC/MNC	SME
"So, tool wise, the changes are done in a system called Team Center. This is what R&D uses for development where they store drawings... So yeah, it's a process where the idea is just to gather all the relevant functions to review them together. Before we send the new drawings to the supplier. So, everybody has an opportunity to understand why and what's changed and can come up with suggestions in case the designer didn't think of that basically." – Informant 3	*"I mean, that's tricky. People do it in different ways (share knowledge formally and informally). Some of my colleagues, they find an article, or they find something about something that has been done in another part of organizations. And they email it to various stakeholders and say, Hey, look at this, what I found, other people, they interact informally by saying, Oh, I found this, and this could be interesting for us. What do you think about this? And other people, they set up a meeting with PowerPoint. Chunking slides say this is how you do and in this part of organization, we should do it as well. So, it's different culture, company culture, and also individual skill sets of change management." – Informant 6* (So, you would say this is how you share the knowledge in a smaller organization?) *"Yes. But the ladder, the PowerPoint chunking (formal KT), that's not very frequent in small organization. It could happen. But to do this in a successful way, you need to do the individual interactions first, because then if you provide in a smaller organization, the level of autonomy is much*

	higher than in a bigger from my perspective. So, you need to anchor what you should do with individuals so that they feel that they could do it. And I think that it's a good idea." *– Informant 6*

Table 4.1.8.7 Quotes from informants in the LC/MNC and SME. Source: developed by book' authors.

There is no secondary data found that can support this finding 41.

In summary, Informant (3 and 6) discuss different approaches to how knowledge is disseminated in the organization, Informant 6 from the SME address how informal knowledge dissemination is more frequent over formal since the level of autonomy of the employees, as you need to address specific individuals that perhaps the knowledge affects. Informant 3 from the LC/MNC approach to it, is by more formal nature, since the knowledge is accessible, but it is not specifically directed for the employees to attend the dissemination of knowledge, it is by their own choice, which also shows that the level of autonomy is high in the LC/MNC, but to a different perspective of dissemination of knowledge.

4.1.9 Primary and secondary data about Aspect: Knowledge characteristics
According to the literature and theoretical framework of this study (section 2.23), tacit and explicit knowledge contains characteristics such as:

1) The understanding of the knowledge in individuals, known as tacitness (Kalling, 2003), when tacitness is deemed high, the understanding of the knowledge is increased (context), resulting in a higher rate of successful knowledge transfer.

2) The uncertainty of knowledge, known as ambiguity (Wijk et al., 2008) depends of the interaction of the context and content of the knowledge that is being transferred, when these aspects are high, the transfer of knowledge is less likely to be successful.

3) The complexity of knowledge (Stock & Tatikonda, 2000), is affected by the how many variables are involved in the context and content of the knowledge, as when these are deemed high, the transfer of knowledge is less likely to be successful.

Sub-aspect: (1) Knowledge tacitness in tacit and explicit knowledge

We have found in this study based on collectively agreed informants that:

Finding 42: LC/MNC as a source in a knowledge transfer experience more difficulties in the transfer when the understanding of the knowledge is deemed low by the partner (e.g., with SME as a recipient side due to the less diversified expertise), while the SME as a source in a knowledge transfer experience less difficulties in the transfer when the understanding of the knowledge is deemed high by the partner (e.g., with LC/MNC as a recipient side due to the more diversified expertise).

This was concluded based on the following information:

Primary data:

Regarding tacitness of knowledge, Informants (1, 2, 3, 7 and 9) from the LC/MNC and Informants (4 and 5) from the SME support that when tacitness is low on the recipients side the knowledge transfer becomes difficult. The reasoning behind this is as both organizations are aware of that the understanding of the knowledge that is being transferred is affected by the context of which it is applied.

As an illustration on the type of quotes we analysed and their connection with the theory, we show table 4.1.9.1 that explain the quotes from informants in both the LC/MNC and SME. Remaining Informants (6 and 8) from the SME did not provided opinions on this perspective.

LC/MNC	**SME**
"Yeah. Well, it's both during the development phase, I would say it's always better to discuss when the partner or the customer or the partners person, because they understand the partner or the customer, we understand like the developmental phase. So, it's much easier to if they kind of explain what it is, they want, they need it easier for the developers to understand that. Instead of like sitting in developing something, get it into a manual, documented those GTS and they were supposed to get customer installed." – *Informant 1*	*"Normally, I think it's much easier to work with large organizations because then you have you have specialists. Normally you have more knowledge... often long experience of things and they can make quick summaries and so on."* – *Informant 4*
	"I think the technical questions are more like. How do you how do you manoeuvre? How do you handle the machine or the technical things or Microsoft or you learned something. But the softer things, example change management if you teach someone, it's not easy because the people, it's the soft questions. It's not easy all the time because you need to work with yourself and how you communicate, how you talk to people, how you do things, how you understand that people acting depending on what kind of information they have." – *Informant 5*
"You need to see how they can apply it in their everyday work. So, you need to kind of transpose. So, whatever you were using these skills or whatever skills you were using before, you need to remember what they will be used for in the new context. So, don't try to explain what you use it before. Like just tried to see what are people doing now and give them concrete examples." – *Informant 2*	

"In your experience when you shared knowledge, what was the most difficult to transfer to the partner? Did you notice any difference when knowledge was transferred? Absolutely experience. And I mean, not my experience but their experience. So recently we require a smaller company and then we are onboarding them, want to transfer the knowledge in regard to there are certain processes they have to follow. It's not up for discussion. This is the way we work. And by doing so then this is the KPI that will need to put in place and so on. And then if they don't see the benefits immediately, that is mostly because of the maturity level as they don't see the bigger picture. And also when we acquire smaller companies, what I often call them garage companies. This is that those innovative people, they are not structured at all. But maybe there's the strong side because innovation is their focus and telling them that you need to have a system that will handle your revision, handling of drawings are just waste for them, you know, this paperwork, it doesn't bring value because they don't see value in that. But when you start sending by mistake, you send an old design or drawing to a supplier and that he may manufacturers 50000 units, wrong design. And then it's a huge cost. And if in addition, push that to the market and you find out it's wrong after it has been installed, then the cost is 20 to 30 times higher. Then you see the benefit. But they don't see because they didn't have experience with that. They do not come that far." – Informant 3

"I mean, of course I would listen into I kind of have a mental picture. I know that the receiving end what they are doing, right? So, I don't try to teach to building company about textile design or something. So, of course, if you try to form the way you communicate with the receiver's context, maybe I should put it like that. Why and how? Why, of course is I would be aiming bit beside of the target if I don't try to shape it into something that they can relate to. Because I guess the pickup rate or their understanding will be much better if

I fight for me to something that they can relate to in their daily work." – Informant 7	
"Another example, and then in Oslo here though is there's a big campus. It has like a 20 thousand student rooms. It's run by a small organization. They have a small IT department. They have also external IT consultants. And I don't understand. They have had some limited access control before. And at this point, they didn't understand our concept of access control, which is quite different from regular access control, in the way we solve it, which is part of our success. That we have this unique approach to how we do access control. And that's bringing over that concept to persons that already had perceived concept of how it should work. That was quite difficult actually, and we spent a lot of time actually discussing, had to go back to discuss the architecture of the system without revealing any secrets in order for them to buy into this. And probably we underestimated the task. Normally, we will make this interface into property management type of system that already are used to do all this type of integration. This time, they were building the system themselves and this was the only system built special purpose for this campus. They didn't buy the standard product. They bought a lot of engineers to build on the system. So that was difficult to change that concept actually, how to do this, and we had to go back to the basic and actually start, start from scratch." – Informant 9	

Table 4.1.9.1 Quotes from informants in the LC/MNC and SME. Source: developed by book' authors.

There is no secondary data found that can support this finding 42.

Overall, tacitness varies for both organizations, while both the LC/MNC and SME are aware of the difficulties that arise from when tacitness is deemed low, both organizations attempt to reduce it by relating the knowledge to how the partner organization could apply it to their operations. As all informants from the LC/MNC and two out of four from the SME focused on tacitness, this could be since the LC/MNC have more niche roles, their individual tacitness of the knowledge could be higher, while the roles in the SME are broader, dampening the tacitness of the knowledge received, which ultimately, changes the scope and focus of knowledge characteristics depending on the size of the recipient and source organization in the knowledge transfer.

Sub-aspect: (2) Knowledge ambiguity in tacit and explicit knowledge

We have found in this study based on collectively agreed informants that:

Finding 43: LC/MNC as a source in a knowledge transfer experience more difficulties in the transfer when the understanding of the interactions and components of the knowledge is deemed low by the partner (e.g., with SME as a recipient side due to their maturity level and experience), while SME as a source in a knowledge transfer experience less difficulties in the transfer when the understanding of the interactions and components of the knowledge is deemed high by the partner (e.g., with LC/MNC as a recipient side due to their maturity level and experience).

This was concluded based on the following information:

Primary data:

Informants (2 and 7) from the LC/MNC and Informant 5 from the SME support that when ambiguity is high, transferring tacit knowledge to a recipient partner becomes difficult. As the underlaying context interacts differently with content it is applied to, both organizations are to some extent aware of the ambiguity involved when transferring tacit knowledge, while Informant (2 and 5) focus more on how they communicate the knowledge to reduce the ambiguity, Informant 7 simply state that the ambiguity rise from the tacitness since it becomes diluted when obtained.

As an illustration on the type of quotes we analysed and their connection with the theory, we show table 4.1.9.2 that explain the quotes from informants in both the LC/MNC and SME. Remaining Informants (4, 6 and 8) from the SME and Informant (1, 3 and 9) did not provided opinions on this perspective.

LC/MNC	SME
"You need to see how they can apply it in their everyday work. So, you need to kind of transpose. So, whatever you were using these skills or whatever skills you were using before, you need to remember what they will be used for in the new context. So, don't try to explain what you use it before. Like just tried to see what are people doing now and give them concrete examples." – Informant 2	*"I think the technical questions are more like. How do you how do you manoeuvre? How do you handle the machine or the technical things or Microsoft or you learned something. But the softer things, example change management if you teach someone, it's not easy because the people, it's the soft questions. It's not easy all the time because you need to work with yourself and how you communicate, how you talk to people, how you do things, how you understand that people acting depending on what kind of information they have." – Informant 5*
"I think just getting the getting all the details and communicating all of the like positive energy that I had. Getting it, it's kind of hard to convey that to someone who weren't there and heard it from the same professional source that I did. Right? So, this is become	

like an second-hand knowledge and then that person it becomes a third hand knowledge, so it gets diluted." – Informant 7	

Table 4.1.9.2 Quotes from informants in the LC/MNC and SME. Source: developed by book' authors.

Secondary data:

This secondary data shows that the LC/MNC attempts to codify tacit knowledge to reduce ambiguity of the knowledge, which makes transferring it easier.

As an illustration on the type of secondary data we analysed and its connection with the theory, we show table 4.1.9.3 that describes information obtained from the internal website of the case companies.

LC/MNC	SME
"The elements of reusable knowledge ensure that the person receiving the knowledge has all the pieces to decide whether or not the knowledge applies and how far it can be trusted, understood, believable, actionable and generalized. If any of these elements are missing, the knowledge won't be reused to its full potential. The company culture needs to support the effort it takes to generate reusable knowledge and reward those who actively seek the benefits of knowledge reuse. This can be achieved through policies, processes and performance expectations that reward people for capturing and using reusable knowledge to save the company's time, money and capacity." Source: The internal website of the LC/MNC: "Avenue"	None.

Table 4.1.9.3 Secondary data from the LC/MNC and SME. Source: developed by book' authors.

Overall, knowledge ambiguity were not the focus of the informants from the LC/MNC and the SME, as a small portion of the interviewees had the focus on ambiguity, they showed awareness of how the components and interactions are affected, while secondary data from the LC/MNC confirms that transferring tacit knowledge is affected by ambiguity and to reduce it, they attempt to codify the knowledge, as none of the informants from either organization mentioned knowledge ambiguity for explicit knowledge and no secondary data for the SME, it still shows the importance of understanding the ambiguity and how to reduce it.

Sub-aspect: (3) Knowledge complexity in tacit and explicit knowledge

We have found in this study based on collectively agreed informants that:

Finding 44: LC/MNC as a source in a knowledge transfer experience less difficulties in the transfer when there are low amount of variables involved in the interactions and components of the knowledge that is transferred to the partner (e.g., with SME as a recipient side due to their organizational demands), while SME as a source in a knowledge transfer experience more difficulties in the transfer when there are high amount of variables involved in the interactions and components of the knowledge that is transferred to the partner (e.g., with LC/MNC as a recipient side due to their organizational demands).

This was concluded based on the following information:

Primary data:

Informant (1, 2, 3 and 7) from the LC/MNC and Informant (6 and 8) from SME support that when complexity is deemed high, the transfer of knowledge becomes difficult to a recipient partner.

As an illustration on the type of quotes we analysed and their connection with the theory, we show table 4.1.9.4 that explain the quotes from informants in both the LC/MNC and SME. Remaining Informants (4 and 5) from the SME and Informant 9 did not provided opinions on this perspective.

LC/MNC	SME
"It's the way they use the system. And also, how many rooms there are, how they kind of build up the hotel. Because you can build in the access management system, you can build it in different ways. And for small enterprises are smaller. There's not that much complexity. You have certain number of floors. You have a certain amount of rooms, corridors, maybe half an hour later. You have like two entrance doors maybe. And large enterprise you might 20 elevators, you might have 5000 rooms. You have conference rooms, you have spa's, you have gyms. So, it's like the, the bigger the hotel, the more complexities." – Informant 1	*"So, we've done some knowledge transfer activities to hospitality when it comes to testing and development methods and techniques. And we've done that as a mutual sharing experience that we've had conferences with them, discussing and workshopping with them. And that's been very good. I think the most value comes from networking and inspiration. So, I actually am quite hesitant to say that we actually did something on that. We were inspired, but we didn't really apply the same things as they had. And I would say that to answer this question, the main blocker for us to share knowledge to the bigger organization is time. And we don't get the benefits from it." – Informant 6*
"You need to see how they can apply it in their everyday work. So, you need to kind of transpose. So, whatever you were using these skills or whatever skills you were using before, you need to remember what they will be used for in the new context. So, don't try to explain what you use it before. Like just tried	(If you go about transferring knowledge to a larger organization, what would you say is, the more difficult in regard to the time constraints?)

to see what are people doing now and give them concrete examples. " – Informant 2	*"Because they had another organization. They have niche roles. So, they have one role that work with just this part of their business. And I'm constantly saying to my colleagues, that's okay, please have network with the larger organizations do that for inspiration or anything. But do not let them to eat your time too much because they had much more time than you have on your area of expertise." – Informant 6*
"So, we can take one of our readers, for instance, it's in RFID and there is a card in each lock and someone reads the card. And we've had for like 6-7 years. This is when you have failure rate that is not too high, but it's not low enough. So, you know, you just keep it there. You've prioritized other things. But at some point, we had to decide. Okay. We need to identify why are they failing as they do. And then we brought on board, Flex who was our biggest supplier for electronics and our R&D department, together with Flex, made an analysis. What can be done? Flex came up with proposals. So did our R&D team and they got together and discussed it. And it was a combination of changing the design, the layout, thickness of the PCBA board, and then production process, which is actually one of those that had the most impact. And based on their experience and result, we took that and share with our other Hannah, who is our second supplier. Because we have dual sourcing for electronics. This is the other less mature company. It's much, much smaller. Flex is one of the biggest electronic companies in the world. So, then we used what we experienced there and shared with Hannah and gotten to implement what they could. They couldn't implement everything because they didn't, they didn't have the advanced tools, but they did implement some of it. And this is like 1.5 years ago. And now we see that they need to do full implementation, because we see still a higher, higher failure rate on their products. And then, so we took it directly from what we experience at Hannah and taught what we learned from Flex to Hannah to how to teach them what to do." – Informant 3	
"I think just getting the getting all the details and communicating all of the like positive energy that I had. Getting it, it's kind of hard to convey that to someone who weren't there and heard it from the same professional source that I did. Right? So, this is become like an second-hand knowledge and then that	*"Yeah, I would say that this goes as well with a larger organization that it's harder to reach the right recipient of the knowledge. Because you think that you might talk the right person in transferring the knowledge, but then it might be someone else who does actually execute it in the end. So I think that's the hardest part to figure out who you actually*

person it becomes a third hand knowledge, so it gets diluted." – Informant 7	*should have involved to get things." – Informant 8*
(You would also say that the complexity of the knowledge is more difficult to transfer since your perspective is not the same as the perspective that was given to you?)	
"Sure. I mean, the source material for the talk could be like a 300-page book, condensed into one-hour speech. And then I get 20 minutes to tell my colleagues (partners) about it. So, the funnel gets narrower and narrower." – Informant 7	

Table 4.1.9.4 Quotes from informants in the LC/MNC and SME. Source: developed by book' authors.

There is no secondary data found that can support this finding 44.

Overall, while Informant (6 and 8) from the SME found the complexity of the recipient partner's organization was the reason complexity arises, due to their organizational size, Informant 1 from the LC/MNC also agreed to the perspective, as the complexity is less when transferring knowledge to a partner based on the organizational size. Informant (2 and 3) found that complexity arise from the partner's experience and their maturity level, as an organization. Informant 7 state that the complexity arises from tacitness, as it dilutes the knowledge according to the individuals understanding of the knowledge. This shows that complexity comes from many different variables that changes when it comes to different organizational sizes and the need of awareness of this for both organizations, which a large portion of informants shows to achieve a successful knowledge transfer.

5. Discussion

The purpose of the research is to extend the understanding and relevance of various aspects that affect KT in terms of different firm's sizes from an inter-organizational perspective. As knowledge transfer is a vital source of a firm sustainable competitive advantage (Easterby-Smith et al., 2008; Cardoni et al., 2019) because merely knowledge creation and retention within the organization are not sufficient. Knowledge must be transferable and shareable both within and across an organization for leveraging competitive advantage and improving business performance (Minbaeva et al., 2003; Levine & Prietula, 2012).

This study found various aspects that affect interfirm knowledge transfer and helps to extend the current literature in terms of what aspects are more relevant to a particular firm size. In table 5.1 all findings from the primary and secondary data is summarized, from both the LC/MNC and the SME.

This section is organized as follows, (1) for each aspect why it is important in inter-organizational knowledge transfer, (2) how it was measured according to the literature and theoretical framework, (3) a discussion on what the study presented in section 4 regarding findings that were found to be inconsistent and consistent with the literature, on e.g., why it was found to be in one way or another, what it means for inter-organizational knowledge transfer and what it means for the different organizations that were involved in the case study.

Dimension	Theoretical definition	LCs/MNC	SMEs
Knowledge governance mechanisms	Formal governance mechanisms consist of management policies and procedures, organizational structure (Foss, 2007).	*Finding 1: Formal knowledge governance mechanisms are used for KT coordination between partners.*	
		Finding 2: Formal is more appropriate for KT with external organizations whereas informal within the company for LCs/MNC.	*Finding 2: Formal knowledge governance helps to show SME that a company is larger than it is to its partners.*
		This viewpoint was not mentioned by informants from the larger organizations probably because the size of the firm is naturally large as a result it was not applicable.	*Finding 3: As firm size increases the more formal knowledge governance mechanisms should be applied in SME.*
		Finding 4: Formal knowledge governance procedures are needed for LC/MNCs to manage their cost structure.	*Finding 4: SMEs are focused on growth and customer base expansion that needs more informal procedures.*
	Informal includes corporate culture, trust, social ties and networks (Foss, 2007).	*Finding 5: Informal knowledge governance mechanisms are essential for LC/MNCs for easier navigation.*	*Finding 5: Informal knowledge governance mechanisms are essential for SMEs because they help to enhance adaptability and flexibility capabilities.*
Trust	Trust is based on a social judgements in terms of (1) evaluation of other partners' benevolence, (2) expertise and competence coupled with (3) assessment of the risk related to untrustworthy behaviour (Inkpen & Tsang, 2005). In addition, to this, the literature proposes to assess	*Finding 6: Trust for the knowledge transfer activities between partners is based on formal documents.*	
		Finding 7: Larger sized companies have the power to enforce the fulfillment of formal trust agreements.	*Finding 7: SMEs have less power and considers enforcement of the fulfillment of formal trust agreements as costly and ineffective.*
		Finding 8: Mutual trust for KT between partners is built in a form of mutual benefit and reciprocity.	

	interfirm trust based on (1) fulfillment of commitments agreed between partners; (2) honest negotiation; and (3) avoidance of receiving excessive benefits of ally organizations (Cummings & Bromiley, 1996). Moreover, mutual trust is also deemed to be an important aspect, since exchange of favours or similar transactions (e.g., reciprocity behaviour) tend to be met in return as "If we helped them today, they would help us later" was a response regarding trust in Al-Jabri & Al-Busaidi (2018) article.	*Finding 9: Mutual business benefits evaluation is more prevalent for KT rather than an assessment of the partner's benevolence.*	
		Finding 10: Trust in partnerships for knowledge transfer is based on an assessment of the competence and expertise of partners.	
		Finding 11: The degree of the trust for KT depends on a company's size: the greater the size of the partner the higher risks are associated with KT activities for LC/MNCs.	*Finding 11: The degree of the trust for KT depends on a company's size: the larger organization the higher credibility and trust for KT for SME.*
Cultural distance	The national cultural distance can serve as barrier for the KT among international partners regardless the firm size such as (1) higher cost of entry and hinders the transmitting the core competencies to the international markets; (2) facing operational challenges in terms of lacking common understanding of norms, values and motivation; (3) misunderstanding between foreign partners can limit sharing of critical organizational knowledge (Wijk et al., 2008).	*Finding 12: Language barrier can impede knowledge transfer between partners.*	*There is no information provided by informants of SME regarding the language barrier probably because they do not often interact with partners at the international level due to smaller size business.*
		This viewpoint was not mentioned by informants from the larger organization probably because they are already aware about decision-making process of their international partners due to operations are inherently in the global context.	*Finding 13: A lack of understanding of the decision-making process in different cultures can negatively affect KT activities between partners.*
		Finding 14: High power distance in terms of hierarchical subordination can hinder KT.	

Geographical distance/proximity	The geographical distance negatively influences inter-organizational KT especially for tacit knowledge exchange but can be mediated in terms of (1) developing trust-based relations between distant partners (e.g., Capaldo & Petruzzelli, 2014); (2) organizational proximity including common corporate culture and values (Korbi & Chouki, 2017); and (3) bilateral way of utilizing modern communication technologies (Choi & Contractor, 2016) in large or multinational companies as well as in SMEs.	*Finding 15: Physical proximity is essential to build trust between partners.*	
		Finding 16: Adaptability capacity between partners is important for geographical distant partners.	*SME did not provide any opinions on this perspective probably because SME has already possessed adaptability capability as an inherent feature due to the smaller scale of the firm.*
		Finding 17: Geographical distance/proximity does not influence interfirm KT.	
		Finding 18: Digital communication tools improve KT between physically distant organizations but as a complementary technique.	
Disseminative capacity	Knowledge source's motivation to teach (Minbaeva et al., 2018)	*Finding 19: Mutual experience sharing, and learning are motivation drivers for KT between partners.*	
		Finding 20: Mutual business performance enhancement motivates partners for KT activities.	
		This viewpoint is supported by LC/MNC, who mentioned they are very selective in the learning process and mainly they want to acquire knowledge due to concrete business purpose. As a result, this finding is applicable for SME only.	*Finding 21: SME as a source of knowledge is less motivated to exchange knowledge and it depends on size partner.*
	Knowledge source's ability to teach (Easterby-Smith et al., 2008).	*Finding 22: The ability to transfer knowledge to partners depends on the knowledge recipient's competence and maturity level.*	

		Finding 23: The ability to teach depends on an assessment of risks related to knowledge leakage and protecting valuable knowledge.	
		Finding 24: The ability to codify the knowledge facilitates knowledge transfer between partners.	
Social ties	Social ties according to Tangaraja et al. (2016) is a relationship based concept, its either conducted directly or indirectly between two parties and differentiated by (1) informal or formal ways. Liu et al. (2015) argues that there are three steps to establishing a high-quality relationship, (2) first step involves locating a partner firm and identify its key contacts, secondly to develop and (3) maintain the relationship established, last stage involves dispatching of knowledge from the source along with knowledge absorption of the recipient, which is how the knowledge is exchanged. Martín Cruz et al. (2009) state that motivation consists of intrinsic and extrinsic motives. Organizational	*Findings 25: LC/MNCs prefer both informal and formal knowledge transfers in social ties with partners but leaning more towards formal social ties.*	*Findings 25: SMEs prefer informal knowledge transfer in social ties with partners.*
		Finding 26: LC/MNCs utilize a role of socialization to establish new social ties when engaging in knowledge transfers with partners.	*Finding 26: SME attempts to establish social ties with the role of socialization when engaging in knowledge transfer with the partner.*
		Finding 27: Both SMEs and LC/MNCs rely on the potential partner to engage in a knowledge transfer to establish a social tie with them.	

	motivation relates to the firms motivation of acquiring knowledge and how to obtain it, the organization must invest time and effort to the process of knowledge transfer (Bengoa & Kaufmann, 2014).	*Finding 28: LC/MNCs rely on informal knowledge transfer in meetings, digital or physical to establish new social ties with partners.*	*Findings 28: SMEs seeks to establish social ties with knowledge transfer via phone calls, emails or meetings to with partners.*
		Finding 29: LC/MNCs utilize a role of socialization to handle their knowledge transfers regarding maintaining social ties with partners.	*Findings 29: SME attempts to maintain social ties by knowledge transfers to the role of socialization in the partner organisation.*
		Finding 30: LC/MNCs utilize more formal knowledge transfer regarding maintenance of social ties with partners.	*Finding 30: SMEs prefers informal knowledge transfer in maintenance of the social tie but often ends up using formal means with their partners.*
Motivation	Martín Cruz et al. (2009) state that motivation consists of intrinsic and extrinsic motives. Organizational motivation relates to the firms motivation of acquiring knowledge and how to obtain it, the organization	*Finding 31: SMEs and LC/MNCs are more incentivises by intrinsic motivation in the form of pleasant work environment and accomplishment to acquire new knowledge from partners and sharing knowledge with them.*	

	must invest time and effort to the process of knowledge transfer (Bengoa & Kaufmann, 2014).	*Finding 32: SMEs and LC/MNCs are not specifically motivated by incentives to acquire and share knowledge with partners, instead it is a part of the job description of the employees.*	
		This viewpoint was not mentioned in this perspective from the LC/MNC, which could be because they are already offered job education.	*Finding 33: SMEs need extrinsic motivation in the form of job education (e.g. training or courses) which give them the understanding (e.g. framework) of how and why they should engage in acquisition and sharing of knowledge, while feel autonomous in KT activities from the partner.*
		Finding 34: LC/MNCs apply intrinsic or/and extrinsic motivation on partners to engage with acquisition and sharing of knowledge to fulfill the LC/MNC criterions of collaboration.	*This viewpoint was not mentioned in this perspective from the SME, which could be because e.g., the viewpoint of the relationship between the organizations where the SME seeks approval.*
		Finding 35: Both SMEs and LC/MNCs attempt to indirectly motivate their employees via the job description to acquire and share knowledge with partners.	

		Finding 36: LC/MNCs offer direct extrinsic motivation in the form of job education and indirect intrinsic motivation in the form of accomplishment.	*Finding 36: SMEs offer direct and indirect intrinsic motivation in the form of accomplishment to acquire and share knowledge with partners.*
Absorptive capacity	The literature of absorptive capacity agrees that it consist of aspects such as acquisition of external knowledge and (1) intra-firm knowledge transmission because after acquiring knowledge it needs to be (2) diffused within the firm, which is how they are these aspects are interrelated (De Luca & Cano Rubio, 2019).	*Finding 37: SMEs and LCs/MNCs rely on implementing acquired knowledge from individuals to the organizational level in systematic data bases.*	
		This viewpoint was not mentioned from the perspective of the LC/MNC, which could be of the natural size difference of the organizations.	*Finding 38: When the SMEs acts as a recipient of knowledge they cannot always fully implement the acquired knowledge from the source partner due to organizational limitations.*
		This viewpoint was not mentioned from the perspective of the LC/MNC, which could be because their organizational structure operates differently compared to the partner.	*Finding 39: SMEs don't implement acquired knowledge to the organizational level, but rather to the individual level.*

Knowledge characteristics			
		Finding 40: LCs/MNCs diffuse acquired knowledge amongst employees with meetings, emails, conferences, workshops and their technology network.	*Finding 40: SMEs diffuse the knowledge by staff meetings, projects, recordings and emails.*
		Finding 41: LCs/MNCs rely on the individuals to access the knowledge from the data systems themselves rather than having it given directly to them.	*Finding 41: SMEs diffuse knowledge more informally in the organization.*
Knowledge characteristics	**Tacit and explicit knowledge:** The literature of knowledge characteristics has identified that tacitness, ambiguity and complexity exist for both tacit and explicit knowledge and has a significant impact in inter-organizational knowledge transfer and have multiple times proven that it affects the success and efficiency of knowledge transfer in LCs or MNCs (Simonin, 2004), SMEs (Corral de Zubielqui et al., 2015) and in alliances (Mazloomi Khamseh & Jolly, 2008). **Ambiguity:** Wijk et al. (2008) claims that knowledge ambiguity is	*Finding 42: LC/MNC as a source in a knowledge transfer experience more difficulties in the transfer when the understanding of the knowledge is deemed low by the partner (e.g., with SME as a recipient side due to the less diversified expertise).*	*Finding 42: SME as a source in a knowledge transfer experience less difficulties in the transfer when the understanding of the knowledge is deemed high by the partner (e.g., with LC/MNC as a recipient side due to the more diversified expertise).*
		Finding 43: LC/MNC as a source in a knowledge transfer experience more difficulties in the transfer when the understanding of the interactions and components of the knowledge is deemed low by the partner (e.g., with SME as a recipient side due to their maturity level and experience).	*Finding 43: SME as a source in a knowledge transfer experience less difficulties in the transfer when the understanding of the interactions and components of the knowledge is deemed high by the partner (e.g., with LC/MNC as a recipient side due to their maturity level and experience).*

defined by the uncertainty of the knowledge, what underlying components and sources are in play (content-dependency), along with how they interact with each other (context-dependency). **Complexity**: Complexity consider how the knowledge that is transferred is affected by different variations as it is comprehended by different skills or experiences of organizations (Stock & Tatikonda, 2000)… complexity can be measured by how many different variables are involved (content and context dependency) in a certain situation (Milagres & Burcharth, 2019). **Tacitness**: Kalling, (2003), state that tacitness exists for both tacit and explicit knowledge, whereas tacitness is affected by the individuals understanding of the knowledge they are handling… to measure tacitness in knowledge to some extent we will analyse the individuals understanding of the context at hand (context dependency).	*Finding 44: LC/MNC as a source in a knowledge transfer experience less difficulties in the transfer when there are low amount of variables involved in the interactions and components of the knowledge that is transferred to the partner (e.g., with SME as a recipient side due to their organizational demands).*	*Finding 44: SME as a source in a knowledge transfer experience more difficulties in the transfer when there are high amount of variables involved in the interactions and components of the knowledge that is transferred to the partner (e.g., with LC/MNC as a recipient side due to their organizational demands).*

5.1 Aspect: Knowledge governance mechanisms

The key challenge that can impede the knowledge transfer between different partners is the opportunistic behaviour of parties in terms of evading their obligations and responsibilities related to the improper allocation of knowledge, efforts, and other knowledge-based assets as a result, partners apply various knowledge governance mechanisms to avoid opportunistic behaviours of their partners (Yang et al., 2015; Mesquita et al., 2008).

According to the literature, knowledge governance mechanisms can take two modes such as formal and informal: (1) the formal governance mechanisms consist of management policies and procedures, organizational structure, incentive schemes, information systems, and other control and coordination systems, whereas (2) informal (also referred to as relational) includes corporate culture, trust, social networks, and communities (Foss, 2007).

Our study found that both the LC/MNC and SME utilize formal knowledge governance mechanisms for establishing common goals and structuring of collaboration projects, which eventually lead to facilitation of the knowledge transfer coordination between partners. This finding shows that is consistent with the literature and both sized companies use formal knowledge governance mechanisms for coordination purpose.

However, the literature analysed in this study does not explain how informal knowledge governance mechanisms are applied by different sized companies. Our study shows that informal knowledge governance mechanisms are used by LC/MNCs to transfer knowledge within the organization including to enhance the implementation and progress of KT projects due to better navigation in a large network, while utilizing the formal knowledge governance mechanisms when knowledge is needed to transfer to the external organizations such as customers wherein the knowledge transferred should be clear and structured, which can help reduce misunderstanding, for example, due to cultural and geographical distance.

In addition, since the larger the firm's size the more cost structure management focus in those organizations, therefore formal knowledge governance mechanisms can facilitate improvements of cost efficiency programs of LC/MNC in terms of systemizing and controlling knowledge transfer activities with their partners. In contrast, SMEs are focused on expanding their customer base and firm growth as a result they more apply informal knowledge governance mechanisms to interact with their external partners.

Moreover, informal knowledge governance mechanisms can contribute to enhancement of adaptability and flexibility capabilities of SME but as a firm size increase the more formal knowledge governance mechanisms needs to be applied rather informal.

In conclusion, our study has contributed to the theory extending the concept of the knowledge governance describing how formal and informal knowledge governance mechanisms are applied by different sized firms, for example, since LC/MNC aim to reduce their cost and improve profitability they utilize in greater extent the formal knowledge governance mechanisms, whereas SMEs are focused on growth and expansion and as a result, they use more informal procedures.

5.2 Aspect: Trust

The interorganizational knowledge transfer is significantly influenced by trust and trustworthy interaction between independent partners including LC/MNC as well as for SMEs, but different size companies exploit different modes of the trust-based partnerships (Martinkenaite, 2011). This is because of trust between organizations affects the success of knowledge transfer, as trust correlates to the source's credibility and trustworthiness to deliver adequate quality of both the knowledge and knowledge transfer process (Al-Salti & Hackney, 2011; Chen et al., 2014). According to the literature, trust can be measured based on social judgements in terms of (1) evaluation of other partners' benevolence, (2) expertise and competence coupled with (3) assessment of the risk related to untrustworthy behaviour (Inkpen & Tsang, 2005). In addition, to this, the literature proposes to assess interfirm trust based on (1) fulfillment of commitments agreed between partners; (2) honest negotiation; and (3) avoidance of receiving excessive benefits of ally organizations (Cummings & Bromiley, 1996). Moreover, mutual trust can be built by means of mutual reciprocity behaviour: "*If we helped them today, they would help us later*" (Al-Jabri & Al-Busaidi, 2018).

Our study found that, like previous scholar findings, trust for inter-organizational knowledge transfer is evaluated in terms of competence and expertise of counterparts that is applicable for both sized forms. Furthermore, our research supports that mutual trust for KT is enhanced by virtue of mutual benefit and reciprocity. This implies that one of the components of trust to transfer knowledge is mutual beneficial behaviour of both partners to achieve mutual goals and business performance rather partner's benevolence.

In contrast to the literature, where trust defined as "*the willingness of a party to be vulnerable to the actions of another party based on the expectation that the other will perform a particular action important to the trustor, irrespective of the ability to monitor or control that other party*" (Mayer et al., 1995). Our research shows that trust for KT activities between partners for LC/MNC and SME is based on signing non-disclosure agreements (NDAs) that means that there is limited trust between partners that is restricted by formal trust agreements. This could be due to risk assessment of knowledge leakage that LC/MNC and SMEs evaluate differently. Particularly, the greater the size of the partner the higher risks are associated with the KT activities for LC/MNC, whereas for SME, the larger organization the higher credibility and trust for KT activities. Furthermore, it was found that LC/MNC rely more on formal trust agreements for knowledge transfer as compared SME this probable because LC/MNC possess higher power to enforce the fulfillment of formal trust agreements, while SMEs have less power and considers enforcement of the fulfillment of formal trust agreements as costly and ineffective.

To summarize, these aspects were not explained by the selected literature, in this sense, our book contributes explaining the concept of trust for interfirm knowledge transfer in the context of LC/MNC and SMEs by supporting that there is limited trust for KT activities between partners.

5.3 Aspect: Cultural distance

The literature agrees that due to an expansion of businesses across different countries cultural factors significantly affect inter-organizational knowledge transfer in terms of higher complexity for knowledge flow because diverse national cultures, cultural characteristics as well as organizational culture influence the employee's ultimate behaviour in the interfirm knowledge transfer process (Milagres & Burcharth, 2019; Cheung et al., 2011; Easterby-Smith et al., 2008). As a result, the national cultural distance can serve as barrier for the KT among international partners regardless the firm size such as (1) higher cost of entry and hinders the transmitting the core competencies to the international markets; (2) facing operational challenges in terms of lacking common understanding of norms, values, and motivation; (3) misunderstanding between foreign partners can limit sharing of critical organizational knowledge (Wijk et al., 2008).

Our study supports previous scholars' findings that the language proficiency ability is essential aspect for successful knowledge transfer in the context of different cultures and LC/MNC that often engage in the global level partnerships support that language barrier is key aspect that can hinder knowledge transfer and can lead to misunderstanding between partners. As a result, the case LC/MNC has special dedicated language courses for its employees to improve their language skills, which eventually can enhance knowledge exchange process with international partners. Furthermore, our research also confirms that high-power distance in terms of strong hierarchical subordination among employees that exists in countries such as China and Japan can lead to misunderstanding due to less opportunity to ask questions or provide opposite opinion those managers who possess higher hierarchical position. As a result, interactions between hierarchical levels tend to be more authoritative and paternalistic, eventually, this can lead to impeding knowledge transfer between different organizations and increases the cost due to mistakes related to KT between different cultural societies. These findings are true regardless the firm size.

However, this study has identified that SME strives to understand the process of decision-making processes of their international partners and our study findings suggests that a lack of understanding of the decision-making process in different cultures can negatively affect KT activities between partners. This aspect was not described in the literature, in this sense, our book contributes explaining that SMEs are more strives to explore the decision-making process and motivation behind decisions taken of their international partners as compared to LC/MNC.

5.4 Aspect: Geographical distance/proximity

Geographical proximity or distance is essential relational aspect that impacts the inter-organizational knowledge flow because the transfer of knowledge slower down and less effective as physical distance increases between source and recipient due to the difficulty, the time, and the cost of communication (Battistella, 2016; Cummings & Teng, 2003).

According to the academics findings, the geographical distance negatively influences inter-organizational KT especially for tacit knowledge exchange but can be mediated in terms of (1) developing trust-based relations between distant partners (e.g., Capaldo & Petruzzelli, 2014); (2) organizational proximity including common corporate culture and values (Korbi & Chouki, 2017); and (3) bilateral way of utilizing modern communication technologies (Choi & Contractor, 2016) in LC/MNCs as well as in SMEs.

Our study found that physical proximity is more important when partners want to build trust for knowledge transfer activities for LC/MNC and SME. Particularly, physical proximity via face-to-face meetings can facilitate to transfer of the complex knowledge such as technology transfer and to solve different emergent situations. This implies that tacit knowledge and related complexity can be better transmitted to other partners only via real physical interaction. Implicitly, this means that trust-based relationships are important aspect that can be built via physical proximity. Moreover, for SME, geographical proximity is essential because it can facilitate to establish trust between partners for better learning capacity. This means that to establish trust first it is needed to have physical proximity in terms of face-to-face interaction and then this collaboration can move to distant partnerships. These findings are extending the previous research on this aspect.

Besides, our research found that LC/MNC due to pandemic situation that the world experiencing at present, need to develop adaptability capacity between partners, which is substantial capability for geographical distant partners. This aspect was not explained in the literature; hence, our book contributes describing what sub-aspects can additionally mitigate negative impacts of geographical distant partnerships for successful KT.

In addition, our study supports the current literature that organizational culture in term of common values and willingness to collaborate is more prevalent than the physical distance for knowledge transfer between partners. Furthermore, our findings support the previous literature findings: digital communication tools also positively affect knowledge transfer between distant partners but as a complementary technique since face-to-face meetings are still important to establish trust between partners.

In conclusion, our study has contributed to the theory extending the concept of geographical distance/proximity describing that physical proximity is more prevalent aspect to establish trust-based relationships between distant partners. In addition, LC/MNC should develop adaptability capacity to manage negative effects of geographical distant partnership for knowledge transfer.

5.4 Aspect: Disseminative capacity

The literature suggest that disseminative capacity is essential aspect for the interfirm knowledge flow because if the knowledge source fails to transfer the knowledge, then recipient cannot successfully assimilate and utilize this knowledge (Kuiken & van der Sijde, 2011; Tang et al., 2010).

According to the selected literature in this study, the disseminative capacity is measured in terms of (1) willingness to transfer knowledge and (2) ability to teach by the source of knowledge (Minbaeva et al., 2018; Easterby-Smith et al, 2008).

Concerning motivation of knowledge source, our study shows that motivation of the source to transfer knowledge operates in similar way that scholars previously identified. Particularly, mutual experience sharing and learning from each other are key drivers for KT between partners. Furthermore, another key motivation for knowledge exchange with partners is mutual business performance enhancement, that eventually can shape mutual trust motivates partners for KT activities that was not explained in the literature, in this sense, our book contributes explaining that one of components for the knowledge source motivation is achieving mutual benefits between partners.

In addition, our study describes that SMEs as a source of knowledge is less motivated to exchange knowledge with partners. For instance, SME mentioned as an example, that they are less motivated to exchange knowledge with LC/MNC as a recipient side due to their inflexibility and rigid processes.

As far as knowledge source's ability to teach is concerned, our research shows that knowledge holder, both LC/MNC as well as SME, assesses knowledge recipient's competence that measured by organizational structure and processes of the partner that help to the knowledge source to understand the needs and maturity level of the recipient, eventually this can facilitate KT activities between different partners.

Moreover, our study found that knowledge holder's ability to transfer knowledge depends on evaluation of risks related to the knowledge leakage and protecting valuable knowledge before engaging KT activities with knowledge recipient. Finally, our research shows that both sized firms support that source's ability to codify the knowledge facilitates knowledge transfer between partners.

These all findings are similar that academics identified previously in the literature.

5.5 Aspect: Social ties

Social ties according to Tangaraja et al. (2016) is a relationship based concept, its either conducted directly or indirectly between two parties and differentiated by informal or formal ways. As when organizations have established strong social ties, it acts as a facilitator for knowledge transfer (Kang & Sauk Hau, 2014) and Liu et al. (2015) argues that there are three steps to establishing a high quality relationship, first step involves to locate a partner firm and identify its key contacts, secondly to develop and maintain the relationship established, last stage involves dispatching of knowledge from the source along with knowledge absorption of the recipient, which is how the knowledge is exchanged. Thus, strong social ties enhance the transfer of knowledge in inter-organizational relationships since it facilitates KT better than formal contracts by per example social events, interactions or team building (Chen & McQueen, 2010) where strong social ties promotes communication between organizations (Kraatz, 1998), which strengthens the sharing of knowledge in terms of interests and goals (De Luca & Cano Rubio, 2019) and aids the pursuit of gaining new knowledge from the relationship (Al-Salti & Hackney, 2011).

Thus, to establish and maintain social ties between organization is essential to successfully transfer knowledge between them, therefor, according to the literature, social ties has been researched on SMEs and LC/MNCs if they prefer (1) formal or informal means, (2) how they establish their social ties and (3) how they maintain them.

The study conducted in this book revealed that LC/MNCs leans more towards formal social ties, while SMEs leans more towards informal social ties. This implies that the organizations attempt different approaches towards the social ties they have established, which may impact the social tie negatively in per example expectations, results or misunderstandings as when one organization approaches the potential partner, they are met by different criterions. This indicates that the knowledge transfer is affected differently and that the organizations needs to find a middle ground to meet, however, the study found that SMEs often end up adopting the formal means of the partner, which could be because of several different reasons such as the partner has more influence in the relationship, because e.g., they have more options in regards of alternative social ties which are equivalent to the SME.

When it comes to establishing new social ties, the presence of a role of socialization in the LC/MNC assists the establishment of the social ties, as in this study, the partner was aware of the role of socialization, this implies that the role of socialization for an organization is essential for knowledge transfer in the early stages of social ties as it assists the organizations to establish the initial connection. Additionally, both the SME and LC/MNC utilize same means of establishing their social ties, per example by digital or physical meetings and phone calls or email. Also, the study found that both the SME and LC/MNC relies on the other potential partner organization to establish the social ties, while it was a small portion of the informants that stated this and the reasoning from the LC/MNC was their major position in the market and the informant from the SME did not specify the reason behind it, but the perspective was more from them being contacted. This implies that both the LC/MNC and SME needs to be aware that when they are being approached, they are the ones that needs to be persuaded to engage in the social tie, which also implies that they are not the one researching the other organization (initially), but rather relies on being persuaded. This means that the social tie might not always the be most optimal one for either organization, as they are not the one that initially does the prior research regarding the other organization.

As for maintaining social ties, the finding from previous paragraph regarding the role of socialization is covered in maintenance as well, this shows that both organizations are aware of the connection point that bridge the knowledge transfer in the social tie. As previously discussed regarding formal or informal social ties, the study revealed that LC/MNCs utilize more formal means of maintenance, while the SME prefer informal means, but they end up adapting to the formal means of the partner. Which one again, could be because of the influence of the LC/MNC, as one informant from the SME stated that the partner utilize performance indicators for the SME to follow, which is how they maintained their relationship with the SME, by strict formal knowledge transfers, this indicates that the SME need to adapt to the partner's means of formal maintenance to keep them satisfied with the relationship, which may impact their organization negatively in terms of resource allocation to meet the needs of the partner. This finding shows that the partner needs to be aware of the informal needs and knowledge transfers ways of the SME and perhaps adjust their maintenance of social ties, as it could be more fruitful in the long-term and yield better results in terms of successful knowledge transfer.

5.6 Aspect: Motivation

Motivation is an important aspect reoccurring in the literature of inter-organizational KT, the aspect occurs on an individual and organizational level, where on the former, motivation is key for example achieving effective networking between firms (Harris, 2008), in the collective context of the organization to promote acquisition of knowledge and distribution of it (Jantunen, 2005) and in alliances where learning intent as a motivational aspect facilitates higher efficiency of knowledge transfer (Mazloomi Khamseh & Jolly, 2008). Organizational motivation relates to the firms motivation of acquiring knowledge and how to obtain it, the organization must invest time and effort to the process of knowledge transfer (Bengoa & Kaufmann, 2014), which correlates to the agenda of motivation, as the transaction must be seen worthwhile for the organization to invest in, along with the recipient to be motivated to gain knowledge (Easterby-Smith et al., 2008) and being motivated to learn (Pérez-Nordtvedt et al., 2008) while possessing learning intent (Mazloomi Khamseh & Jolly, 2008).

Martín Cruz et al. (2009) state that motivation consists of (1) intrinsic and extrinsic motives, as such, to motivate employees this study has researched what motivates the employees and how the organization attempts to motivate their employees to acquire and share knowledge with or from their partners.

This study found that to engage in a knowledge transfer regarding acquisition and sharing of it, employees were more incentivised by intrinsic motivation in both SMEs and LC/MNCs, with the form of pleasant work environment and sense of accomplishment. The reasoning behind this could be because of how individuals operate in their daily basis, as by having a work environment which foster the employee's motivation assists them in inter-organizational knowledge transfer as it could e.g., free up time in the future, by feeling a sense of accomplishment employees are motivated by continuous engaging in knowledge transfers as it leads to e.g., successful projects, collaborations and so forth. While both SMEs and LCs/MNC attempts to indirectly motivate their employees with intrinsic motivation, only the LC/MNC were found to offer direct extrinsic motivation in the form of job education, which could be argued is because of the size of the organization or lack of resources in terms of time or finances.

For incentives, in both SMEs and LC/MNCs, it was found that the employees are not necessarily motivated by intrinsic or extrinsic terms to engage in an inter-organizational knowledge transfer, which was due to the expectations of the employee, whereas if it was not a part of the actual job description they are assigned, it is not within their scope of responsibility to engage in these kinds of activities either. This shows that organizations should be aware of how the job description affects inter-organizational knowledge transfer, whereas an absence of it, the employees might deter from engaging in it. Which were found from the organization as well, as both the SME and LC/MNC attempt to increase the acquisition and sharing of knowledge with partners via the job description and indirectly motivate them via their job description rather than applying incentives towards the employees.

As for extrinsic motivation, employees in SMEs were found to have a larger need of it, in the form of job education, as the employees are not always aware of the framework which they can share knowledge or acquire new knowledge. This implies that SMEs should engage more in education their employees regarding knowledge that is/could be deemed valuable to the organization, which could per example increase generation of new knowledge. This could also be like the view of when the partner is the recipient and the LC/MNC is the source, then the perspective from the LC/MNC, as a larger firm they put their smaller partners in different categories of performance and maturity level. Where in one way, the LC/MNC is indirectly motivating the partner, by either intrinsic or extrinsic motivation to increase their knowledge transfer activities to advance their performance and maturity level in the eyes of the LC/MNC, as it would benefit the partner if they were per example put in a better categorization, which could lead to further knowledge transactions with the LC/MNC.

5.7 Aspect: Absorptive capacity

What is true for all approaches to absorptive capacity is that it facilitates inter-organizational knowledge transfer (Mowery et al.. 1996) and contributes to knowledge learning internally in firms (Szulanski. 1996). Therefore, no matter how one author develop different approaches to absorptive capacity, the KT process itself will always be important for absorptive capacity since it is used for exploitation and commercialization of the acquired knowledge (Easterby-Smith et al.. 2008). Thus, for this book the importance of absorptive capacity is clear, organizations should understand how to manage and transfer knowledge from external sources and diffuse the knowledge internally to increase their innovative capabilities and organizational performance.

As our thesis's scenario is based on two organizations already engaging in knowledge transfer, the book will not consider acquisition of external knowledge for absorptive capacity as the acquisition of knowledge is already established from the scenario that is researched. Therefore, the acquisition of external knowledge is not the focus of this study, rather the implementation and dissemination of knowledge. Thus, as the research is done on an organizational level, to then measure absorptive capacity this book will research how organizations go about implementing external knowledge to the organizational level, along with how the organization diffuse it in the organization from the individual.

This study found that both SMEs and LC/MNCs rely on knowledge from the individual level, to the organizational level is by implementing the acquired external knowledge in systematic data bases, while it exists to a higher degree in LCs/MNC, because of their natural size, they have more systematic ways to implement knowledge as they are more naturally invested in more variety of knowledge. This implies that implementing knowledge to a LC/MNC requires more navigation of where to implement it, while it could be easier to implement the knowledge in the partner organization, where there could be e.g., less diverse data bases. This finding also relates to how SMEs as a recipient, were found to not be able to implement acquired knowledge fully when it was acquired from a source partner, as the knowledge is naturally involving more variables and the SME needed to divide the knowledge partially, to be able to implement the knowledge into the organization. Furthermore, SMEs tended not to implement acquired knowledge as frequently to the organizational level, but rather to the individual level, while unspecified if it remained on that level, it shows that if the SME does not implement the knowledge to the organizational level, the knowledge may lose it value in the future, as it could be lost or misunderstood if it is not codified.

As for dissemination of knowledge in the organizations, both the SME and LC/MNC adopted similar ways, however the findings shows that the LC/MNC attempts to spread the knowledge in a broader sense, in the organization, either by larger gatherings such as conferences or by their technological networks where they store and share knowledge. While the SME also utilize their technological network, it was only confirmed by secondary data and not primary, which shows that the employees might not always be aware of the technological means when disseminating knowledge via their networks, instead the SME relies more on smaller scale of dissemination of knowledge, per example staff meetings, projects, recordings and emails. This implies that the dissemination of knowledge has different scopes of focus in the different organizations, while LC/MNCs attempts to disseminate it more broadly, as their size is naturally larger, it is only natural that they utilize means adapted for the size, the SMEs attempts to disseminate more narrowly, also natural to their size, as the knowledge might not need to be dissemination to everyone in the organization. Additionally, SMEs were found to disseminate knowledge more informally in the organization, since they have a smaller societal structure with less hierarchal structure, while the LC/MNC did it more formally, which could be since they have a larger societal structure and more hierarchal structure in the organization. This shows that the knowledge transfer when disseminating it in the organizations is affected differently by the organizational different sizes.

5.8 Aspect: Knowledge characteristics

The literature has identifies various knowledge characteristics that distinguish the object of the transfer and that can influence the transfer process of knowledge (Battistella, 2016). Furthermore, the knowledge transfer process includes an exchange of the object between two actors, wherein the knowledge as a object is transferred with its nature characteristics (Ajith Kumar & Ganesh, 2009) and depends on context-specific properties (Cohen & Levinthal, 1990). The nature of the knowledge is characterized with two dimensions such as tacit and explicit (Nonaka, 1991), whereas the context of knowledge is described in terms of tacitness, ambiguity and complexity, which exist for both tacit and explicit knowledge because these context-specific properties can serve as main barriers for interfirm KT regardless of the knowledge type exchanged between different entities (Milagres & Burcharth, 2019).

As the knowledge characteristics such as (1) tacitness, (2) ambiguity and (3) complexity exist for both tacit and explicit knowledge and has a significant impact in inter-organizational knowledge transfer and have multiple times proven that it affects the success and efficiency of knowledge transfer in LCs or MNCs (Simonin, 2004), SMEs (Corral de Zubielqui et al., 2015) and in alliances (Mazloomi Khamseh & Jolly, 2008). As such, this study has researched how these characteristics (1-3) are perceived and prioritized by the LC/MNC and SME when it comes to tacit and explicit knowledge.

Regarding tacitness of explicit and tacit knowledge, this study has shown that it is of high priority by LC/MNC as when they engage as a source in a knowledge transfer with a recipient partner, the informant often mentioned the maturity level and the experience is of great importance of the recipient when it comes to both explicit and tacit knowledge, this implies that the knowledge transfer is affected by tacitness of the recipient, in terms of the partners maturity level and experience, which the understanding or context of the knowledge is highly affected by. This means that LC/MNC must develop correct means of transferring both tacit and explicit knowledge when they act as a source in the knowledge transfer to increase the tacitness (understanding) of the recipient, as it affects the knowledge by how it is understood, which implies that if there is an absence of understanding the recipient cannot generate value from said knowledge, which could also become costly for the source.

As for SMEs and tacitness, the study shown it was to some extent prioritized, however when they engage in a knowledge as a source in the transfer, the informant mentioned that tacitness is generally higher in e.g., a larger partner organization due to their diverse access to knowledge and individuals, which implies that SMEs as a source does not need to be as concerned about the tacitness of knowledge characteristics when they engage in a knowledge transfer with a recipient partner of a larger size, as the recipient has a natural advantage when it comes to tacitness due to the organizational size.

The ambiguity in knowledge from the perspective of the LC/MNC and SME as a source in the knowledge transfer, was shown in the study to not be of a high priority, as the informants showed high awareness that they need to adapt the knowledge in terms of how the components and interactions (content) of the knowledge changes when it is applied to a different context, this implies that the ambiguity is well understood in our case study, which could be perceived differently in other case studies.

Complexity in knowledge in the perspective of the LC/MNC acts as a source in a knowledge transfer to a partner recipient was found to be low, which was due to the partner organizational size, this finding was also validated from the perspective of the SME but when the SME acts as a source in the knowledge transfer with their partner recipient, complexity was found to be high due to the organizational size of the partner and time constraints. This implies that there is less complexities when the recipient is a SME and more complexities in a LC/MNC, which means that when the recipient is a LC/MNC, they should be aware of their organizational complexities and attempt to decrease it by per example communication or similar means when the source SME attempts a knowledge transfer with them, to achieve a successful transfer.

6. Conclusions and Theoretical implications

The research conducted aims to provide a higher understanding of the various aspects for inter-organizational knowledge transfer in different firm sizes such as SMEs to their partners and LC/MNCs to their partners. As this setting of analysis provides two scenarios, it assist both sized organizations to adequately leverage KT to generate sustainable competitive advantage (Easterby- Smith et al., 2008; Cardoni et al., 2019), as knowledge itself must be transferable and sharable across organizations for it to contribute competitive advantage for firms and to improve their business performance (Minbaeva et al., 2003; Levine & Prietula, 2012). It is then clear that organizations such as SMEs and LC/MNCs are affected differently due to their different capabilities and priorities, when engaging in an inter-organizational knowledge transfer.

As the existing bult of literature has established the knowledge transfer aspects, with a focus on KT in LC/MNC and SME, this book has found many similar findings compared to the existing literature but also contributed to the theory of inter-organizational knowledge transfer on the organizational level for SMEs and LC/MNCs in the sense of what similarities and differences the organizations experiences regarding knowledge transfer aspects such as knowledge governance mechanisms, trust, cultural distance, geographical distance/proximity, disseminative capacity, social ties, motivation, absorptive capacity and knowledge characteristics when they engage in a knowledge transfer with a partner. In this sense, this book extends the literature of KT for both SMEs and LC/MNCs when they engage with a recipient e.g., of the other size, as it assists the organization to understand the recipient better to achieve a successful transfer of knowledge.

Thus, the identified aspects were found to be most reoccurring during the research and affect the different case selections in terms of:

(1) Knowledge governance mechanisms:

The theoretical framework in this book has defined how trust affect inter-organizational KT (Foss, 2007). The study found that both the LC/MNC and SME utilize formal knowledge governance mechanisms for establishing common goals and structuring collaboration projects, which eventually lead to the facilitation of the knowledge transfer coordination between partners. However, the literature analysed in this study does not explain in detail how formal and informal knowledge governance mechanisms are applied by different-sized companies. Our study has contributed to the theory extending the concept of knowledge governance describing how formal and informal knowledge governance mechanisms are applied by different sized firms, for example, since LC/MNC are aimed to reduce their cost and improve profitability they utilize in greater extent the formal knowledge governance mechanisms, whereas SMEs are focused on growth and expansion and as a result, they use more informal procedures.

(2) Trust:

The theoretical framework in this book has defined how trust affect inter-organizational KT (Al-Jabri & Al-Busaidi, 2018; Inkpen & Tsang, 2005; Cummings & Bromiley, 1996). The research identified that mutual trust for KT is enhanced by virtue of mutual benefit and reciprocity for LC/MNCs and SMEs. This implies that one of the components of trust to transfer knowledge is mutual beneficial behaviour of both partners to achieve mutual goals and business performance rather than partner's benevolence. Nevertheless, our research shows that trust for KT activities between partners for LC/MNC and SME is based on signing non-disclosure agreements (NDAs), which means that there is limited trust between partners that is restricted by formal trust agreements. Overall, our book contributes to explaining the concept of trust for interfirm knowledge transfer in the context of LC/MNC and SMEs by supporting that there is limited trust for KT activities between partners.

(3) Cultural distance:

The theoretical framework in this book has defined how trust affect inter-organizational KT (Wijk et al., 2008). The study supports previous scholars' findings that the language proficiency ability is an essential aspect for successful knowledge transfer in the context of different cultures and LC/MNC that often engage in the global level partnerships support that language barrier is a key aspect that can hinder knowledge transfer and can lead to misunderstanding between partners. However, this study has identified that SMEs strive to understand the process of decision-making processes of their international partners and our study findings suggest that a lack of understanding of the decision-making process in different cultures can negatively affect KT activities between partners. As a result, our book contributes to the literature by explaining that SMEs are focused on examining the decision-making process and motivation behind decisions taken of their international partners as compared to LC/MNC.

(4) Geographical distance/proximity:

The theoretical framework in this book has defined how trust affect inter-organizational KT (Korbi & Chouki, 2017; Choi & Contractor, 2016; Capaldo & Petuzzelli, 2014). The study found that physical proximity is more important when partners want to build trust for knowledge transfer activities for LC/MNC and SMEs. Particularly, physical proximity via face-to-face meetings can facilitate to transfer of complex knowledge such as technology transfer and to solve different emergent situations. Hence, our study has contributed to the theory extending the concept of geographical distance/proximity describing that physical proximity is a more prevalent aspect to establish trust-based relationships between distant partners. In addition, LC/MNC should develop adaptability capacity to manage negative effects of geographical distant partnership for knowledge transfer.

(5) Disseminative capacity:

The theoretical framework in this book has defined how trust affect inter-organizational KT (Minbaeva et al., 2018; Easterby-Smith et al., 2008). The study found that mutual experience sharing and learning, mutual business performance enhancement, knowledge holder's ability to codify the knowledge significantly facilitate knowledge transfer between partners. However, our study found that SMEs as a source of knowledge are less motivated to exchange knowledge with different partners such as LC/MNCs due to their rigid processes and inflexibility.

(6) Social ties:

The theoretical framework in this book has defined how social ties affect inter-organizational KT (Tangaraja et al., 2016; Liu et al., 2015). This study has found that the LC/MNC and SME prefer different types of types of social ties, whereas LC/MNC has both types, they often ended up using formal social ties, while the SME also had both types, they often ended up adapting to the partners need (e.g., a LC/MNC) of formal social ties even though the SME prefer informal social ties. This mismatch of means towards a social tie leverages the SME to adapt to a larger partner needs, as e.g., the SME might face replacement or being less prioritized by the larger partner. In the eye of the LC/MNC, the reason to why they lean towards formal means could be of their natural size as an organization, as it becomes more structured to follow formal means. The similarities in social ties were how the organizations (LC/MNC and SME) establishes social ties, where both were found to be aware of the partners role of socialization and attempted to establish a social tie primarily with this individual to further assist developing the social tie.

(7) Motivation:

The theoretical framework in this book has defined how social ties affect inter-organizational KT (Martín Cruz et al., 2009). The study found that SMEs and LC/MNC have similar incentives in the form of intrinsic ones, where they motivate their employees to acquire and share knowledge by a sense of accomplishment, whereas the difference was that the LC/MNC offer more extrinsic motivation in the form of job education, which was of high absence in the SME, additionally the SME were found to be in need of the extrinsic motivation in the form of job education to give their employees a better understanding of the scope of which they can share knowledge and acquire it (what specifically to acquire).

Moreover, both the LC/MNC and the SME were found to not necessarily be motivated by incentives to acquire and share knowledge, it was rather a responsibility of the job description dedicated towards one employee. This indicates that both organizations must be aware of when specifying a job description to one employee, the employee will not engage in inter-organizational KT unless specified to do so.

(8) Absorptive capacity:

The theoretical framework in this book has defined how social ties affect inter-organizational KT (De Luca & Cano Rubio, 2019; Cohen & Levinthal, 1990; Tripsas, 1997). The study found that SMEs could not always implement the knowledge received from e.g., a LC/MNC partner, due to organizational limitation. This finding shows the importance in the perspective of the LC/MNC that when they are the source in a knowledge transfer to a partner that is e.g., a SME, they must be aware of that the knowledge that is being transferred might not be successfully transferred, due to the limitations of the recipient organization and by highlighting this scenario, it could assist LC/MNCs to successfully adapt the knowledge to the needs of e.g., the SME recipient to achieve a successful knowledge transfer.

(9) Knowledge characteristics:

The theoretical framework in this book has defined how social ties affect inter-organizational KT (Milagres & Burcharth, 2019; Fang et al., 2013; Nonaka & von Krogh, 2009; Kalling, 2003). The study found that regarding ambiguity, complexity and tacitness the LC/MNC perspective as a source in KT and e.g., a partner recipient from a SME, the interviewees often discussed about the maturity level, experience and organizational demands were less in comparison to their organization, where complexity were found to be easier transferred to a partner of e.g., smaller size due to less organizational demands, ambiguity and tacitness were found to be more difficult to transfer to a partner of e.g., a smaller size due to e.g., the maturity level or experience of the organization. In contrast, the perspective of the SME validated the perspective from the LC/MNC, as the interviewees were found to implicitly state the same, but in the other way, where complexity was more difficult due to organizational demands, tacitness and ambiguity were easier to transfer due to more experience and a higher maturity level when the partner was a e.g., LC/MNC.

7. Managerial implications

This book apart from theoretical contribution brings practical implications to senior managers of companies at the strategical level.

First, the findings provide an understanding of aspects that affect inter-organizational knowledge transfer where one organization can prioritize aspects that belong to different sized partners and shape accordingly a strategy on collaboration and interaction with external partners of different sizes. Therefore, when top managers are aware of aspects that affect knowledge exchange between partners, they can identify various risks of success and failure in KT processes according to the list of aspects provided in this study and prioritize actions and opportunities that can arise due to engagement in that partnership: cultural, geographical, motivation, partner's capabilities, ambiguity, complexity and etc.

Secondly, to draw a holistic snapshot of various aspects in the findings that affect KT activities from the inter-organizational perspective for different firm sizes in terms of similarities and differences of KT activities for LC/MNCs and SMEs, managers can establish specific strategies to analyse their own knowledge transfer activities to adjust accordingly when participate in inter-organizational knowledge transfer with partners depending on the firm size.

8. Future research

The field of knowledge transfer researched in this book explores how the inter-organizational knowledge transfer aspects are affected by different sized organizations such as SMEs and LC/MNCs when they act as a source in a transfer of knowledge to a recipient partner on an organizational level, by utilizing a theoretical framework based on existing literature of knowledge transfer and a case study containing the different sized organizations.

This path of research revealed several options to pursuit regarding different phenomena of analysis, to which future research could be done on an inter-organizational level to further research the impact of how knowledge transfer aspects are affected by different sized firms, some future research based on this book could be covering aspects like:

Inter-organizational KT theoretically does not define properly the role of each participant and the size. In the same direction, the literature on KT does not establish aspects that affect the process leading to situations in which the interaction occurs from the same size and different size companies. Studies based on empirical data is needed to establish:

- The perspective of a SME as a source transferring knowledge to a recipient LC/MNC.
- The perspective of a LC/MNC as a source transferring knowledge to a recipient SME.
- The perspective of a SME as a recipient of knowledge from a LC/MNC.
- The perspective of a LC/MNC as a recipient of knowledge from a LC/MNC.

The measurement of the aspects affecting KT in inter-organizational context can be clarified to determine, the level of maturity in the KT practices, the effectiveness of the KT and the best practices to improve KT capabilities.

9. References

Ahammad, M. F., Tarba, S. Y., Liu, Y., & Glaister, K. W. (2016). Knowledge transfer and cross-border acquisition performance: The impact of cultural distance and employee retention. *International Business Review*, *25*(1), 66–75. https://doi.org/10.1016/j.ibusrev.2014.06.015

Ajith Kumar, J., & Ganesh, L. S. (2009). Research on knowledge transfer in organizations: A morphology. *Journal of Knowledge Management*, *13*(4), 161–174. https://doi.org/10.1108/13673270910971905

Ali, I., Ali, M., Salam, M. A., Bhatti, Z. A., Arain, G. A., & Burhan, M. (2020). How international SME's vicarious learning may improve their performance? The role of absorptive capacity, strength of ties with local SMEs, and their prior success experiences. *Industrial Marketing Management*, *88*, 87–100. https://doi.org/10.1016/j.indmarman.2020.04.013

Al-Jabri, H., & Al-Busaidi, K. A. (2018). *Inter-organizational knowledge transfer in Omani SMEs: Influencing factors*. 20.

Al-Salti, Z., & Hackney, R. (2011). Factors impacting knowledge transfer success in information systems outsourcing. *Journal of Enterprise Information Management*, *24*(5), 455–468. https://doi.org/10.1108/17410391111166521

Ambos, T. C., Ambos, B., & Schlegelmilch, B. B. (2006). Learning from foreign subsidiaries: An empirical investigation of headquarters' benefits from reverse knowledge transfers. *International Business Review*, *15*(3), 294–312. https://doi.org/10.1016/j.ibusrev.2006.01.002

Andersson, U., Gaur, A., Mudambi, R., & Persson, M. (2015). Unpacking Interunit Knowledge Transfer in Multinational Enterprises. *Global Strategy Journal*, *5*(3), 241–255. https://doi.org/10.1002/gsj.1100

Apriliyanti, I. D., & Alon, I. (2017). Bibliometric analysis of absorptive capacity. *International*

Business Review, *26*(5), 896–907. https://doi.org/10.1016/j.ibusrev.2017.02.007

Argote, L., & Ingram, P. (2000). Knowledge Transfer: A Basis for Competitive Advantage in Firms. *Organizational Behavior and Human Decision Processes*, *82*(1), 150–169. https://doi.org/10.1006/obhd.2000.2893

Argote, L., McEvily, B., & Reagans, R. (2003). Managing Knowledge in Organizations: An Integrative Framework and Review of Emerging Themes. *Management Science*, *49*(4), 571–582.

Bapuji, H., & Crossan, M. (2005). CO-EVOLUTION OF SOCIAL CAPITAL AND KNOWLEDGE: AN EXTENSION OF THE NAHAPIET AND GHOSHAL (1998) FRAMEWORK. *Academy of Management Proceedings*, *2005*(1), DD1–DD6. https://doi.org/10.5465/ambpp.2005.18778410

Barney, J. (1991). Firm Resources and Sustained Competitive Advantage. *Journal of Management*, *17*(1), 99–120. https://doi.org/10.1177/014920639101700108

Bathelt, H., Malmberg, A., & Maskell, P. (2004). Clusters and knowledge: Local buzz, global pipelines and the process of knowledge creation. *Progress in Human Geography*, *28*(1), 31–56. https://doi.org/10.1191/0309132504ph469oa

Battistella, C. (2015). Inter-organisational technology/knowledge transfer: A framework from critical literature review. *Journal of Technology Transfer*, 41.

Battistella, C., De Toni, A. F., & Pillon, R. (2016). Inter-organisational technology/knowledge transfer: A framework from critical literature review. *The Journal of Technology Transfer*, *41*(5), 1195–1234. https://doi.org/10.1007/s10961-015-9418-7

Becerra, M., Lunnan, R., & Huemer, L. (2008). Trustworthiness, Risk, and the Transfer of Tacit and Explicit Knowledge Between Alliance Partners. *Journal of Management Studies*, *45*(4), 691–713. https://doi.org/10.1111/j.1467-6486.2008.00766.x

Becker, M. C., & Knudsen, M. P. (2006). *Intra and inter-organizational knowledge transfer processes: Identifying the missing links.*

Bell, G. G., & Zaheer, A. (2007). Geography, Networks, and Knowledge Flow. *Organization Science, 18*(6), 955–972.

Bender, S., & Fish, A. (2000). The transfer of knowledge and the retention of expertise: The continuing need for global assignments. *Journal of Knowledge Management, 4*(2), 125–137. https://doi.org/10.1108/13673270010372251

Bengoa, D. S., & Kaufmann, H. R. (2014). Questioning Western Knowledge Transfer Methodologies: Toward a Reciprocal and Intercultural Transfer of Knowledge. *Thunderbird International Business Review, 56*(1), 11–26. https://doi.org/10.1002/tie.21593

Berchicci, L., de Jong, J. P. J., & Freel, M. (2016). Remote collaboration and innovative performance: The moderating role of R&D intensity. *Industrial and Corporate Change, 25*(3), 429–446. https://doi.org/10.1093/icc/dtv031

Blome, C., Schoenherr, T., & Eckstein, D. (2014). The impact of knowledge transfer and complexity on supply chain flexibility: A knowledge-based view. *International Journal of Production Economics, 147*, 307–316. https://doi.org/10.1016/j.ijpe.2013.02.028

Blomkvist, K. (2012). Knowledge management in MNCs: The importance of subsidiary transfer performance. *Journal of Knowledge Management, 16*(6), 904–918. https://doi.org/10.1108/13673271211276182

Bock, G. W., & Kim, Y.-G. (2002). Breaking the Myths of Rewards: An Exploratory Study of Attitudes about Knowledge Sharing. *Information Resources Management Journal, 15*(2), 14–21. https://doi.org/10.4018/irmj.2002040102

Bosch-Sijtsema, P. M., & Postma, T. J. B. M. (2010). Governance factors enabling knowledge transfer in interorganisational development projects. *Technology Analysis & Strategic Management, 22*(5), 593–608. https://doi.org/10.1080/09537325.2010.488064

Bresman, H., Birkinshaw, J., & Nobel, R. (1999). Knowledge transfer in international acquisitions. *Journal of International Business Studies, 30*(3), 439–462. ABI/INFORM Global.

Brown, J. S., & Duguid, P. (2000). *Balancing Act: How to Capture Knowledge Without Killing It. 7*.

Bryman, A., & Bell, E. (2011). *Business research methods* (3rd ed). Oxford University Press.

Cambra-Fierro, J., Florin, J., Perez, L., & Whitelock, J. (2011). Inter-firm market orientation as antecedent of knowledge transfer, innovation and value creation in networks. *Management Decision, 49*(3), 444–467. https://doi.org/10.1108/00251741111120798

Capaldo, A., & Petruzzelli, A. M. (2014). Partner Geographic and Organizational Proximity and the Innovative Performance of Knowledge-Creating Alliances. *European Management Review, 11*(1), 63–84. https://doi.org/10.1111/emre.12024

Cardoni, A., Dumay, J., Palmaccio, M., & Celenza, D. (2019). Knowledge transfer in a start-up craft brewery. *Business Process Management Journal, 25*(1), 219–243. https://doi.org/10.1108/BPMJ-07-2017-0205

Cavusgil, S. T., Calantone, R. J., & Zhao, Y. (2003). Tacit knowledge transfer and firm innovation capability. *The Journal of Business & Industrial Marketing, 18*(1), 6–21. ABI/INFORM Global. https://doi.org/10.1108/08858620310458615

Cerchione, R., Centobelli, P., Zerbino, P., & Anand, A. (2020). Back to the future of Knowledge Management Systems off the beaten paths. *Management Decision, ahead-of-print*(ahead-of-print). https://doi.org/10.1108/MD-11-2019-1601

Cerchione, R., Esposito, E., & Spadaro, M. R. (2017). A literature review on knowledge management in SMEs. *Knowledge Management Research & Practice*. https://doi.org/10.1057/kmrp.2015.12

Chen, C.-J. (2004). The effects of knowledge attribute, alliance characteristics, and absorptive capacity on knowledge transfer performance. *R&D Management, 34*(3), 311–321. https://doi.org/10.1111/j.1467-9310.2004.00341.x

Chen, C.-J., Hsiao, Y.-C., & Chu, M.-A. (2014). Transfer mechanisms and knowledge transfer: The cooperative competency perspective. *Journal of Business Research, 67*(12), 2531–2541. https://doi.org/10.1016/j.jbusres.2014.03.011

Chen, J., & Lovvorn, A. S. (2011). The speed of knowledge transfer within multinational enterprises: The role of social capital. *International Journal of Commerce and Management, 21*(1), 46–62. https://doi.org/10.1108/10569211111111694

Chen, J., & McQueen, R. J. (2010). Knowledge transfer processes for different experience levels of knowledge recipients at an offshore technical support center. *Information Technology & People, 23*(1), 54–79. https://doi.org/10.1108/09593841011022546

Chen, S., Duan, Y., & Edwards, J. S. (2006). Inter-Organisational Knowledge Transfer Process Model. In *Encyclopedia of Communities of Practice in Information and Knowledge Management.* https://doi.org/10.4018/978-1-59140-556-6.ch043

Chen, S., Duan, Y., Edwards, J. S., & Lehaney, B. (2006). Toward understanding inter-organizational knowledge transfer needs in SMEs: Insight from a UK investigation. *Journal of Knowledge Management, 10*(3), 19.

Cheung, M.-S., Myers, M. B., & Mentzer, J. T. (2011). The value of relational learning in global buyer-supplier exchanges: A dyadic perspective and test of the pie-sharing premise. *Strategic Management Journal, 32*(10), 1061–1082. https://doi.org/10.1002/smj.926

Choi, J., & Contractor, F. J. (2016). Choosing an appropriate alliance governance mode: The role of institutional, cultural and geographical distance in international research & development (R&D) collaborations. *Journal of International Business Studies, 47*(2), 210–232. https://doi.org/10.1057/jibs.2015.28

Christensen, C. M. (2013). *The innovator's dilemma: When new technologies cause great firms to fail*. Harvard Business Review Press.

Cohen, W. M., & Levinthal, D. A. (1990). Absorptive Capacity: A New Perspective on Learning and Innovation. *Administrative Science Quarterly*, *35*(1), 128. https://doi.org/10.2307/2393553

Conner, K. R., & Prahalad, C. K. (1996). A Resource-Based Theory of the Firm: Knowledge versus Opportunism. *Organization Science*, *7*(5), 477–501.

Corral de Zubielqui, G., Jones, J., Seet, P.-S., & Lindsay, N. (2015). Knowledge transfer between actors in the innovation system: A study of higher education institutions (HEIS) and SMES. *Journal of Business & Industrial Marketing*, *30*(3/4), 436–458. https://doi.org/10.1108/JBIM-07-2013-0152

Corral de Zubielqui, G., Lindsay, N., Lindsay, W., & Jones, J. (2019). Knowledge quality, innovation and firm performance: A study of knowledge transfer in SMEs. *Small Business Economics*, *53*(1), 145–164. https://doi.org/10.1007/s11187-018-0046-0

Crescentini, A., & Mainardi, G. (2009). Qualitative research articles: Guidelines, suggestions and needs. *Journal of Workplace Learning*, *21*(5), 431–439. https://doi.org/10.1108/13665620910966820

Cummings, J. L., & Teng, B.-S. (2003). Transferring R&D knowledge: The key factors affecting knowledge transfer success. *Journal of Engineering and Technology Management*, *20*(1–2), 39–68. https://doi.org/10.1016/S0923-4748(03)00004-3

Cummings, L. L., & Bromiley, P. (1996). The organizational trust inventory (OTI). *Trust in Organizations: Frontiers of Theory and Research*, *302*(330), 39–52.

Cyert, R. and March, J. (1963) A Behavioral Theory of the Firm. Prentice-Hall, Englewood Cliffs.

De Luca, P., & Cano Rubio, M. (2019). The curve of knowledge transfer: A theoretical model. *Business Process Management Journal*, *25*(1), 10–26. https://doi.org/10.1108/BPMJ-

06-2017-0161

Desouza, K. C., & Awazu, Y. (2006). Knowledge management at SMEs: Five peculiarities. *Journal of Knowledge Management*, *10*(1), 32–43. https://doi.org/10.1108/13673270610650085

Dhanaraj, C., Lyles, M. A., Steensma, H. K., & Tihanyi, L. (2004). Managing Tacit and Explicit Knowledge Transfer in IJVs: The Role of Relational Embeddedness and the Impact on Performance. *Journal of International Business Studies*, *35*(5), 428–442.

Dierickx, I., & Cool, K. (1989). Asset Stock Accumulation and Sustainability of Competitive Advantage. *Management Science*, *35*(12), 1504–1511. https://doi.org/10.1287/mnsc.35.12.1504

Driffield, N., Love, J. H., & Yang, Y. (2016). Reverse international knowledge transfer in the MNE: (Where) does affiliate performance boost parent performance? *Research Policy*, *45*(2), 491–506. https://doi.org/10.1016/j.respol.2015.11.004

Durst, S., & Runar Edvardsson, I. (2012). Knowledge management in SMEs: A literature review. *Journal of Knowledge Management*, *16*(6), 879–903. https://doi.org/10.1108/13673271211276173

Durst, S., & Wilhelm, S. (2012). Knowledge management and succession planning in SMEs. *Journal of Knowledge Management*, *16*(4), 637–649. https://doi.org/10.1108/13673271211246194

Easterby-Smith, M., Lyles, M. A., & Tsang, E. W. K. (2008). Inter-Organizational Knowledge Transfer: Current Themes and Future Prospects. *Journal of Management Studies*, *45*(4), 677–690. https://doi.org/10.1111/j.1467-6486.2008.00773.x

Easterby-Smith, M., Thorpe, R., & Lowe, A. (1991). *Management Research: An Introduction.* SAGE Publications. https://books.google.se/books?id=NSy1QgAACAAJ

Eiriz, V., Gonçalves, M., & Areias, J. S. (2017). Inter-organizational learning within an institutional knowledge network: A case study in the textile and clothing industry. *European Journal of Innovation Management, 20*(2), 230–249. https://doi.org/10.1108/EJIM-11-2015-0117

Eisenhardt, K. M., & Santos, F. M. (2006). Knowledge-Based View: A New Theory of Strategy? In *Handbook of Strategy and Management* (pp. 139–164). SAGE Publications Ltd. https://doi.org/10.4135/9781848608313.n7

Etikan, I. (2016). Comparison of Convenience Sampling and Purposive Sampling. *American Journal of Theoretical and Applied Statistics, 5*(1), 1. https://doi.org/10.11648/j.ajtas.20160501.11

Fang, S.-C., Yang, C.-W., & Hsu, W.-Y. (2013). Inter-organizational knowledge transfer: The perspective of knowledge governance. *Journal of Knowledge Management, 17*(6), 943–957. https://doi.org/10.1108/JKM-04-2013-0138

Farquhar, J. D. (2012). *Case Study Research for Business*. SAGE Publications. http://ebookcentral.proquest.com/lib/halmstad/detail.action?docID=880850

Fernie, S., Green, S. D., Weller, S. J., & Newcombe, R. (2003). Knowledge sharing: Context, confusion and controversy. *International Journal of Project Management, 21*(3), 177–187. https://doi.org/10.1016/S0263-7863(02)00092-3

Filieri, R., & Alguezaui, S. (2014). Structural social capital and innovation. Is knowledge transfer the missing link? *Journal of Knowledge Management, 18*(4), 728–757. https://doi.org/10.1108/JKM-08-2013-0329

Fletcher, M., & Prashantham, S. (2011). Knowledge assimilation processes of rapidly internationalising firms: Longitudinal case studies of Scottish SMEs. *Journal of Small Business and Enterprise Development, 18*(3), 475–501. https://doi.org/10.1108/14626001111155673

Fong Boh, W., Nguyen, T. T., & Xu, Y. (2013). Knowledge transfer across dissimilar cultures.

Journal of Knowledge Management, *17*(1), 29–46. https://doi.org/10.1108/13673271311300723

Foss, N. J. (2007). The Emerging Knowledge Governance Approach: Challenges and Characteristics. *Organization,* *14*(1), 29–52. https://doi.org/10.1177/1350508407071859

Foss, N. J., & Michailova, S. (2009). *Knowledge Governance: Processes and Perspectives.* OUP Oxford.

Foss, N. J., & Pedersen, T. (2002). Transferring knowledge in MNCs: The role of sources of subsidiary knowledge and organizational context. *Journal of International Management,* 19.

Frishammar, J., Ericsson, K., & Patel, P. C. (2015). The dark side of knowledge transfer: Exploring knowledge leakage in joint R&D projects. *Technovation, 41–42,* 75–88. https://doi.org/10.1016/j.technovation.2015.01.001

Garavelli, A. C., Gorgoglione, M., & Scozzi, B. (2002). Managing knowledge transfer by knowledge technologies. *Technovation,* *22*(5), 269–279. https://doi.org/10.1016/S0166-4972(01)00009-8

Garcia, R., Araujo, V., Mascarini, S., Santos, E. G., & Costa, A. (2018). Is cognitive proximity a driver of geographical distance of university–industry collaboration? *Area Development and Policy, 3,* 1–19. https://doi.org/10.1080/23792949.2018.1484669

García-Morales, V. J., Lloréns-Montes, F. J., & Verdú-Jover, A. J. (2007). Influence of personal mastery on organizational performance through organizational learning and innovation in large firms and SMEs. *Technovation,* *27*(9), 547–568. https://doi.org/10.1016/j.technovation.2007.02.013

Gaur, A. S., Ma, H., & Ge, B. (2019). MNC strategy, knowledge transfer context, and knowledge flow in MNEs. *Journal of Knowledge Management, 23*(9), 1885–1900.

https://doi.org/10.1108/JKM-08-2018-0476

Ghauri, P. N., & Grønhaug, K. (2005). *Research Methods in Business Studies: A Practical Guide*. Financial Times Prentice Hall. https://books.google.se/books?id=-sTUDbaefgkC

Giannakis, M. (2008). Facilitating learning and knowledge transfer through supplier development. *Supply Chain Management: An International Journal, 13*(1), 62–72. https://doi.org/10.1108/13598540810850328

Gilbert, M., & Cordey-Hayes, M. (1996). Understanding the process of knowledge transfer to achieve successful technological innovation. *Technovation, 16*(6), 301–312. https://doi.org/10.1016/0166-4972(96)00012-0

Goh, S. C. (2002). Managing effective knowledge transfer: An integrative framework and some practice implications. *Journal of Knowledge Management, 6*(1), 23–30. https://doi.org/10.1108/13673270210417664

Gollwitzer, P. M. (1990). *Action_Phases_MindSets*.

Gollwitzer, P. M., & Oettingen, G. (2011). *Planning Promotes Goal Striving*. 24.

Gollwitzer, P. M., & Oettingen, G. (2015). Motivation: History of the Concept. In *International Encyclopedia of the Social & Behavioral Sciences* (pp. 936–939). Elsevier. https://doi.org/10.1016/B978-0-08-097086-8.03102-0

Golzadeh Kermani, F. (2019). *The Interdependence of Knowledge Absorptive and Disseminative Capacities: An Explanatory Framework* [PhD, University of Sheffield]. http://etheses.whiterose.ac.uk/25472/

Grant, R. M. (1996). Toward a knowledge-based theory of the firm: Knowledge-based Theory of the Firm. *Strategic Management Journal, 17*(S2), 109–122. https://doi.org/10.1002/smj.4250171110

Gray, C. (2006). Absorptive capacity, knowledge management and innovation in entrepreneurial small firms. *International Journal of Entrepreneurial Behavior & Research, 12*(6), 345–360. https://doi.org/10.1108/13552550610710144

Guba, E. G., & Lincoln, Y. S. (1981). *Effective evaluation: Improving the usefulness of evaluation results through responsive and naturalistic approaches.* Jossey-Bass.

Gupta, A. K., & Govindarajan, V. (2000). Knowledge flows within multinational corporations. *Strategic Management Journal, 21*(4), 473–496. https://doi.org/10.1002/(SICI)1097-0266(200004)21:4<473::AID-SMJ84>3.0.CO;2-I

Hamdoun, M., Chiappetta Jabbour, C. J., & Ben Othman, H. (2018). Knowledge transfer and organizational innovation: Impacts of quality and environmental management. *Journal of Cleaner Production, 193*, 759–770. https://doi.org/10.1016/j.jclepro.2018.05.031

Hansen, M. T. (1999). The Search-Transfer Problem: The Role of Weak Ties in Sharing Knowledge across Organization Subunits. *Administrative Science Quarterly, 44*(1), 82. https://doi.org/10.2307/2667032

Harel, R., Schwartz, D., & Kaufmann, D. (2020). Sharing knowledge processes for promoting innovation in small businesses. *European Journal of Innovation Management, ahead-of-print*(ahead-of-print). https://doi.org/10.1108/EJIM-04-2020-0122

Harris, R. J. (2008). *Improving tacit knowledge transfer within SMEs through e-collaboration.* 18.

Hoetker, G., & Mellewigt, T. (2009). Choice and performance of governance mechanisms: Matching alliance governance to asset type. *Strategic Management Journal, 30*(10), 1025–1044. https://doi.org/10.1002/smj.775

Hofstede, G. (1994). The business of international business is culture. *International Business Review, 3*(1), 1–14. https://doi.org/10.1016/0969-5931(94)90011-6

Holstein, J., & Gubrium, J. (1997). *The New Language of Qualitative Method.* https://global.oup.com/ushe/product/the-new-language-of-qualitative-method-

9780195099942?cc=se&lang=en&

House, R. J., Hanges, P. J., Javidan, M., Dorfman, P. W., & Gupta, V. (2004). *Culture, leadership, and organizations: The GLOBE study of 62 societies.* Sage publications.

Huang, M.-C., Chiu, Y.-P., & Lu, T.-C. (2013). Knowledge governance mechanisms and repatriate's knowledge sharing: The mediating roles of motivation and opportunity. *Journal of Knowledge Management, 17*(5), 677–694. https://doi.org/10.1108/JKM-01-2013-0048

Hughes, T., O'Regan, N., & Sims, M. A. (2009). The effectiveness of knowledge networks: An investigation of manufacturing SMEs. *Education + Training, 51*(8/9), 665–681. https://doi.org/10.1108/00400910911005226

Hutchinson, V., & Quintas, P. (2008). Do SMEs do Knowledge Management?: Or Simply Manage what they Know? *International Small Business Journal: Researching Entrepreneurship, 26*(2), 131–154. https://doi.org/10.1177/0266242607086571

Hutzschenreuter, T., & Horstkotte, J. (2010). *Knowledge transfer to partners: A firm level perspective.* 21.

Hyde, K. F. (2000). Recognising deductive processes in qualitative research. *Qualitative Market Research: An International Journal, 3*(2), 82–90. https://doi.org/10.1108/13522750010322089

Inkpen, A. C. (2008). Knowledge transfer and international joint ventures: The case of NUMMI and General Motors. *Strategic Management Journal, 29*(4), 447–453. https://doi.org/10.1002/smj.663

Inkpen, A. C., & Pien, W. (2006). An Examination of Collaboration and Knowledge Transfer: China-Singapore Suzhou Industrial Park. *Journal of Management Studies, 43*(4), 779–811. https://doi.org/10.1111/j.1467-6486.2006.00611.x

Inkpen, A. C., & Tsang, E. W. K. (2005). Social Capital, Networks, and Knowledge Transfer.

The Academy of Management Review, 30(1), 146–165. https://doi.org/10.2307/20159100

Ishihara, H., & Zolkiewski, J. (2017). Effective knowledge transfer between the headquarters and a subsidiary in a MNC: The need for heeding capacity. *Journal of Business & Industrial Marketing, 32*(6), 813–824. https://doi.org/10.1108/JBIM-06-2015-0109

Jantunen, A. (2005). Knowledge-processing capabilities and innovative performance: An empirical study. *European Journal of Innovation Management, 8*(3), 336–349. https://doi.org/10.1108/14601060510610199

Jeong, G.-Y., Chae, M.-S., & Park, B. I. (2017). Reverse knowledge transfer from subsidiaries to multinational companies: Focusing on factors affecting market knowledge transfer. *Canadian Journal of Administrative Sciences / Revue Canadienne des Sciences de l'Administration, 34*(3), 291–305. https://doi.org/10.1002/cjas.1366

Jiménez-Jiménez, D., Martínez-Costa, M., & Sanz-Valle, R. (2014). Knowledge management practices for innovation: A multinational corporation's perspective. *Journal of Knowledge Management, 18*(5), 905–918. https://doi.org/10.1108/JKM-06-2014-0242

Joia, L. A., & Lemos, B. (2010). Relevant factors for tacit knowledge transfer within organisations. *Journal of Knowledge Management, 14*(3), 410–427. https://doi.org/10.1108/13673271011050139

Joshi, K. D., Sarker, S., & Sarker, S. (2007). Knowledge transfer within information systems development teams: Examining the role of knowledge source attributes. *Decision Support Systems, 43*(2), 322–335. https://doi.org/10.1016/j.dss.2006.10.003

Kalling, T. (2003). Organization-internal transfer of knowledge and the role of motivation: A qualitative case study. *Knowledge and Process Management, 10*(2), 115–126. https://doi.org/10.1002/kpm.170

Kaminski, P. C., de Oliveira, A. C., & Lopes, T. M. (2008). Knowledge transfer in product development processes: A case study in small and medium enterprises (SMEs) of the metal-mechanic sector from São Paulo, Brazil. *Technovation*, *28*(1), 29–36. https://doi.org/10.1016/j.technovation.2007.07.001

Kang, M., & Sauk Hau, Y. (2014). Multi-level analysis of knowledge transfer: A knowledge recipient's perspective. *Journal of Knowledge Management*, *18*(4), 758–776. https://doi.org/10.1108/JKM-12-2013-0511

Khamseh, H. M., & Jolly, D. (2014). Knowledge transfer in alliances: The moderating role of the alliance type. *Knowledge Management Research & Practice*, *12*(4), 409–420. https://doi.org/10.1057/kmrp.2012.63

Kogut, B., & Zander, U. (1992). Knowledge of the Firm, Combinative Capabilities, and the Replication of Technology. *Organization Science*, *3*(3), 383–397. https://doi.org/10.1287/orsc.3.3.383

Korbi, F. B., & Chouki, M. (2017). Knowledge transfer in international asymmetric alliances: The key role of translation, artifacts, and proximity. *Journal of Knowledge Management*, 00–00. https://doi.org/10.1108/JKM-11-2016-0501

Kraatz, M. S. (1998). LEARNING BY ASSOCIATION? INTERORGANIZATIONAL NETWORKS AND ADAPTATION TO ENVIRONMENTAL CHANGE. *Academy of Management Journal*, *41*(6), 621–643. https://doi.org/10.2307/256961

Kuiken, J., & van der Sijde, P. (2011). Knowledge Transfer and Capacity for Dissemination: A Review and Proposals for Further Research on Academic Knowledge Transfer. *Industry and Higher Education*, *25*(3), 173–179. https://doi.org/10.5367/ihe.2011.0041

Kumar, N. (2013). Managing reverse knowledge flow in multinational corporations. *Journal of Knowledge Management*, *17*(5), 695–708. https://doi.org/10.1108/JKM-02-2013-0062

Kuzel, A. J., & Like, R. C. (1991). *Standards of trustworthiness for qualitative studies in primary care.*

Lavie, D. (2006). The Competitive Advantage of Interconnected Firms: An Extension of the Resource-Based View. *Academy of Management Review*, *31*(3), 638–658. https://doi.org/10.5465/amr.2006.21318922

Lee, C. Y., & Wu, F. C. (2006). *Factors Affecting Knowledge Transfer And Absorptive Capacity In Multinational Corporations.*

Lee, H., & Choi, B. (2003). Knowledge management enablers, processes, and organizational performance: An integrative view and empirical examination. *Journal of Management Information Systems*, *20*(1), 179–228.

Leonard-Barton, D. (1990). A dual methodology for case studies: Synergistic use of a longitudinal single site with replicated multiple sites. *Organization Science*, *1*(3), 248–266.

Levine, S. S., & Prietula, M. J. (2012). How Knowledge Transfer Impacts Performance: A Multilevel Model of Benefits and Liabilities. *Organization Science*, *23*(6), 1748–1766. https://doi.org/10.1287/orsc.1110.0697

Lewin, A. Y., Massini, S., & Peeters, C. (2011). Microfoundations of Internal and External Absorptive Capacity Routines. *Organization Science*, *22*(1), 81–98. https://doi.org/10.1287/orsc.1100.0525

Li, J. H., Chang, X. R., Lin, L., & Ma, L. Y. (2014). Meta-analytic comparison on the influencing factors of knowledge transfer in different cultural contexts. *Journal of Knowledge Management*, *18*(2), 278–306. https://doi.org/10.1108/JKM-08-2013-0316

Lichtenthaler, U., & Ernst, H. (2006). Attitudes to externally organising knowledge management tasks: A review, reconsideration and extension of the NIH syndrome. *R and D Management*, *36*(4), 367–386. https://doi.org/10.1111/j.1467-9310.2006.00443.x

Lièvre, P., & Tang, J. (2015). SECI and inter-organizational and intercultural knowledge transfer: A case-study of controversies around a project of co-operation between France and China in the health sector. *Journal of Knowledge Management, 19*, 1069–1086. https://doi.org/10.1108/JKM-02-2015-0054

Lin, H. (2007). Knowledge sharing and firm innovation capability: An empirical study. *International Journal of Manpower, 28*(3/4), 315–332. https://doi.org/10.1108/01437720710755272

Liu, X., Gao, L., Lu, J., & Wei, Y. (2015). The role of highly skilled migrants in the process of inter-firm knowledge transfer across borders. *Journal of World Business, 50*(1), 56–68. https://doi.org/10.1016/j.jwb.2014.01.006

Liyanage, C., Elhag, T., Ballal, T., & Li, Q. (2009). Knowledge communication and translation – a knowledge transfer model. *Journal of Knowledge Management, 13*(3), 118–131. https://doi.org/10.1108/13673270910962914

Loebbecke, C., van Fenema, P. C., & Powell, P. (2016). Managing inter-organizational knowledge sharing. *The Journal of Strategic Information Systems, 25*(1), 4–14. https://doi.org/10.1016/j.jsis.2015.12.002

Lucas, L. M., & Ogilvie, dt. (2006). Things are not always what they seem: How reputations, culture, and incentives influence knowledge transfer. *The Learning Organization, 13*(1), 7–24. https://doi.org/10.1108/09696470610639103

Lyles, M. A., & Salk, J. E. (1996). Knowledge Acquisition from Foreign Parents in International Joint Ventures: An Empirical Examination in the Hungarian Context. *Journal of International Business Studies, 27*(5), 877–903. https://doi.org/10.1057/palgrave.jibs.8490155

Maciejovsky, B., & Budescu, D. V. (2013). Markets as a structural solution to knowledge-sharing dilemmas. *Organizational Behavior and Human Decision Processes, 120*(2),

154–167. https://doi.org/10.1016/j.obhdp.2012.04.005

Makino, S., & Delios, A. (1996). Local Knowledge Transfer and Performance: Implications for Alliance Formation in Asia. *Journal of International Business Studies, 27*(5), 905–927.

Martín Cruz, N., Martín Pérez, V., & Trevilla Cantero, C. (2009). The influence of employee motivation on knowledge transfer. *Journal of Knowledge Management, 13*(6), 478–490. https://doi.org/10.1108/13673270910997132

Martinkenaite, I. (2011). Antecedents and consequences of inter-organizational knowledge transfer: Emerging themes and openings for further research. *Baltic Journal of Management, 6*(1), 53–70. https://doi.org/10.1108/17465261111100888

Martins, E. C., & Terblanche, F. (2003). Building organisational culture that stimulates creativity and innovation. *European Journal of Innovation Management, 6*(1), 64–74. https://doi.org/10.1108/14601060310456337

Maurer, I., Bartsch, V., & Ebers, M. (2011). The Value of Intra-organizational Social Capital: How it Fosters Knowledge Transfer, Innovation Performance, and Growth. *Organization Studies, 32*(2), 157–185. https://doi.org/10.1177/0170840610394301

Maxwell, J. (2009). Designing a Qualitative Study. In L. Bickman & D. Rog, *The SAGE Handbook of Applied Social Research Methods* (pp. 214–253). SAGE Publications, Inc. https://doi.org/10.4135/9781483348858.n7

Mayer, R. C., Davis, J. H., & Schoorman, F. D. (1995). An Integrative Model of Organizational Trust. *The Academy of Management Review, 20*(3), 709–734. https://doi.org/10.2307/258792

Mazloomi Khamseh, H., & Jolly, D. R. (2008). Knowledge transfer in alliances: Determinant factors. *Journal of Knowledge Management, 12*(1), 37–50. https://doi.org/10.1108/13673270810852377

Menon, T., & Pfeffer, J. (2003). Valuing Internal vs. External Knowledge: Explaining the Preference for Outsiders. *Management Science, 49*(4), 497–513.

Mesquita, L. F., Anand, J., & Brush, T. H. (2008). Comparing the resource-based and relational views: Knowledge transfer and spillover in vertical alliances. *Strategic Management Journal, 29*(9), 913–941. https://doi.org/10.1002/smj.699

Milagres, R., & Burcharth, A. (2019). Knowledge transfer in interorganizational partnerships: What do we know? *Business Process Management Journal, 25*(1), 27–68. https://doi.org/10.1108/BPMJ-06-2017-0175

Millie, M., & Cheung, P.-K. (2006). The Knowledge Transfer Process: From Field Studies to Technology Development. *Journal of Database Management, 17*, 17.

Minbaeva, D. B. (2007). Knowledge Transfer in Multinational Corporations. *Management International Review, 47*(4), 567–593. ABI/INFORM Global.

Minbaeva, D. B., & Michailova, S. (2004). Knowledge transfer and expatriation in multinational corporations: The role of disseminative capacity. *Employee Relations, 26*(6), 663–679. https://doi.org/10.1108/01425450410562236

Minbaeva, D., Park, C., Vertinsky, I., & Cho, Y. S. (2018). Disseminative capacity and knowledge acquisition from foreign partners in international joint ventures. *Journal of World Business, 53*(5), 712–724. https://doi.org/10.1016/j.jwb.2018.03.011

Minbaeva, D., Pedersen, T., Björkman, I., & Fey, C. F. (2003). MNC knowledge transfer, subsidiary absorptive capacity and HRM. *Journal of International Business Studies, 34*(6), 586–599.

Montazemi, A. R., Pittaway, J. J., Qahri Saremi, H., & Wei, Y. (2012). Factors of stickiness in transfers of know-how between MNC units. *The Journal of Strategic Information Systems, 21*(1), 31–57. https://doi.org/10.1016/j.jsis.2012.01.001

Mowery, D. C., Oxley, J. E., & Silverman, B. S. (1996). Strategic alliances and interfirm knowledge transfer. *Strategic Management Journal*, *17*(S2), 77–91. https://doi.org/10.1002/smj.4250171108

Mu, J., Tang, F., & MacLachlan, D. L. (2010). Absorptive and disseminative capacity: Knowledge transfer in intra-organization networks. *Expert Systems with Applications*, *37*(1), 31–38. https://doi.org/10.1016/j.eswa.2009.05.019

Najafi-Tavani, Z., Zaefarian, G., Naudé, P., & Giroud, A. (2015). Reverse knowledge transfer and subsidiary power. *Industrial Marketing Management*, *48*, 103–110. https://doi.org/10.1016/j.indmarman.2015.03.021

Nakauchi, M., Washburn, M., & Klein, K. (2017). Differences between inter- and intra-group dynamics in knowledge transfer processes. *Management Decision*, *55*(4), 766–782. https://doi.org/10.1108/MD-08-2016-0537

Narteh, B. (2008). Knowledge transfer in developed-developing country interfirm collaborations: A conceptual framework. *Journal of Knowledge Management*, *12*(1), 78–91. https://doi.org/10.1108/13673270810852403

Nelson, R. E. (1989). THE STRENGTH OF STRONG TIES: SOCIAL NETWORKS AND INTERGROUP CONFLICT IN ORGANIZATIONS. *Academy of Management Journal*, *32*(2), 377–401. https://doi.org/10.2307/256367

Newell, S. (2005). KNOWLEDGE TRANSFER AND LEARNING: PROBLEMS OF KNOWLEDGE TRANSFER ASSOCIATED WITH TRYING TO SHORT-CIRCUIT THE LEARNING CYCLE. *Journal of Information Systems and Technology Management*, *2*(3), 17.

Noblet, J.-P., & Simon, E. (2012). SECTION 2. Management in firms and organizations. *Problems and Perspectives in Management*, *10*(3), 10.

Nonaka, I. (1991). The Knowledge-Creating Company. *The Economic Impact of Knowledge*, 175–187. https://doi.org/10.1016/B978-0-7506-7009-8.50016-1

Nonaka, I., & Takeuchi, H. (1995). *The knowledge-creating company: How Japanese companies create the dynamics of innovation*. Oxford university press.

Nonaka, I., & von Krogh, G. (2009). Tacit Knowledge and Knowledge Conversion: Controversy and Advancement in Organizational Knowledge Creation Theory. *Organization Science, 20*(3), 635–652.

O'Dwyer, M., & O'Flynn, E. (2005). MNC–SME strategic alliances—A model framing knowledge value as the primary predictor of governance modal choice. *Journal of International Management, 11*(3), 397–416. https://doi.org/10.1016/j.intman.2005.06.006

OECD. (2019). *OECD SME and Entrepreneurship Outlook 2019*. OECD. https://doi.org/10.1787/34907e9c-en

Patton, M. Q. (2002). Two Decades of Developments in Qualitative Inquiry: A Personal, Experiential Perspective. *Qualitative Social Work, 1*(3), 261–283. https://doi.org/10.1177/1473325002001003636

Paulin, D., & Suneson, K. (2012). *Knowledge Transfer, Knowledge Sharing and Knowledge Barriers – Three Blurry Terms in KM. 10*(1), 12.

Penrose, E. (1959) The Theory of the Growth of the Firm. Basil Blackwell, Oxford.

Pérez-Nordtvedt, L., Kedia, B. L., Datta, D. K., & Rasheed, A. A. (2008). Effectiveness and Efficiency of Cross-Border Knowledge Transfer: An Empirical Examination. *Journal of Management Studies, 45*(4), 714–744. https://doi.org/10.1111/j.1467-6486.2008.00767.x

Phene, A., & Tallman, S. (2014). Knowledge Spillovers and Alliance Formation. *Journal of Management Studies, 51*(7), 1058–1090. https://doi.org/10.1111/joms.12088

Polanyi, M. (1966). *The Tacit Dimension*. Doubleday.

Porter, M. E. (1985). TECHNOLOGY AND COMPETITIVE ADVANTAGE. *Journal of*

Business Strategy, *5*(3), 60–78. https://doi.org/10.1108/eb039075

Qian, Y., & Xu, C. (1998). Innovation and bureaucracy under soft and hard budget constraints. *The Review of Economic Studies*, *65*(1), 151–164.

Qiu, X. (2019). The role of regulatory focus and trustworthiness in knowledge transfer and leakage in alliances. *Industrial Marketing Management*, 12.

Reed, R., & Defillippi, R. J. (1990). Causal Ambiguity, Barriers to Imitation, and Sustainable Competitive Advantage. *The Academy of Management Review*, *15*(1), 88. https://doi.org/10.2307/258107

Robinson, H., Carrillo, P., Anumba, C. J., Patel, M., & Patel, M. (2010). *Governance and Knowledge Management for Public-Private Partnerships*. John Wiley & Sons, Incorporated.

http://ebookcentral.proquest.com/lib/halmstad/detail.action?docID=477896

Robson, C. (2002). *Real World Research: A Resource for Social Scientists and Practitioner-researchers.* Oxford: Blackwell.

Robson, C., & McCartan, K. (2017). *Real World Research, 4th Edition.*

Rogers, M. (2004). Networks, Firm Size and Innovation. *Small Business Economics*, *22*(2), 141–153. https://doi.org/10.1023/B:SBEJ.0000014451.99047.69

Rousseau, D. M., Sitkin, S. B., Burt, R. S., & Camerer, C. (1998). Introduction to Special Topic Forum: Not so Different after All: A Cross-Discipline View of Trust. *The Academy of Management Review*, *23*(3), 393–404.

Saunders, M., Lewis, P., & Thornhill, A. (2009). *Research methods for business students.* Pearson education.

Schamp, E. W., Rentmeister, B., & Lo, V. (2004). Dimensions of proximity in knowledge-based networks: The cases of investment banking and automobile design. *European Planning Studies*, *12*(5), 607–624. https://doi.org/10.1080/0965431042000219978

Schildt, H., Keil, T., & Maula, M. (2012). The temporal effects of relative and firm-level absorptive capacity on interorganizational learning. *Strategic Management Journal*, *33*(10), 1154–1173. https://doi.org/10.1002/smj.1963

Schulz, M., & Jobe, L. A. (2001). Codification and tacitness as knowledge management strategies: An empirical exploration. *The Journal of High Technology Management Research*, *12*(1), 139–165. https://doi.org/10.1016/S1047-8310(00)00043-2

Schulze, A., Brojerdi, G., & Krogh, G. von. (2014). Those Who Know, Do. Those Who Understand, Teach. Disseminative Capability and Knowledge Transfer in the Automotive Industry. *Journal of Product Innovation Management*, *31*(1), 79–97. https://doi.org/10.1111/jpim.12081

Schumpeter, J. A. (1947). The Creative Response in Economic History. *The Journal of Economic History*, *7*(2), 149–159. https://doi.org/10.1017/S0022050700054279

Segarra-Ciprés, M., Roca-Puig, V., & Bou-Llusar, J. C. (2014). External knowledge acquisition and innovation output: An analysis of the moderating effect of internal knowledge transfer. *Knowledge Management Research & Practice*, *12*(2), 203–214. https://doi.org/10.1057/kmrp.2012.55

Shen, H., Li, Z., & Yang, X. (2015). Processes, characteristics, and effectiveness: An integrative framework for successful knowledge transfer within organizations. *Journal of Organizational Change Management*, *28*(3), 486–503. https://doi.org/10.1108/JOCM-12-2013-0251

Shenton, A. K. (2004). Strategies for ensuring trustworthiness in qualitative research projects. *Education for Information*, *22*(2), 63–75.

Sikombe, S., & Phiri, M. A. (2019). Exploring tacit knowledge transfer and innovation capabilities within the buyer–supplier collaboration: A literature review. *Cogent Business & Management*, *6*(1), Article 1.

https://doi.org/10.1080/23311975.2019.1683130

Simonin, B. L. (2004). An empirical investigation of the process of knowledge transfer in international strategic alliances. *Journal of International Business Studies*, *35*(5), 407–427. https://doi.org/10.1057/palgrave.jibs.8400091

Stock, G. N., & Tatikonda, M. V. (2000). A typology of project-level technology transfer processes. *Journal of Operations Management*, *18*(6), 719–737. https://doi.org/10.1016/S0272-6963(00)00045-0

Spender, J.-C. (1996). Making knowledge the basis of a dynamic theory of the firm: Making Knowledge. *Strategic Management Journal*, *17*(S2), 45–62. https://doi.org/10.1002/smj.4250171106

Spraggon, M., & Bodolica, V. (2012). A multidimensional taxonomy of intra-firm knowledge transfer processes. *Journal of Business Research*, *65*(9), 1273–1282. https://doi.org/10.1016/j.jbusres.2011.10.043

Squire, B., Cousins, P. D., & Brown, S. (2009). Cooperation and Knowledge Transfer within Buyer–Supplier Relationships: The Moderating Properties of Trust, Relationship Duration and Supplier Performance*. *British Journal of Management*, *20*(4), 461–477. https://doi.org/10.1111/j.1467-8551.2008.00595.x

Sytrz and Porter, (1372), Motivation and behaviour at work, Translator: Amin Ali Alavi, Published by Administration Training Centre, Page 216.

Sveiby, K.-E. (1996). Transfer of knowledge and the information processing professions. *European Management Journal*, *14*(4), 379–388. https://doi.org/10.1016/0263-2373(96)00025-4

Szulanski, G. (1996). Exploring internal stickiness: Impediments to the transfer of best practice within the firm. *Strategic Management Journal*, *17*(S2), 27–43. https://doi.org/10.1002/smj.4250171105

Szulanski, G. (2000). The Process of Knowledge Transfer: A Diachronic Analysis of Stickiness. *Organizational Behavior and Human Decision Processes, 82*(1), 9–27. https://doi.org/10.1006/obhd.2000.2884

Tang, F. (2011). Knowledge transfer in intra-organization networks. *Systems Research and Behavioral Science, 28*(3), 270–282. https://doi.org/10.1002/sres.1074

Tang, F., Mu, J., & MacLachlan, D. L. (2010). Disseminative capacity, organizational structure and knowledge transfer. *Expert Systems with Applications, 37*(2), 1586–1593. https://doi.org/10.1016/j.eswa.2009.06.039

Tangaraja, G., Mohd Rasdi, R., Abu Samah, B., & Ismail, M. (2016). Knowledge sharing is knowledge transfer: A misconception in the literature. *Journal of Knowledge Management, 20*(4), 653–670. https://doi.org/10.1108/JKM-11-2015-0427

Tangaraja, G., Mohd Rasdi, R., Ismail, M., & Abu Samah, B. (2015). Fostering knowledge sharing behaviour among public sector managers: A proposed model for the Malaysian public service. *Journal of Knowledge Management, 19*(1), 121–140. https://doi.org/10.1108/JKM-11-2014-0449

Teece, D., Helfat, C. E., Finkelstein, S., Mitchell, W., Peteraf, M., Singh, H., & Winter, S. G. (2009). *Dynamic Capabilities Understanding Strategic Change in Organizations.* https://nbn-resolving.org/urn:nbn:de:101:1-201412012281

TheTheoryoftheGrowthoftheFirmPenrose1959.docx. (n.d.).

Thompson, V. A. (1965). Bureaucracy and Innovation. *Administrative Science Quarterly, 10*(1), 1–20. https://doi.org/10.2307/2391646

Tohidinia, Z., & Mosakhani, M. (2010). Knowledge sharing behaviour and its predictors. *Industrial Management & Data Systems, 110*(4), 611–631. https://doi.org/10.1108/02635571011039052

Tripsas, M. (1997). Surviving Radical Technological Change through Dynamic Capability:

Evidence from the Typesetter Industry. *Industrial and Corporate Change*, *6*(2), 341–377. https://doi.org/10.1093/icc/6.2.341

Tsai, W. (2001). KNOWLEDGE TRANSFER IN INTRAORGANIZATIONAL NETWORKS: EFFECTS OF NETWORK POSITION AND ABSORPTIVE CAPACITY ON BUSINESS UNIT INNOVATION AND PERFORMANCE. *Academy of Management Journal*, *44*(5), 996–1004. https://doi.org/10.2307/3069443

Tsang, E. W. K. (1999). A preliminary typology of learning in international strategic alliances. *Journal of World Business*, *34*(3), 211–229. https://doi.org/10.1016/S1090-9516(99)00016-4

van den Hooff, B., & de Ridder, J. A. (2004). Knowledge sharing in context: The influence of organizational commitment, communication climate and CMC use on knowledge sharing. *Journal of Knowledge Management*, *8*(6), 117–130. https://doi.org/10.1108/13673270410567675

Vlajcic, D., Marzi, G., Caputo, A., & Dabic, M. (2019). The role of geographical distance on the relationship between cultural intelligence and knowledge transfer. *Business Process Management Journal*, *25*(1), 104–125. https://doi.org/10.1108/BPMJ-05-2017-0129

Wang, S., & Noe, R. A. (2010). Knowledge sharing: A review and directions for future research. *Human Resource Management Review*, *20*(2), 115–131. https://doi.org/10.1016/j.hrmr.2009.10.001

Wee, J., & Chua, A. (2013). The peculiarities of knowledge management processes in SMEs: The case of Singapore. *Journal of Knowledge Management*, *17*(6), 958–972. https://doi.org/10.1108/JKM-04-2013-0163

Wei Chong, C., Choy Chong, S., & Chew Gan, G. (2011). Inter-organizational knowledge transfer needs among small and medium enterprises. *Library Review*, *60*(1), 37–52. https://doi.org/10.1108/00242531111100568

Wernerfelt, B. (1984). A resource-based view of the firm. *Strategic Management Journal*, *5*(2), 171–180. https://doi.org/10.1002/smj.4250050207

Wijk, R. V., Jansen, J. J. P., & Lyles, M. A. (2008). Inter- and Intra-Organizational Knowledge Transfer: A Meta-Analytic Review and Assessment of its Antecedents and Consequences. *Journal of Management Studies*, *45*(4), 830–853. https://doi.org/10.1111/j.1467-6486.2008.00771.x

Wilkesmann, U., Fischer, H., & Wilkesmann, M. (2009). Cultural characteristics of knowledge transfer. *Journal of Knowledge Management*, *13*(6), 464–477. https://doi.org/10.1108/13673270910997123

Yang, H., Zheng, A., & Zaheer, A. (2015). ASYMMETRIC LEARNING CAPABILITIES AND STOCK MARKET RETURNS. *The Academy of Management Journal*, *58*(2), 356–374.

Yang, Q., Mudambi, R., & Meyer, K. E. (2008). Conventional and Reverse Knowledge Flows in Multinational Corporations†. *Journal of Management*, *34*(5), 882–902. https://doi.org/10.1177/0149206308321546

Yih-Tong Sun, P., & Scott, J. L. (2005). An investigation of barriers to knowledge transfer. *Journal of Knowledge Management*, *9*(2), 75–90. https://doi.org/10.1108/13673270510590236

Yin, R. K. (2014). *Case study research: Design and methods* (5th ed.). Thousand Oaks, CA: Sage.

Zahra, S. A., & George, G. (2002). Absorptive Capacity: A Review, Reconceptualization, and Extension. *The Academy of Management Review*, *27*(2), 185–203. https://doi.org/10.2307/4134351

Zairi, M. (1997). Business process management: A boundaryless approach to modern competitiveness. *Business Process Management Journal*, *3*(1), 64–80.

https://doi.org/10.1108/14637159710161585

Zakova Talpova, S. (2019). *Further Insight into Knowledge Transfer in Postacquisition Integration: A Case Study in a MNE. 17*, 66–78.

Zander, U., & Kogut, B. (1995). Knowledge and the Speed of the Transfer and Imitation of Organizational Capabilities: An Empirical Test. *Organization Science, 6*(1), 76–92. https://doi.org/10.1287/orsc.6.1.76

Zapata, C. L., Rialp, C. J., & Rialp, C. A. (2009). Generation and transfer of knowledge in IT-related SMEs. *Journal of Knowledge Management, 13*(5), 243–256. https://doi.org/10.1108/13673270910988088

Zeng, R., Grøgaard, B., & Steel, P. (2018). Complements or substitutes? A meta-analysis of the role of integration mechanisms for knowledge transfer in the MNE network. *Journal of World Business, 53*(4), 415–432. https://doi.org/10.1016/j.jwb.2018.02.001

Zhao, D., Zuo, M., & Deng, X. (Nancy). (2015). Examining the factors influencing cross-project knowledge transfer: An empirical study of IT services firms in China. *International Journal of Project Management, 33*(2), 325–340. https://doi.org/10.1016/j.ijproman.2014.05.003

10. Appendixes

Appendix 1 – The manuscript for the interviews

Presentation structure:
The one **who begins** the introduction takes slide: **1 + 2 + 4**, the **other person** takes slide **3** so the interviewee hears both voices before we begin the interview.

Slide **6**: Aizhan
Slide **7**: Aizhan
Slide **8**: Claes
Slide **9**: Claes
Slide **10**: Aizhan

When time limit is out for the question:
In order to fill out answers for as many questions as we can, I'm sorry to interrupt you now, but we should proceed to the next questions, thank you for answering this question so thoroughly.

If interviewee stray away from the question too much:
Yes, but to go back to the question at hand what we are looking for is more of (state question).

Opening statement
First of all, we would like to thank you for the support of our book project and welcome you to this interview and I would like to introduce the team behind it, my name is Aizhan (or Claes) and my colleague Claes (or Aizhan).

About the book project
We are doing a study focused on examination of various aspects that affect knowledge transfer between large companies and small and medium enterprises.

As you might be aware, knowledge transfer is considered as the most valuable practice to enhance the competitive advantage of any company. Therefore, we aimed to explore the aspects that affect inter-organizational knowledge from different sized companies that will help the (LC/MNC name or SME name, depending on the origination of the interviewee) to establish holistic knowledge transfer strategies with other partners to the organization.

Guidance for the interview process

- A dialog 60 min between the master students and the expert from the company or 3-4 minutes per question.
- The type of questions is related to your experience in projects and activities about new knowledge.

- Speakers can answer questions best to their ability and skip if they are unwilling to answer
- To facilitate our interview, we kindly ask for permission to record the interview for further use of this data for the research purpose. All recordings will be saved in (LC/MNC name) storage and will not be distributed or shared with third parties. We guarantee that information will be used only for the research purpose and we signed a non-disclosure agreement (NDA).
- Importantly, interviews and data recording are a vital part of any research process and based on the information that we will collect, we will be able to build our analysis in our study.
- If you have any questions, please feel free to ask us. Thanks again for your participation and let's have a look at the questions.

(Now we state the questions described in table 3.4.5, one by one).

Appendix 2 – The PowerPoint used for the interview
Slide 1:

- Purpose:

✓ To examine various aspects that affect inter-organizational knowledge transfer between different sized organizations and partners.

- Key activities:

✓ Interviews with employees from the Hospitality division and Senior Care division of ASSA ABLOY Global Solutions.

✓ Analysis of secondary data such as reports, websites and so on.

- Outcomes:

✓ Research on knowledge transfer at the inter-organizational level (in terms of barriers, key aspects that affect knowledge transfer by exploring large organizations and small and medium enterprises).

Slide 2:

✓ A dialog 60 min between the master student and the expert from the company or 3-4 min per question. The time can be divided if needed.

✓ The type of questions is related to your experience in projects and activities about new knowledge.

✓ Interviewees can answer questions best to their ability and skip if they are unwilling to answer.

✓ The terms of confidentiality: All information will be treated with the utmost confidentiality. All answers that will be collected will not be shared with anybody and will be used only for the purpose of the study. No source, individual, organizational will be identified or comment attributed without written permission of the originator.

✓ To facilitate our interview, we kindly ask for permission to record the interview for further correct transcribing of the answers.

✓ Collected answers will be transcribed and send back again to the interviewees for confirmation of the information collected.

✓ If you have any questions, please feel free to ask us.

Thank you for your cooperation!

Slide 3:

1. Could you provide some information about yourself?

    a) Job position?

    b) How many years of experience in total?

    c) How many years of experience in (the SME or LC/MNC)

2. What are your skills/expertise areas? (e.g., product development, R&D)

(Remaining slides are the questions seen in table 3.4.5.)

Appendix 3 – The coding of primary data

Dimensions/ Characteristics	Aspects that affect KT	Subcategory of the aspect	Primary data source	Quote from the informant	Relating literature	Conclusion/Analysis	Summary
Characteristics of the relationship	Social ties	(**SOCT1**) How to establish new relationships and (**SOCT2**) How to maintain relationships	Informant 1 (LC)	(How does your company establish communication with potential and known business small and medium-sized partners?) *"I would assume. This is actually, I mean, since I'm a little bit deeper down in the organization, I really don't know what the first contact is. I assume that someone sells something. (**SOCT1+2**) We have a sales organization. So, we have some kind of key account managers. And then its kind of drops down to the product management (source of KT) for the product. And then he has contact with the customers (recipient of KT). I think. I really don't know what the answer is otherwise, but that's my assumption."* (If we direct the question to your department, how do R&D communicate with the sales department?)	Strong social ties acts as a facilitator for knowledge transfer (Kang & Sauk Hau, 2014) and Liu et al. (2015) argues that there are three steps to establishing a high quality relationship, (**SOCT1**) first step involves to locate a partner firm and identify its key contacts, (**SOCT2**) secondly to develop and maintain the relationship established, last stage involves dispatching of knowledge from the source along with knowledge	The informant directs (**SOCT1+2**) to be handled by another role of socialization within the organization. While outside of the framework, it is consistent with the literature.	Social ties are established and maintained by a role of socialization in the large company (**SOCT1+2**)

Informant			
			Large companies use informal meetings to establish new social ties with SMEs (**SOCT1+3**)
			Explicitly inconsistent with the literature as the informant relates (**SOCT1**) is achieved by informal council meetings, whereas the literature state its to locate a partner firm and identify its key contacts. However, this could implicitly have been done before hand,
	"That will be the product management organization. I think? Because he is kind of responsible for the entire product for domestic product that I'm sitting on. And actually, he discusses all, I mean, all of them like features and the needs for the customers. And then it kind of drops down for the rest of the organization or not."	absorption of the recipient, which is how the knowledge is exchanged. Social ties according to Tangaraja et al. (2016) is a relationship based concept, its either conducted directly or indirectly between two parties and differentiated by (**SOCT3**) informal or formal ways.	
Informant 2 (LC)	(How does your company establish communication with potential and known business small and medium-sized partners?) *"When we reach out to others, there's no established way (to contact recipient's firm). There's no one way that we establish communication. (**SOCT1**) So, you end up knowing or meeting people through these council meetings. And often this is how we reach out to people (contact a*		Strong social ties acts as a facilitator for knowledge transfer (Kang & Sauk Hau, 2014) and Liu et al. (2015) argues that there are three steps to establishing a high quality relationship, (**SOCT1**) first step involves to locate a

| | | | | *recipient's firm). We get to know people from other partners (other recipients) and we reach out through those. So, it's quite (SOCT3) informal.* | partner firm and identify its key contacts, **(SOCT2)** secondly to develop and maintain the relationship established, last stage involves dispatching of knowledge from the source along with knowledge absorption of the recipient, which is how the knowledge is exchanged.

Social ties according to Tangaraja et al. (2016) is a relationship based concept, its either conducted directly or indirectly between two parties and differentiated by **(SOCT3)** informal or formal ways. | prior to the council meetings.

The informant does not mention **SOCT2**.

The informant mention that SMEs that this relationship is conducted by informal means, which is explicitly consistent with the literature **(SOCT3)**. | |

				Informant 3 (LC)	(How does your company establish communication with potential and known business small and medium-sized partners?) "I mean, the last year has been teams. That's the only thing that we've had. And I will absolutely think that going forward, this the COVID year will absolutely change the future as well. (**SOCT2**) Historically has been a lot of traveling, a lot of face-to-face. And we see now that there is not maybe the most effective way spending internal resources, but same time face-to-face is absolutely important and essential to maintain good relations with external companies especially. And also, I would mention is in regard to communication, what I often say, what happens in a meeting, it's very often that just to formalize what you often decide outside of that meeting. So that the formal meeting is, of course you can't avoid those, but also interacting in other forms is very important as well. If applicable that again, this then comes which part organizational talking we talk about internally. It's the interaction between R and D and a partial	Strong social ties acts as a facilitator for knowledge transfer (Kang & Sauk Hau, 2014) and Liu et al. (2015) argues that there are three steps to establishing a high quality relationship, (**SOCT1**) first step involves to locate a partner firm and identify its key contacts, (**SOCT2**) secondly to develop and maintain the relationship established, last stage involves dispatching of knowledge from the source along with knowledge absorption of the recipient, which is how the knowledge is exchanged.	The informant does not mention **SOCT1**. Explicitly consistent with the literature as the informant relates maintenance of relationships is done previously with face-to-face and lately it has been done via virtual environments (**SOCT2**). The informant mentions that the relationship is both informal and formal, whereas the formal part is only done to establish what informal decisions has been made outside of the meeting, this is implicitly inconsistent with the literature as the relationship is done by ways (**SOCT3**).	Large companies utilize formal and informal to maintain their social ties with SMEs (**SOCT2**).

			Social ties according to Tangaraja et al. (2016) is a relationship based concept, its either conducted directly or indirectly between two parties and differentiated by (**SOCT3**) informal or formal ways.	SMEs research prior to the relationship if the social tie is suitable for them and establish contact via phone calls, emails or digital meetings (**SOCT1**). SMEs maintain social ties
				The informant state that SMEs establish new social ties by researching prior to the other company to find more suitable social ties (**SOCT1**), which is explicitly consistent with the literature. For (**1.SOCT2**) the informant state that they maintain relations via digital meetings and physical meetings if necessary.
	Informant 4 (SME)	*realization over cup of coffee. During workshops, launch. Very important as well. "* (So you would say that the formal and informal both have their role in effective communication alongside face-to-face or online meetings?) *"Absolutely."*	Strong social ties acts as a facilitator for knowledge transfer (Kang & Sauk Hau, 2014) and Liu et al. (2015) argues that there are three steps to establishing a high quality relationship, (**SOCT1**) first step involves to locate a partner firm and identify its key contacts, (**SOCT2**) secondly to develop and	
		(How does your company establish communication with potential and known business partners of large size?) *"(**SOCT1**) I think it's a phone call or email that start the communication or a Teams meeting. First contact (with the recipient) probably via email first."* (Does your organization do any research beforehand establishing the other meeting? About the organization?)		

	"Yes. To find out possible alternatives (of other potentially better relationships)." (And how would you say you maintain your relations with other partners?) "(**1.SOCT2**) Nowadays, It's digital meetings. So, if we have a relationship, it would be regularly Teams meetings or digital video meetings and communication through email. This is very common." (If you say that, scenario before pandemic. How you maintain these contacts. Using e-mail, you said?) "Primarily, I would say email, but also visits. Depending on how far the distance (geographical distance/proximity) in between the companies." (If I go back to maintaining relations, would you say your more informal or formal on maintaining the relationship?)	maintain the relationship established, last stage involves dispatching of knowledge from the source along with knowledge absorption of the recipient, which is how the knowledge is exchanged. Social ties according to Tangaraja et al. (2016) is a relationship based concept, its either conducted directly or indirectly between two parties and differentiated by (**SOCT3**) informal or formal ways.	The informant relates that the social tie is based on formal means which is consistent with the literature (**2.SOCT2**).	with LCs by digital meetings and physical meetings (**1.SOCT2**). SMEs utilize formal means to maintain social ties with LCs (**2.SOCT2**).

		Interview data	Literature	Analysis	Findings
		"(2.SOCT2) Yeah. In my situation, I think we often have meetings, minutes of meeting, and I've written actions and so on. Not so informal, more like documented meetings or minutes of meetings."		(**SOCT1**), explicitly inconsistent with the literature as the informant state that they don't locate other partner firms and their key contacts, however this could implicitly be consistent if one take the perspective of the other side, where they identify the SME and the key contact (the informant). The informant state that they schedule meetings to develop and maintain the relationship established, last stage involves dispatching of knowledge from	SMEs rely on LCs to contact them to establish new social ties (**SOCT1**). SMEs maintain their social ties by scheduling meetings with LCs (**SOCT2**). SMEs maintain their social ties both formally and informally with LCs, depending on
	Informant 5 (SME)	(How does your company establish communication with potential and known business partners of large size?) *"A lot of partners who wants to work with Phoniro. (**SOCT1**) They contact me. I meet with them and see something we can work together on, do we, have any opportunities in the future, then try to learn what are they doing and If it's something we can learn from each other. In ASSA OBLOY I try to find my own network."* (How does your company maintain communication with potential and known business partners of large size?) *"(**SOCT2**) Schedule meetings. Same way I did my previous company. In the previous company I somehow"*	Strong social ties acts as a facilitator for knowledge transfer (Kang & Sauk Hau, 2014) and Liu et al. (2015) argues that there are three steps to establishing a high quality relationship, (**SOCT1**) first step involves to locate a partner firm and identify its key contacts, (**SOCT2**) secondly develop and maintain the relationship established, last stage involves dispatching of knowledge from		

				know that these people that I needed to know could be good to know because they can lead me or help to support me in getting contact with other people. So, I try to establish a good network." (Would you say your communication is more formal or informal when you tried to maintain those relationships?) *"(SOCT2) I think informal. Yeah. Depends on who it is."*	the source along with knowledge absorption of the recipient, which is how the knowledge is exchanged. Social ties according to Tangaraja et al. (2016) is a relationship based concept, its either conducted directly or indirectly between two parties and differentiated by **(SOCT3)** informal or formal ways.	contact is, it then becomes formal or informal, which implicitly is consistent with the literature as it can be either **(SOCT2)**.	the individual connection **(SOCT2)**.
			Informant 6 (SME)	(How does your company establish communication with potential and known business partners of large size?) *"(SOCT3) So I do a lot of storytelling and help my colleagues do storytelling. So if people do something good, I create a story on it and I share that with others. So then they get engaged and they get a thrill*	Strong social ties acts as a facilitator for knowledge transfer (Kang & Sauk Hau, 2014) and Liu et al. (2015) argues that there are three steps to establishing a high quality relationship,	The informant does not specifically mention **SOCT1**. The informant state that the maintenance of the social tie with LCs is done formally (KPI), it should be done more informally (storytelling), while not	SMEs want more informal maintenance of social ties while LCs wants more formal **(SOCT2+3)**.

					the way I want to tell a story as well. I want to do this or I want to do something else. Because it's like, I mean, you want to be in the newspaper with the good news, right? So you want a story from you individually or your team or your company. And the KPIs, well, I think you need KPIs, but you need to be very aware of that. They, they are they can be in the KPIs, can be an illusion... I think direct dialogue is very best. So, I think it comes from building relations and gaining trusts (with the recipient). And I think that you should establish it on a very human level (informal communication). I think you should avoid managers speaking to managers (avoid hierarchical structures of source and recipient's organizations). You should instead encourage teams or individuals to speak to teams or individuals." (How does your company maintain the relationship?) *"(**SOCT2**) I think that this one is kind of tricky because in bigger organizations you work with KPIs (Performance indicator). That's an*	(**SOCT1**) first step involves to locate a partner firm and identify its key contacts, (**SOCT2**) secondly to develop and maintain the relationship established, last stage involves dispatching of knowledge from the source along with knowledge absorption of the recipient, which is how the knowledge is exchanged. Social ties according to Tangaraja et al. (2016) is a relationship based concept, its either conducted directly or indirectly between two parties and differentiated by	specifically specified in **SOCT2**, the relationship can be maintained by either. In the perspective of the informant, the SME establish social ties via storytelling (informal) while the LC communicates more formally (KPIs), this is consistent with the literature in terms of (**SOCT3**), as it can be either formal or informal.	

				established way of getting an illusion of control. Illusional communication on how things go within different parts of the organizations. It's also not, from my perspective, not very strong. I think you should work lots more with storytelling. So, if you have something that's smaller or bigger or whatever size company, if you get a group or individuals to story tell about their achievements (to the recipient). And from this you gain commitment by other part (the recipient), by other parts of the company and say, I want to do this as well. And then you get this interaction between these two individuals. So, these two teams, then you get the full-fledged effects. What you, what you need to do potentially." (So you would say that by KPIs, larger organizations tend to maintain this relationship between, for example, your company by KPIs, but you think they should start using storytelling instead?) *"(**SOCT3**) I think KPIs is very bad way of measuring. What we do good, because KPIs can always be altered.*	(**SOCT3**) informal or formal ways.		

Informant	Interview extract	Theory	Finding / interpretation
			Social ties are established and maintained by a role of socialization in the large company (**SOCT1+2**)
	So if you put a KPI on one part of the organization, say, Hey, you must measure this. And I say, okay, what should we reach? What KPI is our goal. What is I'll go and, and the, and you get a goal and they alter the ways that they work so that they achieve the goal. So, I mean, it's, it's pretty bad way of measuring how we do things in a good way.		The informant directs (**SOCT1+2**) to be handled by a role of socialization within the organization. This is is explicitly consistent with the literature. The informant does not mention **SOCT3**.
Informant 7 (LC)	*(How does your company establish communication with potential and known business partners of a small or medium size?)* "*Yeah, this is kind of outside of my role since I'm more embedded within the development department. It's unfortunately outside my role to interact with new partners or even existing ones... unfortunately a few layers of different roles, which is outside my role. I'm the Product Owner for a product and a team. And then (**SOCT1+2**) I have like a technical contact person and a key account manager. And they in turn have people. Who then interact with the customers.*"	Strong social ties acts as a facilitator for knowledge transfer (Kang & Sauk Hau, 2014) and Liu et al. (2015) argues that there are three steps to establishing a high quality relationship, (**SOCT1**) first step involves to locate a partner firm and identify its key contacts, (**SOCT2**) secondly to develop and maintain the relationship established, last stage involves	

						dispatching of knowledge from the source along with knowledge absorption of the recipient, which is how the knowledge is exchanged. Social ties according to Tangaraja et al. (2016) is a relationship based concept, its either conducted directly or indirectly between two parties and differentiated by (**SOCT3**) informal or formal ways.		
				Informant 8 (SME)	(How does your company establish communication with potential and known business partners of large size?) *"(**SOCT1**) I think that it's important to first, figure out who's the key contact (of recipient firm). And I would guess you could say, even if*	Strong social ties acts as a facilitator for knowledge transfer (Kang & Sauk Hau, 2014) and Liu et al. (2015) argues that there are three steps to establishing a	Explicitly consistent with the literature as the informant state that they figure out the key contact first, preferably the key account manager in the LC to aid in establishing the relationship (**SOCT1**).	To establish new social ties, SMEs attempt to identify the account manager in the LC to assist in

	establishing the relationship (**SOCT1**). The more transactions between the LC and SME in the social ties, the more formal governance structure of maintenance is needed by the SME (**SOCT2+3**).
	The maintenance of the relationship depends on the frequency of transactions according to the informant (**SOCT2**), while not mentioned specifically in the literature it is logical. The social tie is either formal or informal depending on the frequency of transactions, as more transactions requires more formal governance structures on the relationship, this is inconsistent with the literature (**SOCT3**).
not all use that role that you're kind of key account manager who can help you and navigate through the organization (recipient) that you want to start a relationship with or that you might already happened, right?" (How would you say you maintain this relationship with, for example, the Account Manager?) *"(**SOCT2+3**) It depends on the frequency of the transactions with that company. Because if it's a company that you have many and even maybe even daily transactions with then I usually like to set up a more formal governance structure on how you cooperate and how you exchange information and knowledge. And the other one it's more of an ad hoc and on a need basis."*	high quality relationship, (**SOCT1**) first step involves to locate a partner firm and identify its key contacts, (**SOCT2**) secondly to develop and maintain the relationship established, last stage involves dispatching of knowledge from the source along with knowledge absorption of the recipient, which is how the knowledge is exchanged. Social ties according to Tangaraja et al. (2016) is a relationship based concept, its either conducted directly or indirectly between two parties and

| | | | | Informant 9 (LC) | (How does your company establish communication with potential and known business small and medium-sized partners?)

"Yeah. (*SOCT1*) *Very often, the first I mean, very often. It is them contacting us. Since we have this major position in the market The big name, most other players wants to play with us.*"

(How does your company maintain this relationship with potential and known business small and medium-sized partners?)

"*(SOCT2+3) Regular basis, regular talks to follow up the contractors that we have or follow-up agreements (with the recipient). Agreements are signed for one year at a time (with the recipient). Frame Agreements are assigned for one year at a time. Assignments can be shorter. And every frame agreement and every assignment we meet up. So that's basically it, whenever it's needed.*" | differentiated by **(SOCT3)** informal or formal ways.

Strong social ties acts as a facilitator for knowledge transfer (Kang & Sauk Hau, 2014) and Liu et al. (2015) argues that there are three steps to establishing a high quality relationship, **(SOCT1)** first step involves to locate a partner firm and identify its key contacts, **(SOCT2)** secondly to develop and maintain the relationship established, last stage involves dispatching of knowledge from the source along with knowledge absorption of the recipient, which is | **(SOCT1)**, explicitly inconsistent with the literature as the informant state that they don't locate other partner firms and their key contacts, however this could implicitly be consistent if one take the perspective of the other side, where they identify the LC and the key contact for the company.

The informant state that to maintain social ties they have regular talks with follow-up agreements or frame agreements, which is implicitly a very formal means of maintaining a relationship **(SOCT2+3)** | LCs rely on SMEs to contact them to establish new social ties **(SOCT1)**.

LCs maintain their social ties by formal means with SMEs **(SOCT2+3)** |
|---|---|---|---|---|---|---|---|---|
| | | | | | | | | |
| | | | | | | | | |
| | | | | | | | | |

	Characteristics of the relationship	Social ties	Which social tie does the organization prefer, informal or formal relations	Informant 1 (LC)		
					how the knowledge is exchanged. Social ties according to Tangaraja et al. (2016) is a relationship based concept, its either conducted directly or indirectly between two parties and differentiated by **(SOCT3)** informal or formal ways.	LCs prefer both informal and formal social ties with SMEs **(SOCT4)**
					Social ties according to Tangaraja et al. (2016) is a relationship based concept, its either conducted directly or indirectly between two parties and differentiated by **(SOCT4)** informal or formal ways.	The case that the informant is describing is the needs of the recipient, additionally stating that the size of the recipient is not affected by the need's that they are expecting. Thus, implicitly the informant state that informal or formal social ties do not matter **(SOCT4)**, which is not mentioned by the literature.
					(What do you think is effective communication between your organization and potential or known small and medium-sized partners) *"Yeah. I mean, I think it's I think the difference is when it comes to like, for instance, new projects within in the product. The ones that were involved in right now are quite big (the projects). And those are new sets of features for instance in both projects, which means that we don't have the what the customer (recipient) wants in the product as it is right now. Both*

				*of them are quite large. (**SOCT4**) But I don't think it would matter in that case if the size of the partners actually, because it would be new things that have to be developed. Suddenly, I think that the difference is more when it comes to like, customers tend to buy the thing of the rack and are very satisfied with everything that you have in the product. And compare that to, yeah, we want your product, but we will all these kinds of things. You need to develop them for us to buy a product."*			
			Informant 2 (LC)	(What do you think is effective communication between your organization and potential or known small and medium-sized partners) *"Yeah, So, well, I think effective communication. (**SOCT4**) The definition of effective communication is if at the end you get the result you were looking for. So, so that's, that's my answer. What do you think is effective communication? Effective communication regardless whether it's with smaller, medium size is getting the result you were looking for, not the answer, but the result. So,*	Social ties according to Tangaraja et al. (2016) is a relationship based concept, its either conducted directly or indirectly between two parties and differentiated by (**SOCT4**) informal or formal ways.	The case that the informant is describing is the results achieved from the relationship, additionally stating that the size of the recipient is not affected by the need's that they are expecting. Thus, implicitly the informant state that informal or formal social ties do not matter (**SOCT4**), which is not mentioned by the literature.	LCs prefer both informal and formal social ties with SMEs (**SOCT4**)

					if you were looking to solve a problem or find people who can help you, things like that. I think that's effective communication. If you wanted to get a message through, if you were looking to yeah, all these kind of things. So effective communication is, is that."			
				Informant 3 (LC)	(What do you think is effective communication between your organization and potential or known small and medium-sized partners) *"I mean, the last year has been teams. That's the only thing that we've had. And I will absolutely think that going forward, this the COVID year will absolutely change the future as well. Historically has been a lot of traveling, a lot of face-to-face. And we see now that there is not maybe the most effective way spending internal resources, but same time face-to-face is absolutely important and essential to maintain good relations with external companies especially. (SOCT4) And also, I would mention is in regard to communication, what I often say, what happens in a meeting, it's very often that just to formalize what you often decide outside of that meeting.*	Social ties according to Tangaraja et al. (2016) is a relationship based concept, its either conducted directly or indirectly between two parties and differentiated by **(SOCT4)** informal or formal ways.	The informant state that informal and formal communication is necessary for social ties, as the informal means is where decisions are made, while formal is to formalize the decisions on an organizational level **(SOCT4)**, which is inconsistent with the literature.	LCs prefer both informal and formal social ties with SMEs **(SOCT4)**

				So that the formal meeting is, of course you can't avoid those, but also interacting in other forms is very important as well. If applicable that again, this then comes which part in the organizational we are talking about internally. Like, the interaction between R and D over a cup of coffee. During workshops, launch. Very important as well."		
			Informant 4 (SME)	(What do you think is effective communication between your organization and potential or known large-sized partners?) *"I think email is effective. Also, video meetings. Sometimes a real meeting or meeting in real life is more effective. (SOCT4) It depends on if you know each other (sender and recipient). It can be, you can just send a question, I get an answer. If you don't know each other or if you discuss hard problems and so on. It can be more effective to meet, to have a real meeting. But it's also time-consuming. If I should meet you, I have to travel and so on."*	Social ties according to Tangaraja et al. (2016) is a relationship based concept, its either conducted directly or indirectly between two parties and differentiated by (**SOCT4**) informal or formal ways.	The informant state that if there is an existing social tie, informal means are possible, whereas an absence of a strong social ties it becomes informal (**SOCT4**), which is to some extent inconsistent with the literature.	SMEs prefer informal social ties over formal ones with LCs (**SOCT4**)
			Informant 5 (SME)	(What do you think is effective communication between your organization and potential or known large-sized partners?)	Social ties according to Tangaraja et al. (2016) is a	The informant state that depending on who it is the social tie is maintained informally,	SMEs prefer informal social ties over formal

				"Schedule meetings. Same way I did my previous company. In the previous company I somehow know that these people that i needed to know could be good to know because they can lead me or helping to support me to get in contact with other people. So I try to establish a good network." (Would you say your communication is more formal or informal when you tried to maintain those relationships?) "(**SOCT4**) I think informal. Yeah. Depends on who it is."	relationship based concept, its either conducted directly or indirectly between two parties and differentiated by (**SOCT4**) informal or formal ways.	implicitly preferring informal as this was mentioned firsthand. This implies that SMEs rely on both informal and formal social ties (**SOCT4**), which is consistent with the literature.	ones with LCs (**SOCT4**)
			Informant 6 (SME)	(What do you think is effective communication between your organization and potential or known large-sized partners?) "(**SOCT4**) I think that this one is kind of tricky because with bigger organizations you work with KPIs (Performance indicators) (formal, standardized). That's an established way of getting an illusion of control. Illusional communication on how things go within different parts of the	Social ties according to Tangaraja et al. (2016) is a relationship based concept, its either conducted directly or indirectly between two parties and differentiated by (**SOCT4**) informal or formal ways.	The informant state that LCs prefer formal social ties (with KPIs) whereas the SME prefers informal social ties (storytelling) (**SOCT4**), this is to some extent consistent with the literature.	SMEs prefer informal social ties over formal ones with LCs (**SOCT4**)

				organizations. It's also not, from my perspective, not very strong. I think you should work lots more with storytelling (informal communication). So, if you have something that's smaller or bigger or whatever size company, if you, if you get a group or individuals to, to story tell about their achievements. And from this you gain commitment by other part, by other parts of the company and say, I want to do this as well. And then you get this interaction between these two individuals. So, these two teams, then you get the full fledged effects. What you, what you need to do potentially."			
		Informant 7 (LC)	(What do you think is effective communication between your organization and potential or known large-sized partners?) - (It can be, as I said, not only external companies, but also these external division that you will have in ASSA OBLOY for example. Do you interact with other division, I know that you are working in hospitality,	Social ties according to Tangaraja et al. (2016) is a relationship based concept, its either conducted directly or indirectly between two parties and differentiated by (**SOCT4**) informal or formal ways.	The informant does not mention SOCT4.	No conclusion

				yes. But you work or cross teams like cross divisions?) "Yeah. To some degree, but we are kind of isolated silos, unfortunately. But there are organizational initiatives to kind of work horizontally as well. So a product developed for some other branch can be used in the hospitality branch as well. So we don't create two different hardware. Could be the same, right? So then we tried to save money that way. So we use some other hardware system from another part."			
			Informant 8 (SME)	(What do you think is effective communication between your organization and potential or known large-sized partners?) *"I would say that it's probably easier for a larger company (source) going to a small company (recipient), than for a small company (source) going to the larger company (recipient) because it's from my perspective, it's usually harder being the person who runs different areas towards different persons, all in one person rather than on the other side when you only have one person to go to."*	Social ties according to Tangaraja et al. (2016) is a relationship based concept, its either conducted directly or indirectly between two parties and differentiated by (**SOCT4**) informal or formal ways.	The informant state that informal social ties with LCs is better, but difficult to reach as you have many connection points to form the strong social tie with, thus informal are preferred by SMEs (**SOCT4**), this is not mentioned in the literature.	SMEs prefer informal social ties over formal ones (**SOCT4**)

				(Would you say that when you communicate with a larger organization is informal communication or formal communication more efficient?) *"(SOCT4) I think I would say that in formal is more efficient. But to me, that might be more difficult to reach with a large company as well. Because when you then talk to, if you are, for example, in a small organization (source) and talk to someone in the larger organization (recipient), that person might not have the full view of the relationship and they might not have the full knowledge, on the complete setup. So, they might be more hesitant in maybe in taking action or a promising things or changing things within the relation without coordinating it with the rest of the team evolved."*			
			Informant 9 (LC)	(What do you think is effective communication between your organization and potential or known small and medium-sized partners) *"... (SOCT4) Start with a very informal discussion of what we need (the source), informal is good to get*	Social ties according to Tangaraja et al. (2016) is a relationship based concept, its either conducted directly or indirectly	The informant state that they start with informal social ties with the SME, to get an impression of the organization, however to establish a strong social tie they follow	LCs prefer formal social ties over informal ones with SMEs (**SOCT4**)

| | | | | *the impression (of the recipient), ends up with a very formal frame agreements and structure the way how to do that. And also of course we can assess their (recipients) capabilities or competencies and so on. But that's done more the informal way than actually formal ways like NDAs, contract and frame agreements."* | between two parties and differentiated by (**SOCT4**) informal or formal ways. | formal means such as contracts, frame agreements and NDAs (**SOCT4**), which is to some extent consistent with the literature. | |

Dimensions/ Characteristics	Aspects that affect KT	Subcategory of the aspect	Primary data source	Quote from the informant	Relating literature	Conclusion/Analysis	Summary
Organizational characteristics of the recipient and the source	Absorptive capacity	(**ABS1**) Implementation of acquired knowledge	Informant 1 (LC)	(After you have engaging with a knowledge transfer with a small and medium firm, how do you implement the acquired knowledge so that the organization can utilize it?) *"(**ABS1**) I mean, either it comes from tech service and then it goes through a system called service now. And then into us in our R&D through Jira that is basically the tool we use. And then it really depends on if it's like, if it's a bug at the customer side, then the developers should fix it as soon as possible. But if it's like a new*	The literature of absorptive capacity agrees that it consist of aspects such as (**ABS1**) intra-firm knowledge transmission because after acquiring knowledge it needs to be (**ABS2**) diffused within the firm, which is how they are these aspects are interrelated (De	The informant state that they implement the knowledge acquired into different data systems such as Service now, Jira, backlogs or Confluence (**ABS1**), which is consistent with the literature.	LCs rely on implementing acquired knowledge from individuals to the organizational level in systematic data bases (**ABS1**).

					request for something, a new feature or something, then it goes into the backlog in Jira... and if something must be like more documented for the future, we use confluence."	Luca & Cano Rubio, 2019).		
				Informant 2 (LC)	(After you have engaging with a knowledge transfer with a small and medium firm, how do you implement the acquired knowledge so that the organization can utilize it?) *"Yes, but not in a standardized way, unfortunately. So, I think once it's sent out by email, some people will probably either save it in their email, save it in a box folder somewhere. I kind of like to put it in folders. If we do a Teams meeting, (**ABS1**) I like to save the file in the Teams meeting. And in the global quality team, we do have folders. So, and for example, I worked with compliance, sustainability, and stuff like that. So, I have folders for all those topics. But there's no global way of saving that knowledge."*	The literature of absorptive capacity agrees that it consist of aspects such as (**ABS1**) intra-firm knowledge transmission because after acquiring knowledge it needs to be (**ABS2**) diffused within the firm, which is how they are these aspects are interrelated (De Luca & Cano Rubio, 2019).	The informant state that implementation of acquired knowledge is done in different folders according to the topic of it (**ABS1**), while not specifically mentioned in the literature it is still implicitly consistent with the literature.	LCs rely on implementin g acquired knowledge from individuals to the organization al level in systematic data bases (**ABS1**).
				Informant 3 (LC)	(After you have engaging with a knowledge transfer with a small	The literature of absorptive capacity	The informant state that they implement the	LCs rely on implementin

Informant	Interview Excerpt	Theory	Interpretation
	…and medium firm, how do you implement the acquired knowledge so that the organization can utilize it?) *"Okay. So, tool wise, the changes are done in a system called Team Center. this is to let R and D use it for development where they store drawings. And that system also handles the revision handling of it as well."*	agrees that it consist of aspects such as (**ABS1**) intra-firm knowledge transmission because after acquiring knowledge it needs to be (**ABS2**) diffused within the firm, which is how they are these aspects are interrelated (De Luca & Cano Rubio, 2019).	g acquired knowledge from individuals to the organizational level in systematic data bases (**ABS1**). knowledge acquired into different data systems such as Team Center (**ABS1**), while not specifically mentioned in the literature it is still implicitly consistent with the literature.
Informant 4 (SME)	(After you have engaging with a knowledge transfer with a large firm, how do you implement the acquired knowledge so that the organization can utilize it?) *"(**ABS1a**) That's my specialty. It's that you can define a process, or you can write a procedure or instruction, or you can save a checklist habit available for the staff."* (Do you have systematic means of storing it?)	The literature of absorptive capacity agrees that it consist of aspects such as (**ABS1**) intra-firm knowledge transmission because after acquiring knowledge it needs to be (**ABS2**) diffused within the firm, which is how they are these	SMEs implement acquired knowledge from individuals to the organizational level in systematic data bases (**ABS1**). The informant state that they implement the knowledge acquired into (**ABS1a**) instructions or checklists, while not mentioned if in a data system, the informant state that the organization is about to implement a systematic data base where they can visualize the knowledge (**ABS1b**), while not specifically mentioned in the literature it is still implicitly

			aspects are interrelated (De Luca & Rubio, 2019).	consistent with the literature.	SMEs cannot always fully implement the acquired knowledge from LCs due to organization al limitations (**ABS1**).
		*"(**ABS1b**) Yes. We have a really good process-oriented management system where we visualize our processes. And we can connect activities to instructions and checklists and so on. So that's a really good tool that we are about to implement at the moment."*			
	Informant 5 (SME)	(After you have engaging with a knowledge transfer with a large firm, how do you implement the acquired knowledge so that the organization can utilize it?) *"(**ABS1**) We tried to establish a new way of working in (SME source organization). And we learned from (LC recipient organization) and from Safe, a new agile way of working. And I think it's important to try to anchor it with a co-worker, with all the team members who have to make sure they have the they can get their opinion. Do you think this will work? We can't as I said, we can't copy everything and get to work in (the SME), because we need to find our own way... You look at the pain point, you see how*	The literature of absorptive capacity agrees that it consist of aspects such as (**ABS1**) intra-firm knowledge transmission because after acquiring knowledge it needs to be (**ABS2**) diffused within the firm, which is how they are these aspects are interrelated (De Luca & Rubio, 2019).	The informant state that they implement the acquired knowledge that is received from a LC partially, as they cannot fully adopt the knowledge that is acquired (**ABS1**), instead they take the corner stones of the knowledge that is most essential and use it as support when they implement the knowledge into the SME. This is explicitly inconsistent with the literature.	

						SMEs implement acquired knowledge to the organization al level, but more towards the individual level (**ABS1**).
					The informant state that they (**ABS1a**) do not implement the acquired knowledge to the organizational level, but rather to the individuals that will be utilizing the knowledge. This is partially inconsistent with the literature. The informant state (**ABS1b**) that they don't store acquired knowledge from LCs frequently on an organizational level, instead they take inspiration from other external sources when knowledge is needed which is due to time constraints in the SME, this is explicitly
				The literature of absorptive capacity agrees that it consist of aspects such as (**ABS1**) intra-firm knowledge transmission because after acquiring knowledge it needs to be (**ABS2**) diffused within the firm, which is how they are these aspects are interrelated (De Luca & Rubio, 2019).		
		Informant 6 (SME)	*we can utilize the method or the principle that's practical and use it in our organization. And take the framework of it. You take that framework and you have as support and how do we establish and implemented in our own organization."*	(After you have engaging with a knowledge transfer with a large firm, how do you implement the acquired knowledge so that the organization can utilize it?) *"(**ABS1a**) I do not implement it. That's, that's the key difference in smaller organizations. So, the people that are working in the business, they are implementing it. So, in a smaller organization, I do not go about to implement the change. I ensure that the people that are to implement the change, they have the heart to do so. In small organizations you have like a shotgun of things to do every day. So, if you're not motivated to do this specific work task, you will not do it."* (Would you say that your organization tries to maybe store		

Informant	Interview excerpt	Literature	Finding	
		inconsistent with the literature.	The informant state that they implement the knowledge acquired into different data systems such as Jira (**ABS1**), which is consistent with the literature.	LCs rely on implementing acquired knowledge from individuals to the organizational level in systematic data bases (**ABS1**).
Informant 7 (LC)	the information in any way? Say you learned about the method to conduct a process or so. Does your organization store that in a document or something?) *"(**ABS1b**) Yeah, we do. But I mean, we have Google. So small organization, we don't work very much with storage and documentation of things we do, but we don't have time to keep it up to date and look at it. I mean, we when we had something at hand that we need to do, we Google it. And then we get inspired from that and we do some shots and then you say, hey, we do like this. This is much quicker decision process of implementing things."* (After you have engaging with a knowledge transfer with a large firm, how do you implement the acquired knowledge so that the organization can utilize it?) *"I mean, like the database for me is my inbox. That's where I store most of information and then we have the, like. I'm not sure how familiar you are with software development, but we have these*	The literature of absorptive capacity agrees that it consist of aspects such as (**ABS1**) intra-firm knowledge transmission because acquiring knowledge it needs to be (**ABS2**)		

					*kind of (**ABS1**) user stories, virtual post-it notes that we store in a system and we move them between different statuses like to do, doing, done. These three. And our systems called JIRA in our case. And then we have a lot of information there. And that's also kind of very tied to our development process. So that one is, of course, I don't need to carry it in my head. It's on a computer somewhere."*	diffused within the firm, which is how they are these aspects are interrelated (De Luca & Cano Rubio, 2019).		
				Informant 8 (SME)	(After you have engaging with a knowledge transfer with a large firm, how do you implement the acquired knowledge so that the organization can utilize it?) *"(**ABS1**) We have one process tool where we document all processes. And there we also create links to documents, for example, that can be used when you do a specific activity. Can be all from a presentation material to a template, to an instruction. And then we have two different documentation methods you can say so one is just a normal storing document in SharePoint library. And the other one, more of or less*	The literature of absorptive capacity agrees that it consist of aspects such as (**ABS1**) intra-firm knowledge transmission because after acquiring knowledge it needs to be (**ABS2**) diffused within the firm, which is how they are these aspects are interrelated (De Luca & Cano Rubio, 2019).	The informant state that they implement the knowledge acquired into different data systems such as SharePoint library (**ABS1**), which is consistent with the literature.	SMEs rely on implementin g acquired knowledge from individuals to the organization al level in systematic data bases (**ABS1**).

				of a website that can b structured as a knowledge base, you can say, and you can make articles and connect documents and to some extent also create workflows."			
			Informant 9 (LC)	(After you have engaging with a knowledge transfer with a small and medium firm, how do you implement the acquired knowledge so that the organization can utilize it?) *"… The one we get most often ends in a project or product. But sometimes it's the technology that goes across different types of products, different type of projects. And the way we do this is that whoever learned it. (**ABS1**) Internally, they start by actually documenting on it, on our documentation tool… And we also had workshops between key players within our departments where we try to explain the technology and where we try to establish a common knowledge on this. Everyone cannot learn everything. So, we must be careful, of course, not spending too many hours on it, to make sure that we preserve it within our*	The literature of absorptive capacity agrees that it consist of aspects such as (**ABS1**) intra-firm knowledge transmission because after acquiring knowledge it needs to be (**ABS2**) diffused within the firm, which is how they are these aspects are interrelated (De Luca & Cano Rubio, 2019).	The informant state that they implement the knowledge acquired into different data systems such as Confluence (**ABS1**), which is consistent with the literature.	LCs rely on implementin g acquired knowledge from individuals to the organization al level in systematic data bases (**ABS1**).

Organization al characteristic s of the recipient and the source	Absorptive capacity	**(ABS2)** Diffusion of knowledge, internally	Informant 1 (LC)	*organization. So, Confluence is a tool that we use a lot for conferences. It's a wiki for an internal form for us where we post the things that we know in different kind of areas and try to build up knowledge.* (Does your organization attempt to share the acquired knowledge amongst its employees? By what means?) *"I mean, we have in R&D, there's a system architect meeting. For instance, (**ABS2**) we have system architect meetings in the teams, and they gather, I don't remember how often is but they have meetings where they share knowledge there. Within VOSTIO we have what we call tech-bord, which is the little bit smaller. But it is within the VOSTIO products. But there are several teams involved there as well, but they can discuss."*	The literature of absorptive capacity agrees that it consist of aspects such as (**ABS1**) intra-firm knowledge transmission because after acquiring knowledge it needs to be (**ABS2**) diffused within the firm, which is how they are these aspects are interrelated (De Luca & Cano Rubio, 2019).	The informant state that after acquiring knowledge it is diffused in the organization by (**ABS2**) different means such as meetings, this is explicitly consistent with the literature.	LCs diffuse acquired knowledge amongst the employees with meetings (**ABS2**)
			Informant 2 (LC)	(Does your organization attempt to share the acquired knowledge amongst its employees? By what means?)	The literature of absorptive capacity agrees that it consist of aspects such as (**ABS1**)	The informant state that after acquiring knowledge it is diffused in the organization by (**ABS2**) different means such as	LCs diffuse acquired knowledge amongst the employees

				"So usually if I find out something, I love to spread it out right away to everyone who might need it, even to those who might not need it (internally). So, two ways to do that. Send it out by email. (**ABS2**) Send it out by having a meeting with the people around the table. So, in global solutions for each area of interest. So, areas of interests are sustainability, quality, customer success. So, there's different areas that we highlight in our global quality team. So, we have people in each business area. We get them around a table, virtual table because of the pandemic. And then we just disseminate this information." (And how do you share the knowledge during these meetings?) "So, during meetings, depending who's running the meeting, we have minutes of meetings that that are shared and sent out. Not always, maybe less than half the time. And then we do have folders to keep track of documents. We have something called Miro	intra-firm knowledge transmission because after acquiring knowledge it needs to be (**ABS2**) diffused within the firm, which is how they are these aspects are interrelated (De Luca & Cano Rubio, 2019).	email and meetings, this is explicitly consistent with the literature.	with meetings and email (**ABS2**)

				boards, where we plan and track projects and other things like that. Where we do roadmaps."			
			Informant 3 (LC)	(Does your organization attempt to share the acquired knowledge amongst its employees? By what means?) *"So, tool wise, the changes are done in a system called Team Center. This is what R&D uses for development where they store drawings... So yeah, it's a process where the idea is just to gather all the relevant functions to review them together. Before we send the new drawings to the supplier. So, everybody has an opportunity to understand why and what's changed and can come up with suggestions in case the designer didn't think of that basically."*	The literature of absorptive capacity agrees that it consist of aspects such as (**ABS1**) intra-firm knowledge transmission because after acquiring knowledge it needs to be (**ABS2**) diffused within the firm, which is how they are these aspects are interrelated (De Luca & Cano Rubio, 2019).	The informant state that the acquired knowledge is not (**ABS2**) directly shared but rather it is shared in the data system that it is implemented in, which is accessible by the individuals that interact with the knowledge, this is to some extent consistent with the literature.	LCs rely on the individuals to access the knowledge from the data systems themselves rather than having it given directly to them (**ABS2**).
			Informant 4 (SME)	(Does your organization attempt to share the acquired knowledge amongst its employees? By what means?) *"(**ABS2**) Maybe by a project, defining a project and take advantage of this new knowledge. It can be a project or it can also be some staff meetings to share.*	The literature of absorptive capacity agrees that it consist of aspects such as (**ABS1**) intra-firm knowledge transmission because after acquiring	The informant state that the diffusion of knowledge is achieved in the SME by (**ABS2**) projects or staff meetings, which is to some extent consistent with the literature.	SMEs diffuse acquired knowledge amongst the employees with staff meetings and projects (**ABS2**)

					SMEs diffuse acquired knowledge amongst the employees with meetings (**ABS2**)
			knowledge it needs to be (**ABS2**) diffused within the firm, which is how they are these aspects interrelated (De Luca & Cano Rubio, 2019).	The informant state that the diffusion of knowledge is (**ABS2**) only done with the individuals that are affected by it in the organization, which is inconsistent with the literature.	
		Share the knowledge and exchange.”			
	Informant 5 (SME)	(Does your organization attempt to share the acquired knowledge amongst its employees? By what means?) *“(**ABS2**) We have like different kind of meetings, where if someone, for example, have acquired knowledge, that was shared between the team. Though I think it’s important that not everyone needs to learn everything you can share in-between. For example, if someone attending good seminar or good course, they can share the main topics. What’s your takeaway from this occasion? I think its a good method to do.”*	The literature of absorptive capacity agrees that it consist of aspects such as (**ABS1**) intra-firm knowledge transmission after because acquiring knowledge it needs to be (**ABS2**) diffused within the firm, which is how they are these aspects interrelated (De Luca & Cano Rubio, 2019).		
	Informant 6 (SME)	(Does your organization attempt to share the acquired knowledge amongst its employees? By what means?)	The literature of absorptive capacity agrees that it consist of aspects	The informant state that in SMEs the acquired knowledge is can be diffused both (**ABS2a**)	SMEs diffuse acquired knowledge

informally more often than formally in the organization (**ABS2**).

formally and informally, while (**ABS2b**) the former is not very common in SMEs, this is to some extent consistent with the literature as SMEs tend to maintain their relationships with other organizations in a informal way which could also reflect on how they diffuse knowledge informally.

such as (**ABS1**) intra-firm knowledge transmission after because acquiring knowledge it needs to be (**ABS2**) diffused within the firm, which is how they are these aspects are interrelated (De Luca & Cano Rubio, 2019).

"(**ABS2a**) I mean, that's tricky. People do it in different ways (share knowledge formally and informally). Some of my colleagues, they find an article, or they find something about something that has been done in another part of organizations. And they email it to various stakeholders and say, Hey, look at this, what I found, other people, they interact informally by saying, Oh, I found this, and this could be interesting for us. What do you think about this? And other people, they set up a meeting with PowerPoint. Chunking slides say this is how you do and in this part of organization, we should do it as well. So, it's different culture, company culture, and also individual skill sets of change management."

(So, you would say this is how you share the knowledge in a smaller organization?)

"Yes. But the ladder, (**ABS2b**) the PowerPoint chunking (formal KT), that's not very frequent in

				small organization. It could happen. But to do this in a successful way, you need to do the individual interactions first, because then if you provide in a smaller organization, the level of autonomy is much higher than in a bigger from my perspective. So, you need to anchor what you should do with individuals so that they feel that they could do it. And I think that it's a good idea."			
			Informant 7 (LC)	(Does your organization attempt to share the acquired knowledge amongst its employees? By what means?) *"If I take the conference example. Unfortunately, we don't really share that kind of knowledge. Sometimes, I guess we do, but there has not been that many (**ABS2a**) conferences last year."* (But if we go back to the example of the design sprint from the conference. You said when you talk to your team, then that's your ways of sharing that knowledge?) "Yeah."	The literature of absorptive capacity agrees that it consist of aspects such as (**ABS1**) intra-firm knowledge transmission because after acquiring knowledge it needs to be (**ABS2**) diffused within the firm, which is how they are these aspects are interrelated (De Luca & Cano Rubio, 2019).	The informant state that acquired knowledge is diffused in the organization via (**ABS2a**) conferences and (**ABS2b**) meetings, which is consistent with the literature.	LCs diffuse acquired knowledge amongst the employees with meetings and conferences (**ABS2**)

		Quote	Analysis
			SMEs diffuse acquired knowledge amongst the employees with recordings, meetings and email (**ABS2**).
			The informant state that acquired knowledge is diffused in the organization via (**ABS2a**) recordings of the (**ABS2b**) knowledge, (**ABS2c**) email and meetings, which is to some extent consistent with the literature.
		(*Was it conducted by meeting or do send out emails or how did you go about it?*) *"Yeah. I think I mentioned that I was like this conference during our stand-up meeting in the morning. And at least previously there has been in other roles. Some of the companies I've had like maybe a (**ABS2b**) team meeting where I have 20 minutes and talk about this was what the conference was about and it was pretty cool. Here's the link if you want to look at the keynote and so on."*	The literature of absorptive capacity agrees that it consist of aspects such as (**ABS1**) intra-firm knowledge transmission because after acquiring knowledge it needs to be (**ABS2**) diffused within the firm, which is how they are these
	Informant 8 (SME)	(*Does your organization attempt to share the acquired knowledge amongst its employees? By what means?*) *"Yes, I would say so if the evaluation that to be worth to be known more persons than the ones who were involved in the exchange. So, we have a few different ways. We want to get good example now, for example, where we have had customer meetings with some, We call them model customers, not sure what*	

				*the English term for this. But front runners as customers. And we have made in interviews with them to kind of figure out how they work and way they have made the choices they have made in choosing other solution. So maybe also choosing others. (**ABS2a**) And then we recorded this and we should try to spread it within the organization. So that's more people get to know about our customers and how they work, for example. And then spread via the internet."* (How do you spread it to the other employees? By e-mail or additional meetings?) *"I would say that it can either (**ABS2b**) e-mail or via a post that is either written email or that you can watch a recorded content. I would say that in some cases to do (**ABS2c**) additional meetings, but it's I would say that it's usually access to material or distribution by email."*	aspects are interrelated (De Luca & Cano Rubio, 2019).		
			Informant 9 (LC)	(Does your organization attempt to share the acquired knowledge	The literature of absorptive capacity	The informant state that acquired knowledge is	LCs rely on technology

	networks to diffuse the knowledge in the organization along with workshops and meetings (**ABS2**).
diffused in the organization via (**ABS2a**) technology network which includes (**ABS2c**) Confluence, (**ABS2b**) workshops and meetings, which is to some extent consistent with the literature.	
agrees that it consist of aspects such as (**ABS1**) intra-firm knowledge transmission because after acquiring knowledge it needs to be (**ABS2**) diffused within the firm, which is how they are these aspects are interrelated (De Luca & Cano Rubio, 2019).	amongst its employees? By what means?) *"Yes, for instance, there is an example now where we have a network, which (**ABS2a**) technology network where we have been using ZigBee. We are going to upgraded to the latest standard of ZigBee. We buy competence to teach us how to do this from Reno CB consulting company, buy services from them."* (You mentioned from the previous question where you said, after you have acquired knowledge you try to explain the technology with (**ABS2b**) workshops and meetings, would you say these are the means you share the acquired knowledge to the employees?) *"And also, (**ABS2c**) Confluence. I mean, everything that we do on this week, we will put on Confluence what the conference is like. You know, you have one engineer starting without something about this. Next thing is then another guy learned*

				Quote from the informant		
				something, and they go in and edit it, updated and correct it. So, it becomes a common knowledge like in Wiki. But then yes, workshops. But actually, before we do the purchase, before we sign the contract with this company, we have a plan for how to share. Not like a written, very formal plan, but an agreement between the people and needs to know this and the management that buys this of how to actually spread it to the right people."		

Dimensions/ Characteristics	Aspects that affect KT	Subcategory of the aspect	Primary data source	Quote from the informant	Relating literature	Conclusion/Analysis	Summary
Organizational characteristics of the recipient and the source	Motivation	Which facilitate knowledge exchange with partners more (**MOT1a**) intrinsic or (**MOT1b**) extrinsic motivation	Informant 1 (LC)	(What kind of improvements could your company do in order to motivate employees for transferring knowledge to partners and acquiring knowledge from them?) *"Well, I don't really know, actually, to be honest with you. (**MOT1a**) I think the employees know that the more information the partners have regarding our product, the*	Martín Cruz et al. (2009) state that motivation consists of (**MOT1a+2a**) intrinsic and (**MOT1b+2b**) extrinsic motives, where the former takes place inside an individual, affected by per example pleasant work environment,	The informant state that to acquire new knowledge and share it with/from a partner the individual is motivated by (**MOT1a**) intrinsic motivation, whereas they will have less work to do if they share knowledge early on, which is consistent with the literature.	LCs are more incentivises by intrinsic motivation in the form of pleasant work environment to acquire new knowledge from partners and sharing knowledge with

| | | | | | *less they come back to your R&D to ask questions."*

(Would you say that in the R&D department you might be motivated to exchange knowledge in order to acquire more knowledge from the partners? And that might motivate you as an individual, in terms of getting a promotion or other things?)

"Yes... (MOT1a+b) but I think I mean, it's like that's part of the job, to spread the knowledge about the system to stakeholder." | atmosphere of mutual respect, accomplishment, self-respect, prestige, personal values and involvement. As for the latter, extrinsic motivation according to the authors is external motives which affects the individual, by per example monetary rewards that comes in direct or indirect ways, whereas the former relates to wages, incentives or bonuses and the latter relates to time not working, job education, health benefits or allowances. | The informant state that the employees are not specifically **(MOT1a+b)** motivated by incentives to exchange knowledge with partners, rather it is a part of the job description, this is inconsistent with the literature. | them **(MOT1a)**.

LCs are not specifically motivated by incentives to acquire and share knowledge with partners, instead it is a part of the job description of the employees **(MOT1a+b)**. |
| | | | | Informant 2 (LC) | (What kind of improvements could your company do in order to motivate employees for transferring knowledge to partners and acquiring knowledge from them?) | Martín Cruz et al. (2009) state that motivation consists of **(MOT1a+2a)** intrinsic and **(MOT1b+2b)** extrinsic motives, | The informant state that the employees are not specifically **(MOT1a+b)** motivated by incentives to exchange knowledge with | LCs are not specifically motivated by incentives to acquire and share knowledge with |

				*"(**MOT1a+b**) People have different kind of jobs. So, for example, if your job is coming to work and doing your work, you're not going to come to work and try to think, who can I reach out to today, right? So, whereas I come to work, and I have things to do, but then I'm trying to get people together, so that's my job. So, if your job is not trying to get together with people to do knowledge exchange?"* (How would you motivate to share knowledge or acquire knowledge from customers then?) *2) "(**MOT1a**) So, we meet our quality standards, so with ISO or whatever and internally we do continual improvement. Of course, in each business area does its own quality work. And if there's problems, it's with our suppliers that we work, not with our customers, right? We do NPS, which is Net Promoter Score. So, we do poll our customers on how*	where the former takes place inside an individual, affected by per example pleasant work environment, atmosphere of mutual respect, accomplishment, self-respect, prestige, personal values and involvement. As for the latter, extrinsic motivation according to the authors is external motives which affects the individual, by per example monetary rewards that comes in direct or indirect ways, whereas the former relates to wages, incentives or bonuses and the latter relates to time not working, job education, health benefits or allowances.	partners, rather it is a part of the job description, this is inconsistent with the literature. The informant state that to acquire new knowledge and share it with/from a partner the individual is motivated by (**MOT1a**) intrinsic motivation, whereas they get a sense of accomplishment, which is consistent with the literature.	partners, instead it is a part of the job description of the employees (**MOT1a+b**). LCs are more incentivises by intrinsic motivation in the form of accomplishmen t to acquire new knowledge from partners and sharing knowledge with them (**MOT1a**).

				LCs apply intrinsic or/and extrinsic motivation on SMEs to engage with acquisition and sharing of knowledge to fulfill the LCs criterions of collaboration. (**MOT1a+b**).
			The perspective of the informant is regarding their partners rather than the employees inside the organization, however, the LC is motivating their partner to engage in knowledge acquisition and sharing to reach a certain level of expectations, which could include both intrinsic and extrinsic motivation, this is inconsistent with the literature.	
		Martín Cruz et al. (2009) state that motivation consists of (**MOT1a+2a**) intrinsic and (**MOT1b+2b**) extrinsic motives, where the former takes place inside an individual, affected by per example pleasant work environment, atmosphere of mutual respect, accomplishment, self-respect, prestige, personal values and involvement. As for the latter, extrinsic motivation according to the authors is external motives		
Informant 3 (LC)	*they feel about our service or products are offering and our on-site service installation and all this. So, we do ask our customers all the time to fill out surveys and tell us how they feel and whatever. So, we do all that stuff that we do, interact with the customers.”* *“(**MOT1a+b**) So, a few years ago, we initiated this type of assessment of the supplier and then put them in different categories based on their performance and maturity level. And that is one way that we have pushed our suppliers to improve because that is presented to them as well. And if they are a certain level actually, they then have so and so much time to improve. Otherwise they will be exchanged. And also, we always set targets based on*			

			SMEs need extrinsic motivation in the form of job education to be aware of the framework of what they can/should acquire and share with/from partners (**MOT1b**).
	which affects the individual, by per example monetary rewards that comes in direct or indirect ways, whereas the former relates to wages, incentives or bonuses and the latter relates to time not working, job education, health benefits or allowances.		The informant state that to incentive knowledge acquisition and sharing with partners, (**1MOT1b**) extrinsic motivation in the form of job education, whereas a lack of it, the employee does not know in which frame they can share knowledge, Which also implicitly. (**MOT1b**).
Informant 4 (SME)	*the assessment of their performance, both in regard to the development itself, but also how quickly they fix issues that appear, because we measured them at that as well. Both in how many cases and how quickly we always put KPIs (Performance indicators) on our suppliers as well and push for improvement. And then this is where supply development team comes in. We go in and we work with them. But the idea is that you collaborate with your partner and you improve your apartment to achieve the goal that we set."* (What kind of improvements could your company do in order to motivate employees for transferring knowledge to partners and acquiring knowledge from them?) *"(1MOT1b) Yeah, that can be in education, how can we possible share information? What are our rules? So, if you are sure about the frame where you can share*	Martin Cruz et al. (2009) state that motivation consists of (**MOT1a+2a**) intrinsic and (**MOT1b+2b**) extrinsic motives, where the former takes place inside an individual, affected by per example pleasant work environment,	

				information then it's easier to do it. So, I think that's education." (Would you say that it's more of the employee to be self-motivated to acquire new knowledge?) *"(2MOT1b) I don't think that's possible. Because you have your workday which is already full. So, I don't think it's possible. I think you need to make it available and discuss how to do it. I think so."* (Would you say different departments might be more incentivized to acquire knowledge from other company?) *"Yeah, definitely the sales department are already used to cooperate with external companies and external persons."* (Would you say that, for example, the sales personnel	atmosphere of mutual respect, accomplishment, self-respect, prestige, personal values and involvement. As for the latter, extrinsic motivation according to the authors is external motives which affects the individual, by per example monetary rewards that comes in direct or indirect ways, whereas the former relates to wages, incentives or bonuses and the latter relates to time not working, job education, health benefits or allowances.	relate to (**2MOT1b**) the job position at hand, as the person are not engaging in these activities firsthand, whereas e.g. the sales department is, which has a (**MOT1a+b**) specific job task to do so. Thus, this is to some extent consistent with the literature.	SMEs are not specifically motivated by incentives to acquire and share knowledge with partners, instead it is a part of the job description of the employees (**MOT1a+b**).

or more incentivize then to acquire knowledge for maybe personal gains or are there other motives?)

"*(MOT1a+b) I think it's in their position (job role, that they have the availability to learn. I think it's my experience from my former company and also these companies that it would be very good if you had the developers to cooperate more with other companies. But it's, often hard to make that happen, but it would be really good. And also, product managers, product departments, to open up and meet more external organizations.*"

(I assume it's due to time constraints. But how would you say the organization could fix that issue?)

"*(1MOT1b) I think that you need to tell the employees what is important and possibility to benchmark and*

				The informant state that when customers know more, they require less maintenance in form of support, (**MOT1a**) this relates to intrinsic motivation of the SME to acquire and share knowledge with their partners, which is consistent with the literature.	SMEs are more incentivises by intrinsic motivation in the form of pleasant work environment to acquire new knowledge from partners and sharing knowledge with them (**MOT1a**).
get information to our organization. And you also must describe how would this be possible? If I would like to do this, how can I do it? Can I have one hour a week for this, or can I attend in a project?... I think you have to make it a part of the job if you should make it happen. And possibly you can ask employees, for instance, during your one-on-one meeting or development meeting, if they are motivated, if they would like to do this more in their profession, I think that might be a way forward.	Informant 5 (SME)	(What kind of improvements could your company do in order to motivate employees for transferring knowledge to partners and acquiring knowledge from them?) *"(**MOT1a**) With customers the more they know the less support they need so that results in less support issues since they know how to handle it."*	Martín Cruz et al. (2009) state that motivation consists of (**MOT1a+2a**) intrinsic and (**MOT1b+2b**) extrinsic motives, where the former takes place inside an individual, affected by per example pleasant work environment, atmosphere of mutual		

				respect, accomplishment, self-respect, prestige, personal values and involvement. As for the latter, extrinsic motivation according to the authors is external motives which affects the individual, by per example monetary rewards that comes in direct or indirect ways, whereas the former relates to wages, incentives or bonuses and the latter relates to time not working, job education, health benefits or allowances.	The informant state that intrinsic motivation in the form of (**1MOTa**) acknowledgement is key for acquiring knowledge from partners, this is	SMEs are more incentivises by intrinsic motivation in the form of accomplishmen t to acquire new knowledge
			Informant 6 (SME)	(What kind of improvements could your company do in order to motivate employees for transferring knowledge to partners and acquiring knowledge from them?)	Martín Cruz et al. (2009) state that motivation consists of (**MOT1a+2a**) intrinsic and (**MOT1b+2b**) extrinsic motives, where the former		

				"So, I think I mentioned a few improvements for motivation. I think I said that in the smaller organizations you need to motivate people. And maybe I didn't say this by them having a feeling that they, they provide value into the business. And I think more so in small than in bigger (organizations). *(1MOT1a)* But in smaller, the acknowledgement is the personal acknowledgment is very key. Because you are like I mentioned, the tribes (small community). Like a tribe where the social interactions are very important. So, you need to, or I usually, when I tend to acknowledge colleagues, I do it i three levels. First, I do it individually. The team with the, with the member or the individual saying why you did this great work. And this is what value are you added. And this is what the effects that we harvest from your work. And then I raise it one level to the team level. Saying	takes place inside an individual, affected by per example pleasant work environment, atmosphere of mutual respect, accomplishment, self-respect, prestige, personal values and involvement. As for the latter, extrinsic motivation according to the authors is external motives which affects the individual, by per example monetary rewards that comes in direct or indirect ways, whereas the former relates to wages, incentives or bonuses and the latter relates to time not working, job education, health benefits or allowances.	consistent with the literature. The informant state that the SME does not **(MOT1b)** offer monetary (extrinsic) motivation to acquire and share knowledge with partners, which could be consistent with the literature. The informant state that when LCs are sharing knowledge with them, the **(2MOT1a)** SME is only interested in acquiring it when they feel autonomous, which is indirectly intrinsic motivation. The informant state that **(2MOT1b)** extrinsic motivation should not be the primary driver in SMEs, rather intrinsic motivation, this is inconsistent with the literature.	from partners and sharing knowledge with them **(MOT1a)**. SMEs are motivated by intrinsic motivation in the form of autonomy when acquiring knowledge from LCs **(MOT1a)**.

it to the team and saying, listen in this individual and why in this case the knowledge, say a change from the bigger organization, what values are as inputs and that we have tweak it. We have done something good that works for this team. And then I put it to the company level and the tribe level and say to everyone in the company this individual or this team has done this great work by knowledge transferring from this organization and done this improvements and put it very much value based. So, this is what we've done and this is the value that we were harvesting from it. So, for my measures of motivation is very individual, but I tend to do it in this is these levels."

(Do you encourage sharing knowledge with partners then?)

"Yes, but not directly. That's really up to the individual or team to decide to do so. I

*mean, that's has traditionally been a good thing, with (the LC) acquiring smaller companies. You do whatever business you do. (**2MOT1a**) We (the large organization) will only be here to support you (the small organization). I think that's very good. But some trends are showing now that the organization says, you should do like this. You should use this system to mention is this process. And that's not very good for the smaller companies because they must choose themself. What tools, what processes, what knowledge to acquire from the bigger part."*

(And this affects the motivation to be engaged with the large organization?)

"Yes, definitely. So, if you feel autonomous, it makes your own decision what parts to use, then the adoption is significantly increased."

(Does your organization offer incentives to employees to learn new knowledge from other companies or share knowledge with them?)

*"(**MOT1b**) No. If incentives are monetary that's a no. Today we only offer soft incentives, meaning the acknowledgements... So, in a smaller organization, I would say it's much more key. So, it's like, I don't know where it is, but it's a way for success, is related to have good leaders that are good in motivating people and making them achieve more."*

(Maybe those incentives that go along with how you motivate people to elevate them that you previously described?)

"Yes. So, in a smaller organization, it's much I would say it's much more key. So, it's like, I don't know where it is, but it's a way for success, is related to have

			LCs are more incentivises by extrinsic motivation in the form of job education to acquire new knowledge from partners and sharing knowledge with them (**MOT1b**).
			The informant state that the LC offer (**MOT1b**) extrinsic motivation in the form of job education, which is consistent with the literature.
	*good leaders that are good in motivating people and making them achieve more... (2**MOT1b**) But as a small organization, money should not be the main driving force. There are people working in a small organization, they have the other driving forces not connected to the wallet.”*		Martín Cruz et al. (2009) state that motivation consists of (**MOT1a+2a**) intrinsic and (**MOT1b+2b**) extrinsic motives, where the former takes place inside an individual, affected by per example pleasant work environment, atmosphere of mutual respect, accomplishment, self-respect, prestige, personal values and involvement. As for the latter, extrinsic motivation according
Informant 7 (LC)	*(Does your organization offer incentives to employees to learn new knowledge from other companies or share knowledge with them?)* *“I mean, not sure. We have like a budget each year which you could use to (**MOT1b**) buy courses or attend conferences. But, if you don't want to take any courses, that's also fine. No real like money motivation for that... I want to sometime in the future I would like to become a manager. And I know that there are like many managerial courses. So, for my personal motivation is to then be ready to take the next*		

				step, I would like to take those courses." (Would you say, the incentives are more of personal growth for the employees?) *"Yeah, Definitely."* (Would you say that this personal growth might incentivize the employees to acquire these training sessions to maybe advance in the organization further?) *"For some, I want to sometime in the future I would like to become a manager. And I know that there are like my (MOT1b) managerial courses. So, for my personal motivation to then be ready to take the next step, I would like to take those courses, but I'm not sure if that answers your question."*	to the authors is external motives which affects the individual, by per example monetary rewards that comes in direct or indirect ways, whereas the former relates to wages, incentives or bonuses and the latter relates to time not working, job education, health benefits or allowances.		
			Informant 8 (SME)	(What kind of improvements could your company do in order to motivate employees	Martín Cruz et al. (2009) state that motivation consists	The informant state that the SME should motivate employees	SMEs are more incentivises by intrinsic

of (**MOT1a+2a**) intrinsic and (**MOT1b+2b**) extrinsic motives, where the former takes place inside an individual, affected by per example work pleasant environment, atmosphere of mutual respect, accomplishment, self-respect, prestige, personal values and involvement. As for the latter, extrinsic motivation according to the authors is external motives which affects the individual, by per example monetary rewards that comes in direct or indirect ways, whereas the former relates to wages, incentives or bonuses and the latter relates to time not working, job education, health	with intrinsic motivation in the (**MOT1a**) form of involvement to acquire and share knowledge with partners, this is consistent with the literature. However, the reasoning of the informant is because the social tie between the organization will be better facilitated if more people are involved in the acquiring and sharing of knowledge from the organization which helps the other organization understand the SME in a better way.	motivation in the form of involvement to acquire new knowledge from partners and sharing knowledge with them (**MOT1a**). SMEs are more motivated to acquire and share knowledge when more employees are involved in the social tie as it aids in the knowledge transfer to the SME (**MOT1a+b**).

for transferring knowledge to partners and acquiring knowledge from them?)

"I would say maybe two things. It's usually very common for an employee to have a very internal perspective. And maybe not always to reflect on what the customer sees or wants or experiences. So I think that we could maybe try to establish a culture (corporate) where more people are (**MOT1a**) *involved in talking to customers, for example, on a more informal basis that doesn't need to be so formerly talking to customers and figuring out and what they want and what they expect. So that you can use that knowledge within the organization and to build the processes and operations to best support that in a way And when it comes to business partner in the other end with suppliers, I think that we could do a better job in having more employees involved in the*

				actual relationship with the supplier in talking to them so that you don't Channel everything through a few persons. And because my experience is also that this makes the understanding in smaller (organizations) of how things work and why they work they do. Because it's I think these ties I spoke about earlier that I think that to make a successful cooperation, you need to be clear on what to expect and how things are working and how you want them to work. And I think that's it. There are employees that just makes an assumption that the supplier knows what we want and how we want it. So I think it's a culture (corporate culture) question in motivating employees to talk to both, customers and suppliers, and to understand the full chain and how they themselves are part of making that work.'	benefits or allowances.		
			Informant 9 (LC)	(Does your organization offer incentives to employees to learn new knowledge from	Martín Cruz et al. (2009) state that motivation consists	The informant state that is it more a part of (**MOT1a+b**) the job	LCs are not specifically motivated by

other companies or share knowledge with them?) *"No, we have not incentives like this to learn knowledge from other companies. We have other types of incentives, know when it comes to patents and these kinds of things and IP there are a lot of incentives, but not, not to learn new knowledge from other companies."* (Are there other incentives? Like promotions or such?) *"We do put it up as goals in the yearly appraisal tool for so on. And we do talk about if they want to have a promotion, either position or, you know, competence steps. Becoming, you know, more senior person. And that we do agree in this patient stores and we do set goals like that. You should if you want. I mean, it has to be it has to be an interest and the other end of course. And sometimes it doesn't have to be 100 percent*	of (**MOT1a+2a**) and intrinsic (**MOT1b+2b**) extrinsic motives, where the former takes place inside an individual, affected by per example pleasant work environment, atmosphere of mutual respect, accomplishment, self-respect, prestige, personal values and involvement. As for the latter, extrinsic motivation according to the authors is external motives which affects the individual, by per example monetary rewards that comes in direct or indirect ways, whereas the former relates to wages, incentives or bonuses and the latter relates to time not working, job education, health	incentives to acquire and share knowledge with partners, instead it is a part of the job description of the employees (**MOT1a+b**).
	description to acquire and share knowledge, rather than being motivated to do so, this is inconsistent with the literature.	

				match with the company also because that's a motivation in itself. But yeah, we do with that way." (When you said that you don't have incentives for knowledge exchange with other companies. And we relate to the question when you said that. If they're like with business partners, they have business decision to engage in special project, the task, and therefore they're engaging in exchange. So, in this question, Can we say that you don't have incentives but it's like a part of job?) *"(**MOT1a+b**) Yeah, definitely."*	benefits or allowances.		
Organizational characteristics of the recipient and the source	Motivation	(**MOT2a**) Intrinsic and (**MOT2b**) extrinsic motivation from the organization	Informant 1 (LC)	(Does your organization offer incentives to employees to learn new knowledge from other companies or share knowledge with them?) *"I don't know if we don't have any special incentives for it, (**MOT2b**) but it's like a part of our job. Meaning, my direct*	Martín Cruz et al. (2009) state that motivation consists of (**MOT1a+2a**) intrinsic and (**MOT1b+2b**) extrinsic motives, where the former takes place inside an individual, affected	The informant state that the LC do not offer any direct incentives but rather attempts to (**MOT2b**) indirectly motivate people by their job role as it benefits the organization if the employees knows	LCs attempt to indirectly motivate their employees via the job description to acquire and share knowledge with

					colleagues tried to do this well within the teams to share knowledge both within teams and for them to learn more (colleagues motivates the team to acquire knowledge and share it). And the manager is very prone to, I mean, he wants people to know more to learn. I think its kind of more to try to get the teams to understand the benefits of knowing more. Tried to motivate them that way. It also depends on how much they must do to be honest. But the more they must do, the less they might want to kind of share the knowledge. But if they share the knowledge, they might have less to do. So, it's kind of a Catch-22. So, I think that's the motivator."	by per example pleasant work environment, atmosphere of mutual respect, accomplishment, self-respect, prestige, personal values and involvement. As for the latter, extrinsic motivation according to the authors is external motives which affects the individual, by per example monetary rewards that comes in direct or indirect ways, whereas the former relates to wages, incentives or bonuses and the latter relates to time not working, job education, health benefits or allowances.	more, this is inconsistent with the literature.	partners (**MOT2a+b**)
					Informant 2 (LC)	(Does your organization offer incentives to employees to learn new knowledge from	Martín Cruz et al. (2009) state that motivation consists of (**MOT1a+2a**)	The informant state that the LC offer (**MOT2b**) direct extrinsic motivation in	LCs offer direct extrinsic motivation in the form of job

				other companies or share knowledge with them?) *"So, we do have access to something called (**MOT2b**) LinkedIn Learning. And that's in our HR system. And you can sign up for as many as may well (training programs), probably not as many because we're kind of busy with regular work, but we can sign up to trainings and stuff like that. But it's up from well, I guess it's from another company, it's from LinkedIn. And we do also have some trainings that are obligatory, like we have to do these trainings. And everybody must do those every so often every two years, every three years. We must renew these trainings."*	intrinsic and (**MOT1b+2b**) extrinsic motives, where the former takes place inside an individual, affected by per example pleasant work environment, atmosphere of mutual respect, accomplishment, self-respect, prestige, personal values and involvement. As for the latter, extrinsic motivation according to the authors is external motives which affects the individual, by per example monetary rewards that comes in direct or indirect ways, whereas the former relates to wages, incentives or bonuses and the latter relates to time not working, job education, health	the form of job education to acquire and share knowledge with partners, this is consistent with the literature.	education to the employees to acquire and share knowledge (**MOT2b**)

		benefits or allowances.	LCs attempt to indirectly motivate their employees via the job description to acquire and share knowledge with partners **(MOT2a+b)**
Informant 3 (LC)	(Does your organization offer incentives to employees to learn new knowledge from other companies or share knowledge with them?) *"No, there are no incentive directly. No, I mean we, the collaboration is something that is very essential, we always strive for. So, we have all these different councils and forums where we bring people together that have the same functions across either business areas within one division, but also across divisions. So, there's a lot of a lot of focus on that. And then we learn from each other's, you know, so if one division have done something that is great, then you just copy directly. So we don't need to do the same thing, global solutions. So we have that type of collaboration a lot. **(MOT2a+b)** But there are no incentives money-wise or*	Martín Cruz et al. (2009) state that motivation consists of **(MOT1a+2a)** intrinsic and **(MOT1b+2b)** extrinsic motives, where the former takes place inside an individual, affected by per example pleasant work environment, atmosphere of mutual respect, accomplishment, self-respect, prestige, personal values and involvement. As for the latter, extrinsic motivation according to the authors is external motives which affects the individual, by per example monetary rewards that comes in direct or indirect ways, whereas the	The informant state that there are **(MOT2a+b)** no direct incentives from the organization to motivate acquisition and sharing knowledge with partners, instead indirectly it is a part of the job description, this is inconsistent with the literature.

				something. It's just the way we work."	former relates to wages, incentives or bonuses and the latter relates to time not working, job education, health benefits or allowances.		
			Informant 4 (SME)	(Does your organization offer incentives to employees to learn new knowledge from other companies or share knowledge with them?) *"I haven't seen any initiatives about this. During my year, I'm not sure."* (Would you say that it's more of the employee to be self-motivated to acquire new knowledge?) *"(MOT2a) I don't think that's possible. Because you have your workday which is already full. So, I don't think it's possible. I think you need to make it available and discuss how to do it. I think so."*	Martín Cruz et al. (2009) state that motivation consists of **(MOT1a+2a)** intrinsic and **(MOT1b+2b)** extrinsic motives, where the former takes place inside an individual, affected by per example pleasant work environment, atmosphere of mutual respect, accomplishment, self-respect, prestige, personal values and involvement. As for the latter, extrinsic motivation according to the authors is external motives	The informant state that the SME does not offer direct incentives in any form, but rather **(MOT2a+b)** indirectly through the job role of the employee, this is inconsistent with the literature.	SMEs attempt to indirectly motivate their employees via the job description to acquire and share knowledge with partners **(MOT2a+b)**

					(Would you say different departments might be more incentivized to acquire or share knowledge with or from other companys?) *"Yeah, definitely the sales department are already used to cooperate with external companies and external persons."*	which affects the individual, by per example monetary rewards that comes in direct or indirect ways, whereas the former relates to wages, incentives or bonuses and the latter relates to time not working, job education, health benefits or allowances.		
				Informant 5 (SME)	(Does your organization offer incentives to employees to learn new knowledge from other companies or share knowledge with them?) "We don't offer incentives." (Say the employee shows tremendous personal growth and are constantly motivated to learn new knowledge inside a company or outside the company and share it as well. Would you say that the organization would favour	Martín Cruz et al. (2009) state that motivation consists of **(MOT1a+2a)** intrinsic and **(MOT1b+2b)** extrinsic motives, where the former takes place inside an individual, affected by per example pleasant work environment, atmosphere of mutual respect, accomplishment, self-respect, prestige,	The informant state that the organization does not offer direct incentives to acquire and share knowledge with partners, it is rather indirectly motivated by intrinsic motivation in the form of accomplishment, this is consistent with the literature.	SMEs offer indirect intrinsic motivation in the form of accomplishmen t to acquire and share knowledge with partners **(MOT2a)**.

				that person for maybe a promotion or other things?) "(**MOT2a**) *Yeah. I think so. If you show that you are willing to learn and put in extra effort.*" (You would say its more indirect motivation then?) "*Yes.*"	personal values and involvement. As for the latter, extrinsic motivation according to the authors is external motives which affects the individual, by per example monetary rewards that comes in direct or indirect ways, whereas the former relates to wages, incentives or bonuses and the latter relates to time not working, job education, health benefits or allowances.		
			Informant 6 (SME)	(Does your organization offer incentives to employees to learn new knowledge from other companies or share knowledge with them?) "*No. (**MOT2b**) If incentives are monetary that's a no. Today we only offer (**MOT2a**) soft incentives, meaning the acknowledgements... So, in a*	Martín Cruz et al. (2009) state that motivation consists of (**MOT1a+2a**) intrinsic and (**MOT1b+2b**) extrinsic motives, where the former takes place inside an individual, affected by per example	The informant state that the SME does not offer any form of (**MOT2b**) extrinsic motivation, rather intrinsic motivation in the form of (**MOT2a**) acknowledgement, which is implicitly accomplishment, this	SMEs offer direct intrinsic motivation in the form of accomplishmen t to acquire and share knowledge with partners (**MOT2a**).

					smaller organization, I would say it's much more key. So, it's like, I don't know where it is, but it's a way for success, is related to have good leaders that are good in motivating people and making them achieve more."	pleasant work environment, atmosphere of mutual respect, accomplishment, self-respect, prestige, personal values and involvement. As for the latter, extrinsic motivation according to the authors is external motives which affects the individual, by per example monetary rewards that comes in direct or indirect ways, whereas the former relates to wages, incentives or bonuses and the latter relates to time not working, job education, health benefits or allowances.	is consistent with the literature.	
				Informant 7 (LC)	(Does your organization offer incentives to employees to learn new knowledge from other companies or share knowledge with them?)	Martín Cruz et al. (2009) state that motivation consists of (**MOT1a+2a**) intrinsic and	The informant state that the LC offer direct extrinsic motivation in the form of job (**MOT2b**) education to	LCs offer direct extrinsic incentives in the form of job education to the

		employees to acquire and share knowledge (**MOT2b**)
		motivate acquisition of and sharing knowledge with partners, this is consistent with the literature.
(**MOT1b+2b**) extrinsic motives, where the former takes place inside an individual, affected by per example pleasant work environment, atmosphere of mutual respect, accomplishment, self-respect, prestige, personal values and involvement. As for the latter, extrinsic motivation according to the authors is external motives which affects the individual, by per example monetary rewards that comes in direct or indirect ways, whereas the former relates to wages, incentives or bonuses and the latter relates to time not working, job education, health benefits or allowances.	*"I mean, not sure. We have like a (**MOT2b**) budget each year which you could use to buy courses or attend conferences. But, if you don't want to take any courses, that's also fine. No real like money motivation for that... I want to sometime in the future I would like to become a manager. And I know that there are like many managerial courses. So, for my personal motivation is to then be ready to take the next step, I would like to take those courses."*	

	Informant 8 (SME)	(Does your organization offer incentives to employees to learn new knowledge from other companies or share knowledge with them?) *"I don't think that we have any incentives as such. But I think that we tried to encourage people to actually to reach out to other companies and ask for knowledge sharing experiences. But I don't think that we have in a specific incentive, monetary or other related to that."* (It is just a part of your job?) *"(MOT2a+b) Yeah."*	Martín Cruz et al. (2009) state that motivation consists of **(MOT1a+2a)** intrinsic and **(MOT1b+2b)** extrinsic motives, where the former takes place inside an individual, affected by per example pleasant work environment, atmosphere of mutual respect, accomplishment, self-respect, prestige, personal values and involvement. As for the latter, extrinsic motivation according to the authors is external motives which affects the individual, by per example monetary rewards that comes in direct or indirect ways, whereas the former relates to wages, incentives or	The informant state that the SME offer indirect incentives via the **(MOT2a+b)** job description of the employee to acquire and share knowledge, this is inconsistent with the literature.	SMEs attempt to indirectly motivate their employees via the job description to acquire and share knowledge with partners **(MOT2a+b)**

					bonuses and the latter relates to time not working, job education, health benefits or allowances.		
			Informant 9 (LC)	(Does your organization offer incentives to employees to learn new knowledge from other companies or share knowledge with them?) *"No, we have not incentives like this to learn knowledge from other companies. We have other types of incentives, know when it comes to patents and these kinds of things and (**MOT1b**) IP there are a lot of incentives, but not, not to learn new knowledge from other companies."*	Martín Cruz et al. (2009) state that motivation consists of (**MOT1a+2a**) intrinsic and (**MOT1b+2b**) extrinsic motives, where the former takes place inside an individual, affected by per example pleasant work environment, atmosphere of mutual respect, accomplishment, self-respect, prestige, personal values and involvement. As for the latter, extrinsic motivation according to the authors is external motives which affects the individual, by per	The informant state that the LC offer (**MOT1b**) extrinsic motivation for specific intellectual properties, which is consistent with the literature in terms of extrinsic motivation. However, this is excluding knowledge acquisition from partners as the informant state this is outside of the role.	LCs offer direct extrinsic motivation for specific job roles (e.g. IP owner) to share knowledge with partners (**MOT2b**)

Dimensions/ Characteristics	Aspects that affect KT	Subcategory of the aspect	Primary data source	Quote from the informant	Relating literature	Conclusion/Analysis	Summary
				example monetary rewards that comes in direct or indirect ways, whereas the former relates to wages, incentives or bonuses and the latter relates to time not working, job education, health benefits or allowances.			
Characteristics of knowledge	Knowledge charact eristics	**(KNOTACIT)** Tacit knowledge	Informant 1 (LC)	(And you explained that every little detail you need to transfer, it's very difficult to document for every system. But you said that you have to sit with the customer. It mean that when you develop some product, then you said that it's very difficult to explain everything to customer. Then you need to sit and like show or like because cannot document everything.)" *"Yeah. Well, it's both during the development phase, (**KNOTACIT**) I*	The literature of knowledge characteristics has identified that tacitness, ambiguity and complexity exist for both **(KNOTACIT)** tacit and explicit knowledge and has a significant impact in inter-organizational knowledge transfer and have multiple times proven that it affects the success and efficiency of knowledge transfer in	The informant states that transfer **(KNOTACIT)** tacit knowledge is affected by **(KNOTAC)** tacitness, in terms of the recipients understanding. This is consistent with the literature. The informant does not mention	When tacitness is low it becomes difficult as a source LC/MNC to transfer tacit knowledge to a recipient SME. **(KNOTACIT, KNOTAC)**

	*would say it's always better to discuss when the partner or the customer or the partners person, because they understand the partner or the customer, (**KNOTAC**) we understand like the developmental phase. So, it's much easier to if they kind of explain what it is, they want, they need it easier for the developers to understand that. Instead of like sitting in developing something, get it into a manual, documented those GTS and they were supposed to get customer installed. "*	LCs or MNCs (Simonin, 2004), SMEs (Corral de Zubielqui et al., 2015) and in alliances (Mazloomi Khamseh & Jolly, 2008). Wijk et al. (2008) claims that knowledge ambiguity is defined by the uncertainty of the knowledge, (**KNOAMB**) what underlying components and sources are in play (content-dependency), along with how they interact with each other (context-dependency). Complexity consider how the knowledge that is transferred (**KNOCOM**) is affected by different variations as it is comprehended by different skills or experiences of organizations (Stock & Tatikonda, 2000)… complexity can be measured by how many different variables are involved (content and context dependency) in a	**KNOCOM** or **KNOAMB** regarding tacit knowledge.	

						certain situation (Milagres & Burcharth, 2019). Kalling, (2003), state that tacitness exists for both tacit and explicit knowledge, whereas tacitness is affected by (**KNOTAC**) the individuals understanding of the knowledge they are handling... to measure tacitness in knowledge to some extent we will analyze the individuals understanding of the context at hand (context dependency).		
				Informant 2 (LC)	(By what criteria do you customize the knowledge you have learnt when you apply to another firm? Why and how?) - (I can provide a short example. Say, you have learned some of your expertise in quality management. And when you transfer or maybe some criteria or something to another firm, how do you temper that knowledge for them to make it comprehensible?)	The literature of knowledge characteristics has identified that tacitness, ambiguity and complexity exist for both (**KNOTACIT**) tacit and explicit knowledge and has a significant impact in inter-organizational knowledge transfer and have multiple times proven that it affects the success and efficiency of knowledge transfer in LCs or MNCs (Simonin, 2004), SMEs (Corral de	The informant state that transferring (**KNOTACIT**) tacit knowledge depends on the recipients understanding (**KNOTAC**) of the context. This is consistent with the literature. The informant state that transferring tacit knowledge is affected by the (**KNOAMB**)	When tacitness is low it becomes difficult as a source LC/MNC to transfer tacit knowledge to a recipient SME. (**KNOTACIT, KNOTAC**) When ambiguity is high it becomes difficult as a source LC/MNC to transfer tacit

				"(KNOAMB+KNOCOM) You need to see how they can apply it in their everyday work. So, you need to kind of transpose. So, (KNOTACIT) whatever you were using these skills or whatever skills you were using before, you need to remember what they will be used for in the (KNOTAC) new context. So, don't try to explain what you use it before. Like just tried to see what are people doing now and give them concrete examples."	Zubielqui et al., 2015) and in alliances (Mazloomi Khamseh & Jolly, 2008). Wijk et al. (2008) claims that knowledge ambiguity is defined by the uncertainty of the knowledge, **(KNOAMB)** what underlying components and sources are in play (content-dependency), along with how they interact with each other (context-dependency). Complexity consider how the knowledge that is transferred **(KNOCOM)** is affected by different variations as it is comprehended by different skills or experiences of organizations (Stock & Tatikonda, 2000)... complexity can be measured by how many different variables are involved (content and context dependency) in a certain situation (Milagres & Burcharth, 2019).	ambiguity of the knowledge, as the skills (content) and how they are used in the new context. This is consistent with the literature. The informant state that transferring tacit knowledge is affected by the **(KNOCOM)** compleixty of the knowledge, as the variables of skills (content) and the context of how they are used in the new context affects the knowledge. This is consistent with the literature.	knowledge to a recipient SME. **(KNOTACIT, KNOAMB)** When compleixty is high it becomes difficult as a source LC/MNC to transfer tacit knowledge to a recipient SME. **(KNOTACIT, KNOAMB)**

						Kalling, (2003), state that tacitness exists for both tacit and explicit knowledge, whereas tacitness is affected by (**KNOTAC**) the individuals understanding of the knowledge they are handling... to measure tacitness in knowledge to some extent we will analyze the individuals understanding of the context at hand (context dependency).		
					Informant 3 (LC)	(By what criteria do you customize the knowledge you have learnt when you apply to another firm? Why and how?) *"So, we can take one of our readers, for instance, it's in RFID and there is a card in each lock and someone reads the card. And we've had for like 6-7 years. This is when you have failure rate that is not too high, but it's not low enough. So, you know, you just keep it there. You've prioritized other things. But at some point we had to make a decision. Okay. We need to identify why are they failing as they do. And then we brought on board, Flex who was our biggest supplier for*	Wijk et al. (2008) claims that knowledge ambiguity is defined by the uncertainty of the knowledge, (**KNOAMB**) what underlying components and sources are in play (content-dependency), along with how they interact with each other (context-dependency). Complexity consider how the knowledge that is transferred (**KNOCOM**) is affected by different variations as it is	The informant state that as the tacit knowledge contained too many complexities it was difficult to transfer the knowledge, this is consistent with the literature.	When complexity is high it becomes difficult as a source LC/MNC to transfer tacit knowledge to a recipient SME. (**KNOTACIT, KNOCOM**)

comprehended by different skills or experiences of organizations (Stock & Tatikonda, 2000)... complexity can be measured by how many different variables are involved (content and context dependency) in a certain situation (Milagres & Burcharth, 2019). Kalling, (2003), state that tacitness exists for both tacit and explicit knowledge, whereas tacitness is affected by (**KNOTAC**) the individuals understanding of the knowledge they are handling... to measure tacitness in knowledge to some extent we will analyze the individuals understanding of the context at hand (context dependency).	*electronics and our R&D department, together with Flex, made an analysis. What can be done? Flex came up with proposals. So did our R&D team and they got together and discussed it.* (**KNOCOM**) *And it was a combination of changing the design, the layout, thickness of the PCBA board, and then production process, which is actually one of those that had the most impact. And based on their experience and result, we took that and share with our other Hannah, who is our second supplier. Because we have dual sourcing for electronics. This is the other less mature company. It's much, much smaller. Flex is one of the biggest electronic companies in the world. So, then we used what we experienced there and shared with Hannah and gotten to implement what they could. They couldn't implement everything because they didn't, they didn't have the advanced tools, but they did implement some of it. And this is like 1.5 years ago. And now we see that they need to do full implementation, because we see still a higher, higher failure rate on their products. And then, so we took it directly from what we experience at Hannah and taught*

				what we learned from Flex to Hannah to how to teach them what to do."			
			Informant 4 (SME)	(In your experience, when you shared knowledge what was the most difficult to transfer to the partner? Did you notice any difference when knowledge was transferred to large or small and medium enterprises?) *"I'm thinking about communication. You say something or you inform something, but the one that received the information or communication. Don't really see what you mean or don't really understand the way that you wanted to. So, I think that it's communication. That is difficult and it can be misunderstandings. "* (By what criteria would you attempt to reduce the barrier of uncertainty?) *"Yeah. I think you can reduce it by follow-up. You communicate and then you can ask follow-up questions to check that you are thinking about the same situation"* "Would you say there was a noticeable difference depending on the size of the partner?)	Wijk et al. (2008) claims that knowledge ambiguity is defined by the uncertainty of the knowledge, **(KNOAMB)** what underlying components and sources are in play (content-dependency), along with how they interact with each other (context-dependency). Complexity consider how the knowledge that is transferred **(KNOCOM)** is affected by different variations as it is comprehended by different skills or experiences of organizations (Stock & Tatikonda, 2000)... complexity can be measured by how many different variables are involved (content and context dependency) in a certain situation (Milagres & Burcharth, 2019).	The informant state that transferring **(KNOTACIT)** tacit knowledge to a LC/MNC recipient is easier since they have **(KNOTAC)** better understanding of knowledge in general, implicitly this is consistent with the literature.	When tacitness is low it becomes easier as a source SME to transfer tacit knowledge to a recipient LC/MNC. **(KNOTACIT, KNOTAC)**

	*"Normally, I think it's much easier to work with large organizations because then you have you have specialists. Normally you have (**KNOTACIT**) more knowledge… often (**KNOTAC**) long experience of things and they can make quick summaries and so on."*	Kalling, (2003), state that tacitness exists for both tacit and explicit knowledge, whereas tacitness is affected by (**KNOTAC**) the individuals understanding of the knowledge they are handling… to measure tacitness in knowledge to some extent we will analyze the individuals understanding of the context at hand (context dependency).	When tacitness is low it becomes easier as a source SME to transfer tacit knowledge to a recipient LC/MNC. (**KNOTACIT, KNOTAC**) When ambiguty is high it becomes difficult as a source SME to transfer tacit knowledge to a recipient LC/MNC.
Informant 5 (SME)	(In your experience, when you shared knowledge what was the most difficult knowledge to transfer to the partner? Did you notice any difference when knowledge was transferred to large or small and medium enterprises?) *"The technical knowledge. It's the easiest. This is how you do this. This is how you manoeuvre or what kind of buttons or how you connect it. (**KNOTACIT**) It's that I think is the soft question that's more difficult to transfer because depends on situation, depends on the customer, depends on the priority, that depends on so many things that comes around. You can't do like that because you need to be*	Wijk et al. (2008) claims that knowledge ambiguity is defined by the uncertainty of the knowledge, what (**KNOAMB**) underlying components and sources are in play (content-dependency), along with how they interact with each other (context-dependency). Complexity consider how the knowledge that is transferred (**KNOCOM**) is affected by different variations as it is comprehended by different	The informant state that (**KNOTACIT**) transferring tacit knowledge is difficult when the (**KNOAMB**) ambiguity is high, since the content and context of the knowledge interacts differently. Additionally, the same applies when (**KNOTAC**) tacitness is high, as the recipients understanding of the tacit knowledge

					flexible. You need to understand the sense of urgency in that kind of situation. So difficult to say do like that because it's not the same way every time." (When you transfer it to a smaller organization or larger, Did you notice any difference In the context of if it was easier to smaller or easier to the larger?) *"Think it's almost the same, I think."* (Could you clarify, for example, technical questions. You mean, for example, what that relates to your products?) *"I think the technical questions are more like. How do you how do you manoeuvre? How do you handle the machine or the technical things or Microsoft or you learned something. But the softer things, example change management if you teach someone, it's not easy because the people, it's the soft questions. It's not easy all the time because you need to work with* **(KNOAMB+KNOTAC)** *yourself and how you communicate, how you talk to people, how you do things, how you*	skills or experiences of organizations (Stock & Tatikonda, 2000)... complexity can be measured by how many different variables are involved (content and context dependency) in a certain situation (Milagres & Burcharth, 2019). Kalling, (2003), state that tacitness exists for both tacit and explicit knowledge, whereas tacitness is affected by **(KNOTAC)** the individuals understanding of the knowledge they are handling... to measure tacitness in knowledge to some extent we will analyze the individuals understanding of the context at hand (context dependency).	affects the transfer. This is consistent with the literaeture.	**(KNOTACIT, KNOTAC)**

					understand that people acting depending on what kind of information they have."			
				Informant 6 (SME)	(In your experience, when you shared knowledge what was the most difficult to transfer to the partner? Did you notice any difference when knowledge was transferred to large or small and medium enterprises?) *"So we've done some knowledge transfer activities to hospitality when it comes to testing and development methods and techniques. And we've done that as a mutual sharing experience that we've had conferences with them, discussing and workshopping with them. And that's been very good. I think the most value comes from networking and inspiration. So, I actually am quite hesitant to say that we actually did something on that. We were inspired, but we didn't really apply the same things as they had. And I would say that to answer this question, the main blocker for us to share knowledge to the bigger organization is time. And we don't get the benefits from it."*	Wijk et al. (2008) claims that knowledge ambiguity is defined by the uncertainty of the knowledge, (**KNOAMB**) what underlying components and sources are in play (content-dependency), along with how they interact with each other (context-dependency). Complexity consider how the knowledge that is transferred (**KNOCOM**) is affected by different variations as it is comprehended by different skills or experiences of organizations (Stock & Tatikonda, 2000)... complexity can be measured by how many different variables are involved (content and context dependency) in a certain situation (Milagres & Burcharth, 2019).	The informant state that transferring knowledge to a LC/MNC (**KNOCOM**) is difficult due to their niche roles in the organization (more variables) and this process takes a lot of the time, this is to some extent consistent with the literature. The informant does not mention **KNOAMB** and **KNOTAC**.	When complexity is high it becomes difficult as a source SME to transfer tacit and explicit knowledge to a recipient LC/MNC. (**KNOTACIT, KNOEXP, KNOCOM**)

	Informant		Literature	Interpretation	Codes
		(If you go about transferring knowledge to a larger organization, what would you say is, the more difficult in regard to the time constraints?) *"Because they had another organization. They have niche roles. So, (**KNOCOM**) they have one role that work with just this part of their business. And I'm constantly saying to my colleagues, that's okay, please have network with the larger organizations do that for inspiration or anything. But do not let them to eat your time too much because they had much more time than you have on your area of expertise."*	Kalling, (2003), state that tacitness exists for both tacit and explicit knowledge, whereas tacitness is affected by (**KNOTAC**) the individuals understanding of the knowledge they are handling… to measure tacitness in knowledge to some extent we will analyze the individuals understanding of the context at hand (context dependency).		
	Informant 7 (LC)	(In your experience, when you shared knowledge what was the most difficult to transfer to the partner? Did you notice any difference when knowledge was transferred to large or small and medium enterprises?) *"I think just getting the getting all the details and communicating all of the like positive energy that I had. Getting it, it's kind of hard to convey that to someone who weren't there and heard it from the same professional source*	Wijk et al. (2008) claims that knowledge ambiguity is defined by the uncertainty of the knowledge, (**KNOAMB**) what underlying components and sources are in play (content-dependency), along with how they interact with each other (context-dependency).	The informant states that transferring (**KNOTACIT**) tacit knowledge is difficult since it becomes second hand knowledge, whereas the (**KNOAMB**) ambiguity of the knowledge in terms of how the content and context interacts with each other is difficult	When complexity is high it becomes difficult as a source LC/MNC to transfer tacit knowledge to a recipient SME. (**KNOTACIT, KNOCOM**) When ambiguity is high it becomes difficult as a

					*that I did. Right? (**KNOTACIT+KNOAMB+KNOCOM**) So, this is become like an second-hand knowledge and then that person it becomes a third hand knowledge, so it gets diluted."* (You would also say that the complexity of the knowledge is more difficult to transfer since your perspective is not the same as the perspective that was given to you?) *"Sure. I mean, the source material for the talk could be like a 300-page book, condensed into one hour speech. And then I get 20 minutes to tell my colleagues (partners) about it. So, the funnel gets narrower and narrower."*	Complexity consider how the knowledge that is transferred (**KNOCOM**) is affected by different variations as it is comprehended by different skills or experiences of organizations (Stock & Tatikonda, 2000)... complexity can be measured by how many different variables are involved (content and context dependency) in a certain situation (Milagres & Burcharth, 2019). Kalling, (2003), state that tacitness exists for both tacit and explicit knowledge, whereas tacitness is affected by (**KNOTAC**) the individuals understanding of the knowledge they are handling... to measure tacitness in knowledge to some extent we will analyze the individuals understanding of the context at hand (context dependency).	to convey, (**KNOCOM**) additionally knowledge complexity affects the knowledge by having the involved variables in the transfer that needs to be understood by the recipient, is also difficult to transfer. This is consistent with the literature.	source LC/MNC to transfer tacit knowledge to a recipient SME. (**KNOTACIT, KNOAMB**)

Informant			
Informant 8 (SME)	(In your experience, when you shared knowledge what was the most difficult knowledge to transfer to the partner? Did you notice any difference when knowledge was transferred to large or small and medium enterprises?) *"Yeah, I would say that this goes as well with a larger organization that it's harder to reach the right recipient of the knowledge. (**KNOCOM**) Because you think that you might talk the right person in transferring the knowledge, but then it might be someone else who does actually execute it in the end. So I think that's the hardest part to figure out who you actually should have involved to get things."*	Wijk et al. (2008) claims that knowledge ambiguity is defined by the uncertainty of the knowledge, what (**KNOAMB**) what underlying components and sources are in play (content-dependency), along with how they interact with each other (context-dependency). Complexity consider how the knowledge that is transferred (**KNOCOM**) is affected by different variations as it is comprehended by different skills or experiences of organizations (Stock & Tatikonda, 2000)… complexity can be measured by how many different variables are involved (content and context dependency) in a certain situation (Milagres & Burcharth, 2019). Kalling, (2003), state that tacitness exists for both tacit and explicit knowledge,	The informant state that transferring knowledge is difficult when the (**KNOCOM**) complexity of the recipient organization is high, this is consistent with the literature.	When complexity is high it becomes difficult as a source SME to transfer tacit and explicit knowledge to a recipient LC/MNC. (**KNOTACIT, KNOEXP, KNOCOM**)

						whereas tacitness is affected by (**KNOTAC**) the individuals understanding of the knowledge they are handling... to measure tacitness in knowledge to some extent we will analyze the individuals understanding of the context at hand (context dependency).	
				Informant 9 (LC)	Same as explicit.	Wijk et al. (2008) claims that knowledge ambiguity is defined by the uncertainty of the knowledge, (**KNOAMB**) what underlying components and sources are in play (content-dependency), along with how they interact with each other (context-dependency). Complexity consider how the knowledge that is transferred (**KNOCOM**) is affected by different variations as it is comprehended by different skills or experiences of organizations (Stock & Tatikonda, 2000)...	Same as explicit.

			(KNO EXP) Explicit knowledge	Informant 1 (LC)			
					complexity can be measured by how many different variables are involved (content and context dependency) in a certain situation (Milagres & Burcharth, 2019). Kalling, (2003), state that tacitness exists for both tacit and explicit knowledge, whereas tacitness is affected by (**KNOTAC**) the individuals understanding of the knowledge they are handling... to measure tacitness in knowledge to some extent we will analyze the individuals understanding of the context at hand (context dependency).		
			(**KNO EXP**) Explicit knowledge	Informant 1 (LC)	(In your experience, when you shared knowledge what was the most difficult to transfer to the partner? Did you notice any difference when knowledge was transferred to large or small and medium enterprises?) *"(**KNOEXP**) You can see it's the same way because it's hard to document or like write the manual for the entire*	The literature of knowledge characteristics has identified that tacitness, ambiguity and complexity exist for both tacit and (**KNOEXP**) explicit knowledge and has a significant impact in inter-organizational knowledge transfer and have multiple	The informant state that it is easier to transfer (**KNOEXP**) explicit knowledge when (**KNOCOM**) there are less complexities (variables) involved in the transfer. This is	When complexity is low it becomes easier as a source LC/MNC to transfer explicit knowledge to a recipient SME. (**KNOEXP, KNOCOM**)

				system in every little detail in the system. And we have to kind of sit with tech service. So, kind of have, we have demos where we present the new function out of difference. But there are always things that are missed." (If I may say that you want to meet the needs of the customer in a new feature or existing feature. Some product that you need to adapt basically, and they come to the R&D department. And if you compare that scenario, depending on if the partner is a small or medium enterprise compared to a large organization, would you see a difference in meeting those needs?) *"Yeah, the needs of a bigger organization or an enterprise is much bigger company."* (More complex systematically?) *"More complex, yes. It's always more complex and it's very, and it's also quite difficult for us to understand the complexity of their enterprise."* (If you try to identify some of the complexity, what would you say they are?)	times proven that it affects the success and efficiency of knowledge transfer in LCs or MNCs (Simonin, 2004), SMEs (Corral de Zubielqui et al., 2015) and in alliances (Mazloomi Khamseh & Jolly, 2008). Wijk et al. (2008) claims that knowledge ambiguity is defined by the uncertainty of the knowledge, (**KNOAMB**) what underlying components and sources are in play (content-dependency), along with how they interact with each other (context-dependency). Complexity consider how the knowledge that is transferred (**KNOCOM**) is affected by different variations as it is comprehended by different skills or experiences of organizations (Stock & Tatikonda, 2000)... complexity can be measured by how many	consistent with the literature. The informant did not **KNOAMB** or **KNOTAC** regarding **KNOEXP**.	

					*"(**KNOCOM1**) It's the way they use the system. And also, how many rooms there are, how they kind of build up the hotel. Because you can build in the access management system, you can build it in different ways. And for small enterprises are smaller. There's not that much complexity. You have certain number of floors. You have a certain amount of rooms, corridors, maybe half an hour later. You have like two entrance doors maybe. And large enterprise you might 20 elevators, you might have 5000 rooms. You have conference rooms, you have spa's, you have gyms. So, it's like the, the bigger the hotel, the more complexities."*	different variables are involved (content and context dependency) in a certain situation (Milagres & Burcharth, 2019). Kalling, (2003), state that tacitness exists for both tacit and explicit knowledge, whereas tacitness is affected by (**KNOTAC**) the individuals understanding of the knowledge they are handling... to measure tacitness in knowledge to some extent we will analyze the individuals understanding of the context at hand (context dependency).		
				Informant 2 (LC)	-	The literature of knowledge characteristics has identified that tacitness, ambiguity and complexity exist for both tacit and (**KNOEXP**) explicit knowledge and has a significant impact in inter-organizational knowledge transfer and have multiple times proven that it affects the success and efficiency	The informant does not mention anything regarding knowledge characteristics in explicit knowledge.	-

of knowledge transfer in LCs or MNCs (Simonin, 2004), SMEs (Corral de Zubielqui et al., 2015) and in alliances (Mazloomi Khamseh & Jolly, 2008).

Wijk et al. (2008) claims that knowledge ambiguity is defined by the uncertainty of the knowledge, (**KNOAMB**) what underlying components and sources are in play (content-dependency), along with how they interact with each other (context-dependency).

Complexity consider how the knowledge that is transferred (**KNOCOM**) is affected by different variations as it is comprehended by different skills or experiences of organizations (Stock & Tatikonda, 2000)... complexity can be measured by how many different variables are involved (content and

				When tacitness is low it becomes easier as a source LC/MNC to transfer explicit knowledge to a recipient SME. **(KNOEXP, KNOTAC)**
			context dependency) in a certain situation (Milagres & Burcharth, 2019). Kalling, (2003), state that tacitness exists for both tacit and explicit knowledge, whereas tacitness is affected by (**KNOTAC**) the individuals understanding of the knowledge they are handling… to measure tacitness in knowledge to some extent we will analyze the individuals understanding of the context at hand (context dependency).	The informant state that the (**KNOTAC**) recipients understanding affects the transfer of (**KNOEXP**) explicit knowledge, this is consistent with the literature.
	Informant 3 (LC)	*"In your experience when you shared knowledge, what was the most difficult knowledge to transfer to the partner? Did you notice any difference when knowledge was transferred? Absolutely experience. (**KNOTAC**) And I mean, not my experience but their experience. So recently we require a smaller company and then we are onboarding them, want to transfer the knowledge in regard to there are (**KNOEXP**) certain processes they have to follow. It's not up for discussion. This is the way we work. And by doing so then this is the*	The literature of knowledge characteristics has identified that tacitness, ambiguity and complexity exist for both tacit and (**KNOEXP**) explicit knowledge and has a significant impact in inter-organizational knowledge transfer and have multiple times proven that it affects the success and efficiency of knowledge transfer in the LCs or MNCs (Simonin,	

				KPI that will need to put in place and so on. And then if they don't see the benefits immediately, that is mostly because of the maturity level as they don't see the bigger picture. And also when we acquire smaller companies, what I often call them garage companies. This is that those innovative people, they are not structured at all. But maybe there's the strong side because innovation is their focus and telling them that you need to have a system that will handle your revision, handling of drawings are just waste for them, you know, this paperwork, it doesn't bring value because they don't see value in that. But when you start sending by mistake, you send an old design or drawing to a supplier and that he may manufacturers 50000 units, wrong design. And then it's a huge cost. And if in addition, push that to the market and you find out it's wrong after it has been installed, then the cost is 20 to 30 times higher. Then you see the benefit. But they don't see because they didn't have experience with that. They do not come that far."	2004), SMEs (Corral de Zubielqui et al., 2015) and in alliances (Mazloomi Khamseh & Jolly, 2008). Wijk et al. (2008) claims that knowledge ambiguity is defined by the uncertainty of the knowledge, (**KNOAMB**) what underlying components and sources are in play (content-dependency), along with how they interact with each other (context-dependency). Complexity consider how the knowledge that is transferred (**KNOCOM**) is affected by different variations as it is comprehended by different skills or experiences of organizations (Stock & Tatikonda, 2000)... complexity can be measured by how many different variables are involved (content and context dependency) in a	

					certain situation (Milagres & Burcharth, 2019). Kalling, (2003), state that tacitness exists for both tacit and explicit knowledge, whereas tacitness is affected by (**KNOTAC**) the individuals understanding of the knowledge they are handling... to measure tacitness in knowledge to some extent we will analyze the individuals understanding of the context at hand (context dependency).	
			Informant 4 (SME)	-	The literature of knowledge characteristics has identified that tacitness, ambiguity and complexity exist for both tacit and (**KNOEXP**) explicit knowledge and has a significant impact in inter-organizational knowledge transfer and have multiple times proven that it affects the success and efficiency of knowledge transfer in LCs or MNCs (Simonin, 2004), SMEs (Corral de	The informant does not mention anything regarding knowledge characteristics in explicit knowledge.	-

Zubielqui et al., 2015) and in alliances (Mazloomi Khamseh & Jolly, 2008).

Wijk et al. (2008) claims that knowledge ambiguity is defined by the uncertainty of the knowledge, (**KNOAMB**) what underlying components and sources are in play (content-dependency), along with how they interact with each other (context-dependency).

Complexity consider how the knowledge that is transferred (**KNOCOM**) is affected by different variations as it is comprehended by different skills or experiences of organizations (Stock & Tatikonda, 2000)... complexity can be measured by how many different variables are involved (content and context dependency) in a certain situation (Milagres & Burcharth, 2019).

						Kalling, (2003), state that tacitness exists for both tacit and explicit knowledge, whereas tacitness is affected by (**KNOTAC**) the individuals understanding of the knowledge they are handling... to measure tacitness in knowledge to some extent we will analyze the individuals understanding of the context at hand (context dependency).		
				Informant 5 (SME)	(In your experience, when you shared knowledge what was the most difficult to transfer to the partner? Did you notice any difference when knowledge was transferred to large or small and medium enterprises?) *"(**KNOEXP**) The technical knowledge. It's the easiest. This is how you do this. This is how you manoeuvre or what kind of buttons or how you connect it. It's that I think is the soft question that's more difficult to transfer because depends on situation, depends on the customer, depends on the priority, that depends on so many things that comes around. You can't do*	The literature of knowledge characteristics has identified that tacitness, ambiguity and complexity exist for both tacit and (**KNOEXP**) explicit knowledge and has a significant impact in inter-organizational knowledge transfer and have multiple times proven that it affects the success and efficiency of knowledge transfer in LCs or MNCs (Simonin, 2004), SMEs (Corral de Zubielqui et al., 2015) and	The informant state that transferring (**KNOEXP**) explicit knowledge is difficult when (**KNOTAC**) tacitness is high, this is consistent with the literature.	When tacitness is low it becomes difficult as a source SME to transfer explicit knowledge to a recipient LC/MNC. (**KNOEXP, KNOTAC**)

					like that because you need to be flexible. You need to understand the sense of urgency in that kind of situation. So difficult to say do like that because it's not the same way every time." (When you transfer it to a smaller organization or larger, Did you notice any difference In the context of if it was easier to smaller or easier to the larger?) *"Think it's almost the same, I think."* (Could you clarify, for example, technical questions. You mean, for example, what that relates to your products?) *"I think the technical questions are more like. How do you how do you manoeuvre? (**KNOTAC**) How do you handle the machine or the technical things or Microsoft or you learned something. But the softer things, example change management if you teach someone, it's not easy because the people, it's the soft questions. It's not easy all the time because you need to work with yourself and how you communicate, how you talk to people,*	in alliances (Mazloomi Khamseh & Jolly, 2008). Wijk et al. (2008) claims that knowledge ambiguity is defined by the uncertainty of the knowledge, (**KNOAMB**) what underlying components and sources are in play (content-dependency), along with how they interact with each other (context-dependency). Complexity consider how the knowledge that is transferred (**KNOCOM**) is affected by different variations as it is comprehended by different skills or experiences of organizations (Stock & Tatikonda, 2000)... complexity can be measured by how many different variables are involved (content and context dependency) in a certain situation (Milagres & Burcharth, 2019).		

				how you do things, how you understand that people acting depending on what kind of information they have."	Kalling, (2003), state that tacitness exists for both tacit and explicit knowledge, whereas tacitness is affected by (**KNOTAC**) the individuals understanding of the knowledge they are handling... to measure tacitness in knowledge to some extent we will analyze the individuals understanding of the context at hand (context dependency).	
			Informant 6 (SME)	Same as tacit.	Wijk et al. (2008) claims that knowledge ambiguity is defined by the uncertainty of the knowledge, (**KNOAMB**) what underlying components and sources are in play (content-dependency), along with how they interact with each other (context-dependency). Complexity consider how the knowledge that is transferred (**KNOCOM**) is affected by different variations as it is comprehended by different	Same as tacit.

			When tacitness is high it becomes easier as a source LC/MNC to transfer tacit and explicit knowledge to a
		skills or experiences of organizations (Stock & Tatikonda, 2000)… complexity can be measured by how many different variables are involved (content and context dependency) in a certain situation (Milagres & Burcharth, 2019). Kalling, (2003), state that tacitness exists for both tacit and explicit knowledge, whereas tacitness is affected by (**KNOTAC**) the individuals understanding of the knowledge they are handling… to measure tacitness in knowledge to some extent we will analyze the individuals understanding of the context at hand (context dependency).	The informant state that transferring knowledge is easier when (**KNOTAC**) tacitness of the recipient is high as their understanding of
		The literature of knowledge characteristics has identified that tacitness, ambiguity and complexity exist for both tacit and explicit (**KNOEXP**) knowledge and has a
	Informant 7 (LC)	(By what criteria do you customize the knowledge you have learnt when you apply to another firm? Why and how?) *"I mean, of course I would listen into I kind of have a mental picture. I know that the receiving end what they are*

					*doing, right? So, I don't try to teach to building company about textile design or something. So, of course, (**KNOTAC**) if you try to form the way you communicate with the receiver's context, maybe I should put it like that. Why and how? Why, of course is I would be aiming bit beside of the target if I don't try to shape it into something that they can relate to. Because I guess the pickup rate or their understanding will be much better if I fight for me to something that they can relate to in their daily work."*	significant impact in inter-organizational knowledge transfer and have multiple times proven that it affects the success and efficiency of knowledge transfer in LCs or MNCs (Simonin, 2004), SMEs (Corral de Zubielqui et al., 2015) and in alliances (Mazloomi Khamseh & Jolly, 2008). Wijk et al. (2008) claims that knowledge ambiguity is defined by the uncertainty of the knowledge, (**KNOAMB**) what underlying components and sources are in play (content-dependency), along with how they interact with each other (context-dependency). Complexity consider how the knowledge that is transferred (**KNOCOM**) is affected by different variations as it is comprehended by different skills or experiences of organizations (Stock &	the knowledge is high, this is consitent with the literature.	recipient LC/MNC. (**KNOTACIT, KNOEXP, KNOTAC**)

					Tatikonda, 2000)... complexity can be measured by how many different variables are involved (content and context dependency) in a certain situation (Milagres & Burcharth, 2019). Kalling, (2003), state that tacitness exists for both tacit and explicit knowledge, whereas tacitness is affected by (**KNOTAC**) the individuals understanding of the knowledge they are handling... to measure tacitness in knowledge to some extent we will analyze the individuals understanding of the context at hand (context dependency).	
			Informant 8 (SME)	Same as tacit.	The literature of knowledge characteristics has identified that tacitness, ambiguity and complexity exist for both tacit and (**KNOEXP**) explicit knowledge and has a significant impact in inter-organizational knowledge	Same as tacit.

transfer and have multiple times proven that it affects the success and efficiency of knowledge transfer in LCs or MNCs (Simonin, 2004), SMEs (Corral de Zubielqui et al., 2015) and in alliances (Mazloomi Khamseh & Jolly, 2008).

Wijk et al. (2008) claims that knowledge ambiguity is defined by the uncertainty of the knowledge, (**KNOAMB**) what underlying components and sources are in play (content-dependency), along with how they interact with each other (context-dependency).

Complexity consider how the knowledge that is transferred (**KNOCOM**) is affected by different variations as it is comprehended by different skills or experiences of organizations (Stock & Tatikonda, 2000)... complexity can be

					measured by how many different variables are involved (content and context dependency) in a certain situation (Milagres & Burcharth, 2019). Kalling, (2003), state that tacitness exists for both tacit and explicit knowledge, whereas tacitness is affected by (**KNOTAC**) the individuals understanding of the knowledge they are handling... to measure tacitness in knowledge to some extent we will analyze the individuals understanding of the context at hand (context dependency).		
			Informant 9 (LC)	(In your experience, when you shared knowledge what was the most difficult to transfer to the partner? Did you notice any difference when knowledge was transferred to large or small and medium enterprises?) *"Another example, and then in Oslo here though is there's a big campus called the S i o. It has like a 20 thousand student rooms. It's run by a*	The literature of knowledge characteristics has identified that tacitness, ambiguity and complexity exist for both tacit and (**KNOEXP**) explicit knowledge and has a significant impact in inter-organizational knowledge transfer and have multiple times proven that it affects	The informant state that the (**KNOTAC**) tacitness of the recipient affects the transfer of knowledge, this is consistent with the literature.	When tacitness is low it becomes difficult as a source LC/MNC to transfer tacit and explicit knowledge to a recipient LC/MNC. (**KNOTACIT,**

				*small organization. They have a small IT department. They have also external IT consultants. And I don't understand. They have had some limited access control before. (**KNOTAC**) And at this point, they didn't understand our concept of access control, which is quite different from regular access control, in the way we solve it, which is part of our success. That we have this unique approach to how we do access control. And that's bringing over that concept to persons that already had perceived concept of how it should work. That was quite difficult actually, and we spent a lot of time actually discussing, had to go back to discuss the architecture of the system without revealing any secrets in order for them to buy into this. And probably we underestimated the task. Normally, we will make this interface into property management type of system that already are used to do all this type of integration. This time, they were building the system themselves and this was the only system built special purpose for this campus. They didn't buy the standard product. They bought a lot of engineers to build on the system. So that was difficult to change that concept actually, how to do this,*	the success and efficiency of knowledge transfer in LCs or MNCs (Simonin, 2004), SMEs (Corral de Zubielqui et al., 2015) and in alliances (Mazloomi Khamseh & Jolly, 2008). Wijk et al. (2008) claims that knowledge ambiguity is defined by the uncertainty of the knowledge, (**KNOAMB**) what underlying components and sources are in play (content-dependency), along with how they interact with each other (context-dependency). Complexity consider how the knowledge that is transferred (**KNOCOM**) is affected by different variations as it is comprehended by different skills or experiences of organizations (Stock & Tatikonda, 2000)... complexity can be measured by how many different variables are	**KNOEXP, KNOTAC)**

				and we had to go back to the basic and actually start, start from scratch."	involved (content and context dependency) in a certain situation (Milagres & Burcharth, 2019). Kalling, (2003), state that tacitness exists for both tacit and explicit knowledge, whereas tacitness is affected by (**KNOTAC**) the individuals understanding of the knowledge they are handling... to measure tacitness in knowledge to some extent we will analyze the individuals understanding of the context at hand (context dependency).		

Dimensions/ Characteristics	Aspects that affect KT	Subcategory of the aspect	Primary data source	Quote from the informant	Relating literature	Analysis	Summary
Organizational characteristics	**Disseminative capacity**	**Motivation to teach**	Informant 1 (LC)	Question: Question: In case you taught your business partners, what motivates you to do so? *"...So, motivation for me to educate if it is a business partner or if it is like an internal stakeholder in some way, it would be for them to know (more clear*	The following aspects drives motivation to teach by knowledge source: (MTDC1)Direct interaction and mutual experience sharing (Squire et al., 2009).	Informant 1 from R&D perspective is motivated to share knowledge to avoid any misunderstanding. Implicitly, this means via mutual experience	**Motivation to teach is to avoid misunderstanding due to transferring complex knowledge**

Informant	Quote	Description (source)	Analysis	Category
	...understanding of the product) the product that they're supposed to use"	(MTDC2) to build a mutual trust between partners (Minbaeva et al., 2018)	sharing with partners, the organization wants to avoid misunderstanding due to transferring complex knowledge related to the product development (MTDC1). Aspect related to MTDC2 was not explicitly mentioned by informant but to avoid misunderstanding between partners can be interpreted as way to build mutual trust between partners.	**related to the product development (MTDC2)**
Informant 2 (LC)	Question: Question: In case you taught your business partners, what motivates you to do so? *"...What motivates me is that we will go home, and we will both be able to do something better. Because when you teach someone or when you show somebody had to do something, you will do it better as well afterwards. Because you learn something when you teach".* *"...I'm going to learn something for sure. And they're going to learn and*	The following aspects drives motivation to teach by knowledge source: (MTDC1)Direct interaction and mutual experience sharing (Squire et al., 2009). (MTDC2) to build a mutual trust between partners (Minbaeva et al., 2018)	Informant 2 is motivated to teach its partners due to mutual experience sharing and learning from each other (MTDC1). Aspect related to MTDC2 was not mentioned by informant.	**Mutual experience sharing and learning from each other (MTDC1)**

		we're probably both going to elevate our level on that topic. So it's win-win" Question: In case you taught your business partners, what motivates you to do so? *" ...If you're looking into a business partner, then you're thinking about a long-term relationship that will benefit both. So, achieving that in addition, absolutely to see some results out of what you are teaching away. And see that bringing them either successful prosperity absolutely motivates me "* *"...if suppliers are not performing and possess higher risks, you want to go out there, build again the relationship and you start to teach them what we can teach them what we can teach them to bring them up to a level where we want them"*	The following aspects drives motivation to teach by knowledge source: (MTDC1)Direct interaction and mutual experience sharing (Squire et al., 2009). (MTDC2) to build a mutual trust between partners (Minbaeva et al., 2018)	Informant 3 discusses about building the long-term relationship to achieve mutual performance results. Motivation for KT is to enhance mutual performance through mutual sharing knowledge (MTDC1) that will enhance partner's competence that needed for mutual performance. MTDC2 was not mentioned explicitly by informant but achieving mutual business performance can lead to establishing mutual trustworthy relation between partners.	**Motivation for KT is to enhance mutual business performance through knowledge transfer (MTDC2)**
	Informant 3 (LC)				
		Question: In case you taught your business partners, what motivates you to do so? *"...from previous experience then so what motivates me to kind of teach or so, I mean, for me it's a way to make sure that we actually get what we*	The following aspects drives motivation to teach by knowledge source: (MTDC1)Direct interaction and mutual experience sharing (Squire et al., 2009).	Informant 7 suggests that motivation to transfer knowledge is considered in terms of mutual experience sharing process with partners rather	**Mutual experience sharing and learning from each other rather teaching that is considered as one**
	Informant 7 (LC)				

Informant	Quote	Coding	Category
	ordered or what we wanted the other part to actually do for us. I have I think a few examples where we thought we ordered something; we get something else back. I think that's learning from that. Of course, the good motivation that to make it as clear as possible to explain what we need to be solved. …I'm just bouncing a bit on the teach word because as a partner we are on equal levels, right? And teaching is I tell someone from an elevated position like you need to do that and this"	teaching that is regarded as unidirectional way to share the knowledge (MTDC1) MTDC2 was not directly mentioned by informant but to be clearer with partners can be interpreted as way to build mutual trust between partners. (MTDC2) to build a mutual trust between partners (Minbaeva et al., 2018)	**way interaction with partners. (MTDC1)** **Transparency with partners is motivation to teach other organizations (MTDC2)**
Informant 9 (LC)	Question: In case you taught your business partners, what motivates you to do so? *" …So, there are different type of partners and there's different motivations for me why I partner up with them. But it's always a cause, a business reason or technology reason. We are very seldom to select the path and just to learn something, we always try to learn doing actual projects or products or components even. So, that is actually our main motivation is always behind some kind of a business case of business relations that we want to turn into a product"*	Informant 9 argues that motivation for interfirm KT activities is mutual learning (MTDC1) in order to create a value for the organization. The main motivation is a business purpose to share knowledge that eventually can be turned in a form of product to the partnering organization (e.g., customer). Implicitly, this means that motivation for The following aspects drives motivation to teach by knowledge source: (MTDC1)Direct interaction and mutual experience sharing (Squire et al., 2009). (MTDC2) to build a mutual trust between partners (Minbaeva et al., 2018)	**Motivation for interfirm KT is to create business value for the knowledge source through mutual sharing knowledge (MTDC2)**

				interfirm KT is to create business value through mutual sharing knowledge that can lead to a mutual trust (MTDC2)	
			Informant 4 (SME)	Question: In case you taught your business partners, what motivates you to do so? *"…It would be to help the situation or improve our relationship or make them grow as our, if it is a supplier, and by that way, also improve the incoming quality or product deliveries"*	The following aspects drives motivation to teach by knowledge source: (MTDC1)Direct interaction and mutual experience sharing (Squire et al., 2009). (MTDC2) to build a mutual trust between partners (Minbaeva et al., 2018)	Informant 4 discusses about improving relationship to achieve mutual performance results in terms of the quality of deliverables. Motivation for KT is to enhance mutual performance through mutual sharing knowledge (MTDC1) that will enhance partner's (supplier) competence that needed for mutual performance. MTDC2 was not mentioned explicitly by informant but achieving mutual business performance can lead to establishing mutual trustworthy relation between partners.	**Motivation for KT is to enhance mutual business performance through knowledge transfer (MTDC2)**

Informant	Question / Quote	Knowledge source	Analysis	Theme / Code
Informant 5 (SME)	Question: In case you taught your business partners, what motivates you to do so? *"...Now, but I think to grow together with your partners, it's a good thing. I mean, you should learn from each other and they tell their experience, we have experience from different people coming into the company ...I think establishing culture of learning and experience is needed. In projects where you evaluate what went good, bad, how can we do differently next time And if you have a good structure of that and a good way of you think the experience for the next one, then you have a learning organization. I think the leadership team and the management need to work with this value at that learning from each other. ...For example, you should copy things you can learn from each other instead of just sitting there and just inventing everything"*	The following aspects drives motivation to teach by knowledge source: (MTDC1)Direct interaction and mutual experience sharing (Squire et al., 2009). (MTDC2) to build a mutual trust between partners (Minbaeva et al., 2018)	Informant 5 gives opinion that mutual experience sharing is one of the important aspects that motivates the knowledge holder (MTDC1). But informant suggests that is more important to establish the learning culture values at the company that structured and it can drive motivation of knowledge holder to share knowledge (MTDC1) MTDC2 was not mentioned by informant.	**Mutual experience sharing and learning from each other (MTDC1)** **Learning culture can drive motivation to transfer knowledge between partners (MTDC1)**
Informant 6 (SME)	Question: In case you taught your business partners, what motivates you to do so? *"... It's to see that that people and businesses thrive by working more efficiently. So, it's more of a personal reason for me that I am glad to see that*	The following aspects drives motivation to teach by knowledge source: (MTDC1)Direct interaction and mutual experience sharing (Squire et al., 2009).	Informant 6 suggest mutual experience sharing allows enhancing the learning capacity of partners (MTDC1) and to enhance mutual	**Motivation for KT is to enhance mutual business performance through mutual knowledge**

	exchange (MTDC1)	business performance through knowledge transfer that can enhance mutual trust between partners (MTDC2)	**Motivation for KT is to enhance mutual business performance through mutual knowledge exchange (MTDC2)** **SME as a source of knowledge is less motivated to exchange knowledge with large organization as recipient side due**
		(MTDC2) to build a mutual trust between partners (Minbaeva et al., 2018)	The following aspects drives motivation to teach by knowledge source: (MTDC1)Direct interaction and mutual experience sharing (Squire et al., 2009). (MTDC2) to build a mutual trust between partners (Minbaeva et al., 2018)
		other people can use my knowledge and my skill sets to work more efficiently *... I mean, not just teaching, I will learn something myself. So, if I have a coaching session where I also invite them to learn something, then I will have the personal win on this as well.* *...My answer to that you could be more efficient and make more money by learning people within my business, things that I do, or we do in a better way* *... And I will be able to recruit more talented people if we work efficiently and make things at a lower cost or better velocity"*	
	Informant 8 (SME)	Question: In case you taught your business partners, what motivates you to do so? *"...It's mainly for the reason in improving our delivery from a qualitative perspective or making it more efficient so that it becomes more efficient in handling on our side* *... it's harder to work with larger companies due to the size of the company, their rigid processes make it difficult to teach new things because there are more set in their own ways"*	

their	**to inflexibility**			**Understanding the recipient's expertise and competence that can facilitate KT (ATDC3)** **Ability to codify the knowledge facilitates KT due to needs for avoiding misunderstandin g between**
		lead to mutual trust between partners. Informant mentions that it is more difficult to share knowledge with larger sized company due to their rigid processes. Implicitly, this means that SME less motivated to exchange knowledge with recipient large organizations due to their inflexibility.		a) This quote of Informant 1 relates to understanding of the extent of the partner's capacity to absorb the knowledge delivered and is consistent with the literature related to recipient competence to absorb the more complex or basic knowledge (ATDC3) b) Informant 1 suggests motivation of
				The following aspects influence ability to teach of the knowledge source: (ATDC1) attractiveness of source companies in terms of competence and knowledge holding value (Pérez-Nordtvedt et al., 2008) (ATDC2) the degree of the source firm's ability to codify knowledge by developing common
	Ability to teach (AT)	Informan t 1 (LC)		Question: What do you think you need to know to be able to teach the other partners? *a) "…So, for me would be to know what the partner need is at the hotel. For instance. I mean, do you need all these fancy schmancy features that we have, or do you just need a smaller package or what's the setup of the hotel, stuff like that. So basically, the partners needs"* *b) "…So, motivation for me to educate if it is a business partner or if it is like an internal stakeholder in some way, it*
Organiz ational characte ristics	**Disse minat ive capac ity (DC)**			

					would be for them to know (more clear understanding of the product) the product that they're supposed to use"	glossaries that can facilitate mutual understanding (Schulze et al., 2014) (ATDC3) the degree of partner's expertise and competence (e.g Schulze et al., 2014)	the knowledge source to provide more clear information to the partners implies that ability to codify the knowledge by the source to avoid misunderstanding is driven by motivation for KT (ATDC2). (ATDC1) was not mentioned.	**partners (ATDC2)**
				Informant 2 (LC)	Question: What do you think you need to know to be able to teach the other partners? *"…I think you just need to know a little bit more than the other person or you just need to know. You just need to have done it. And then you can show them what you know. And then, then that's a teaching moment"*	The following aspects influence ability to teach of the knowledge source: (ATDC1) attractiveness of source companies in terms of competence and knowledge holding value (Pérez-Nordtvedt et al., 2008) (ATDC2) the degree of the source firm's ability to codify knowledge by developing common glossaries that can facilitate mutual understanding (Schulze et al., 2014) (ATDC3) the degree of partner's expertise and competence (e.g Schulze et al., 2014)	Informant 2 supports that knowledge recipient should possess enough competence and valuable knowledge to share knowledge with other organizations (ATDC1). But informant did not specify what type of knowledge should be owned by the knowledge source. Aspects related to (ATDC 2,3) were not mentioned by informant.	**Knowledge source should possess enough competence and valuable knowledge to be able to share the knowledge with other organizations (ATDC1)**

| | | | | Informant 3 (LC) | Question: What do you think you need to know to be able to teach the other partners?
"...*First of all, you would need to know the current state of the company. If you are to fit that information and approach it and a way to teach and show. You need to know their ability to understand and the maturity level*" | The following aspects influence ability to teach of the knowledge source:
(ATDC1) attractiveness of source companies in terms of competence and knowledge holding value (Pérez-Nordtvedt et al., 2008)
(ATDC2) the degree of the source firm's ability to codify knowledge by developing common glossaries that can facilitate mutual understanding (Schulze et al., 2014)
(ATDC3) the degree of partner's expertise and competence (e.g Schulze et al., 2014) | In this perspective, this quote is consistent with this literature in terms of assessing expertise of the recipient that measured by maturity level of the firm (ATDC3), while (ATDC1, 2) were not specified by informant. | **Knowledge recipient should possess enough competence and maturity (ATDC3)** |
| | | | | Informant 7 (LC) | Question: What do you think you need to know to be able to teach the other partners?
"...Needs to know "*in structure and in my experience, larger companies, more structured, more processes. And maybe to do things repeatable*. Smaller company, maybe more adaptable and kind of *listening into our processes and needs*" | The following aspects influence ability to teach of the knowledge source:
(ATDC1) attractiveness of source companies in terms of competence and knowledge holding value (Pérez-Nordtvedt et al., 2008) | Informant 7 suggests that it is important to assess expertise of the recipient that measured by organizational structure and processes of the firm (ATDC3). Informant 7 suggests that larger | **Knowledge recipient should possess enough competence and maturity (ATDC3)**

Large organization have ability to |

	codify the knowledge via formal procedures (ATDC2), while SMEs possess adaptability and flexibility might imply that they more rely on informal KT activities (ATDC)	**Ability to teach is based on assessment risks on knowledge leakage and protecting holding knowledge value (ATDC1)** **Knowledge recipient should possess enough competence and maturity (ATDC3)**
	organizations have repeatable processes, which implies more formal procedures are in place that are related to ability to codify of the knowledge transferred, while SMEs possess adaptability and flexibility that might imply that they more rely on informal KT activities. (ATDC2)	Informant suggests that formal non-disclosure agreements (NDAs) are needed to transfer knowledge, especially if counterpart organization is large because they can bring KT threats during process. Implicitly, this imply that knowledge source due to possessing valuable knowledge wants to protect by signing
	(ATDC2) the degree of the source firm's ability to codify knowledge by developing common glossaries that can facilitate mutual understanding (Schulze et al., 2014) (ATDC3) the degree of partner's expertise and competence (e.g Schulze et al., 2014)	The following aspects influence ability to teach of the knowledge source: (ATDC1) attractiveness of source companies in terms of competence and knowledge holding value (Pérez-Nordtvedt et al., 2008) (ATDC2) the degree of the source firm's ability to codify knowledge by developing common glossaries that can facilitate mutual understanding (Schulze et al., 2014)
Informant 9 (LC)		Question: What do you think you need to know to be able to teach the other partners? "…I really don't share much unless we really have the basics in place as an NDA, which covers the area. And sometimes also a contract that explains something of what is required. But then you can say if there was a very small company, we normally would require more knowledge about their operation before we proceed. Then, for instance, if it's a large company like Disney or someone. Apple is a pretty large company. And then also we need agreements, very early on because they

		(ATDC3) the degree of partner's expertise and competence (e.g Schulze et al., 2014)	NDAs with the partner (ATDC1) In contrast, If the partner is smaller organization, larger organizations as source wants to know more about SMEs learning capacity Informant also suggests that before sharing knowledge with the partners they need information the partners' operations and experience (ATDC3). ATDC2 was not mentioned by the informant.	**Ability to teach is based on assessment risks on knowledge leakage and protecting holding knowledge value (ATDC1)** **Ability to codify of large**
	Informant 4 (SME)	*can act like a bulldozer. To be frank. Therefore, you need to be very careful"* Question: What do you think you need to know to be able to teach the other partners? a) "*… I think I will focus on what we are cooperating around. And I'm also thinking about NDA, like confidentiality and so on. So, if I should share something that is not known by everyone, I would like to do now that we have a secrecy agreement in place*"	The following aspects influence ability to teach of the knowledge source: (ATDC1) attractiveness of source companies in terms of competence and knowledge holding value (Pérez-Nordtvedt et al., 2008) (ATDC2) the degree of the source firm's ability to codify knowledge by	a) Informant suggests that formal non-disclosure agreements (NDAs) are needed to transfer knowledge. Implicitly, this imply that knowledge source due to possessing valuable knowledge wants to protect by signing NDAs with the partner (ATDC1)

				b) *"...The large company knows what they can share and what not They used to have that kind of knowledge. And it can be harder to share information with smaller companies. Also get information from smaller companies. That's my experience... I think it's harder because they (SMEs) often don't know the frames where they are able to share information and what would be effective without risk to share. So, it is simpler to work with larger companies in that way.* ... *They (SMEs) are not so aware of the actual risk of sharing information"*	developing common glossaries that can facilitate mutual understanding (Schulze et al., 2014) (ATDC3) the degree of partner's expertise and competence (e.g Schulze et al., 2014)	b) This viewpoint is inconsistent with the literature: KT exchange with SMEs is more difficult since *"they are not so aware of the actual risk of sharing information"* in contrast KT with larger organization is easier because they know what they can share. This implies that formalized procedures of larger organizations facilitate KT that is related to ability to codify the knowledge, while informal approach that can imply less ability to codification of SMEs impedes KT (ATDC2). (ATDC3) was not mentioned.	**organizations allows facilitating KT, while inability of SME to codify brings difficulties to KT (ATDC2)**
			Informant 5 (SME)	Question: What do you think you need to know to be able to teach the other partners? *"You must know a little bit more about the structure and how other companies work*	The following aspects influence ability to teach of the knowledge source: (ATDC1) attractiveness of source companies in terms of competence and	Informant suggests that it is important to understand expertise and learning capacities of the recipient which are assessed in terms	**Knowledge recipient should possess enough competence and maturity (ATDC3)**

Informant	Quote	Literature / Theory	Analysis	Findings
	... Larger companies are not all going to in the forefront because they are large, things take a lot of time. I have seen that where I come from, in smaller companies, it is easier to do changes. Is not easier to change in a certain way in large companies. So, I think you need to know where the company is headed, how it's built, and why things like they are. And then you can start changing some of the ways they are working	knowledge holding value (Pérez-Nordtvedt et al., 2008) (ATDC2) the degree of the source firm's ability to codify knowledge by developing common glossaries that can facilitate mutual understanding (Schulze et al., 2014) (ATDC3) the degree of partner's expertise and competence (e.g Schulze et al., 2014)	of company structure and processes. Moreover, Informant suggests that with the large organizations is more difficult to share knowledge due to their size and rigidness for changes while is much easier with SME due to flexibility (ATDC3) (ATDC1,2) were not described by informant.	**Knowledge recipient should possess enough competence and maturity (ATDC3)** **Coaching approach is valuable for ability to teach (ATDC2)** **Time is essential resource needed should be considered for**
Informant 6 (SME)	Question: What do you think you need to know to be able to teach the other partners? *"...There are two sides of that coins. I think one side is that yes, you need to get an understanding of what they what they do to give the correct advice. But on the other side of that is that if you put the time and efforts into actually working with the people, working close with them that actually do the work. You will be able to coach them and learn and work with them so that it's more fruitful. So I think time is it's an essential part because change needs time. And you will need to have to that very important dialogue, not just the*	The following aspects influence ability to teach of the knowledge source: (ATDC1) attractiveness of source companies in terms of competence and knowledge holding value (Pérez-Nordtvedt et al., 2008) (ATDC2) the degree of the source firm's ability to codify knowledge by developing common glossaries that can facilitate mutual understanding (Schulze et al., 2014)	Informant confirms that he needs to understand the recipient side expertise (ATDC3); In addition, informants suggests that for better knowledge transfer, coaching approach is applied that is form of frequent interaction rather ability to codify of knowledge (ATDC2) Informant from SME suggests that time as a resource not only	

Dimensions/	Aspects that	Subcategory of	Primary data source	Quote from the informant	Relating literature	Analysis	Summary
				PowerPoint slide to fulfill that change need"	(ATDC3) the degree of partner's expertise and competence (e.g Schulze et al., 2014)	competence and knowledge value owned should be considered for KT activities (ATDC1)	**ability to teach (ATDC1)**
			Informant 8 (SME)	Question: What do you think you need to know to be able to teach the other partners? *"I need to know more about their processes and their people and who to involve them because that might be more widespread than when will we work with a smaller organization where it might be that you can talk to one person who can solve many things. You need to do more navigation and with the larger companies, to figure out who to speak to them when you need to do knowledge transfer so that you reach the right target"*	The following aspects influence ability to teach of the knowledge source: (ATDC1) attractiveness of source companies in terms of competence and knowledge holding value (Pérez-Nordtvedt et al., 2008) (ATDC2) the degree of the source firm's ability to codify knowledge by developing common glossaries that can facilitate mutual understanding (Schulze et al., 2014) (ATDC3) the degree of partner's expertise and competence (e.g Schulze et al., 2014)	Informant confirms that he needs to understand the recipient's expertise including to navigate in organizational structure and find a right contact person to be able to transfer knowledge to larger organizations (ATDC3) (ATDC1,2) were not mentioned.	**Knowledge recipient should possess enough competence and maturity (ATDC3)**

Characteristics	affect KT	the aspect		Data	Analysis	Analysis	Theme
						a) The informant supports the (CD2) and (CD3) aspects in the literature that cultural differences can hinder KT due to misunderstanding including language barrier. Therefore, meetings and interactions should be predefined and well organized to establish common understanding among partners. The aspect related to (CD1) is not specified.	**Language skills in different cultures are main barrier for KT between partners (CD2,3).**
Relational characteristics	**Cultural distance**		Informant 1 (Large company)	Question: In your experience, when you have interacted with people from different countries, what success (hits) and mistakes did you learn? What did you learn? a) *"I know that having like, online meetings… I mean, it's easy to misunderstand actually. I dont know if people are very good at English or not. But it's easy to misunderstand. But it's, it's really important that someone drives meetings and you keep track on what both of you are talking about. And if you're making decisions in the meeting…* ... *I mean, it's on the other hand, if we are like in a meeting with them, with the customer in USA, for instance, you can't really go back and forth all the time either, that kind of cost a lot of money. But and that's what I mean if you had like online meetings, it's important that someone is meeting organizer and keep track on what topics you're supposed to discuss and also what you decide during these meetings…"*	The national cultural distance arises different barriers for the KT among international partners regardless the firm size such as (CD1) higher cost of entry and hinders the transmitting the core competencies to the international markets; (CD2) facing operational challenges in terms of lacking common understanding of norms, values and motivation; (CD3) misunderstanding between foreign partners can limit sharing of critical organizational knowledge (Wijk et al., 2008).		
Characteristics	affect KT	the aspect	Informant 2	Question: In your experience, when you have interacted with people from	The national cultural distance arises different barriers for the KT among	a) The informant supports the (CD2) and (CD3) aspects in	**Language skills in different cultures are main**

				(Large company)	different countries, what success (hits) and mistakes did you have? a) *"So, when you communicate it's important that you remember <u>who you are communicating with and that you adjust your communications with the people (from different culture).</u> So when you communicate, for example, with <u>colleagues in China, you need to make your communications very clear and not that you need to have.</u> For example, short emails. Sentences, very simple. And like almost 1, 2, 3, 4, 5. If you have five questions, actually number them and they will answer. And you write your question and if it's a yes, no? And if it's a number you're looking for, you will write what number, you know, things like that, <u>because you need to remember this. English is not their native language. And you need to make everything easy for them.</u> And this will help you not be frustrated with their answer when, because sometimes people send a very long and complicated e-mail with like five questions that are embedded in the email. And people, one thing, and then they're like sending it back in. And this is not how you communicate with people when it's not their native*	international partners regardless the firm size such as (CD1) higher cost of entry and hinders the transmitting the core competencies to the international markets; (CD2) facing operational challenges in terms of lacking common understanding of norms, values and motivation; (CD3) misunderstanding between foreign partners can limit sharing of critical organizational knowledge (Wijk et al., 2008).	the literature that cultural differences can hinder KT due to misunderstanding including language barrier, which implies that language barrier can be reduced if organization transmits the knowledge in simple form. Informant also suggests that understanding of uniqueness of cultures through adjusting communication can facilitate KT activities. The aspect related to (CD1) is not discussed by informant. b) Informant provided examples that consistent with the literature regarding cultural aspects of China and Japan, where strong hierarchical subordination exists that can lead to misunderstanding and	**barrier for KT between foreign partners (CD2,3)** **Strong hierarchical subordination in organizational structure of companies that exists in various societies can be a barrier for KT (CD1)**

					language. So you need to remember who you're talking to and make it super simple. <u>And I always had very successful communications with my colleagues in China because I made it very simple.</u> *…<u>So, these are things you need to remember, and different countries have different ways of communicating. And we work in an international situation… And people forget that they always think, you know, English is king. It's not true.</u>* Follow up question: Would say culture affects your communication? *b) "Well, I think you need to consider it in the same way, <u>I think the language and the culture as well. So for example, in Japan, they always CC their boss. So if somebody is late in sending you stuff, I do reporting and things like that. I will always be careful not to hound the person too much because I don't want them to get in trouble. I just need my stuff but I don't want them to be in trouble, you know. Yeah. So I think it's, all about cooperating with people"</u>*		other operational challenges and eventually can hinder KT activities (CD2).	
				Informan t 3	Question: In your experience, when you have interacted with people from different countries, what success (hits) and mistakes did you have?	The national cultural distance arises different barriers for the KT among international partners	a) The informant confirms that different cultures can be the barrier for KT between	**Training cultural awareness workshops facilitate**

| | | | (Large company) | a) "_They (Human Resource unit) actually did courses regarding this topic for people that are not experienced to work with different cultures. Absolutely will hit some walls a few times. Because, I mean, we are different and you can absolutely start discussing what's wrong and right, but you just need to accept that the cultures are different_"
b) "...So, I think actually this is one of the exciting parts of being a global company to be honest. Mistakes, so mistakes that I personally have encountered. So, for instance, we hired a consultancy company from India to develop a solution for us. So, they can provide you literally whatever expertise you want. I think they had like 5000 consultants in different areas. And when we started the project, what I'm used to combine an idea as a product manager, and _you want to concept; you have a vision. And then the details are done by the teams._ And this is where it was a bit different here because what my experiences is with people from India in general, they work a lot. They have really can take on a lot. _But you need to be very detailed in what them to do. And if you want to get their ideas, you need to ask for them._ They | regardless the firm size such as (CD1) higher cost of entry and hinders the transmitting the core competencies to the international markets; (CD2) facing operational challenges in terms of lacking common understanding of norms, values and motivation; (CD3) misunderstanding between foreign partners can limit sharing of critical organizational knowledge (Wijk et al., 2008) | organizations (CD2,3) as a result to reduce cultural barriers for KT activities their organization is conducting cultural awareness trainings that helps to increase common understanding, values, and motivation of KT actors. This is in line with the literature. Aspect to (CD1) was not mentioned.
b) The informant provided supportive information of the existing literature that different cultural operational challenges can appear in various cultures during KT activities (CD2,3): in societies with higher power distance, which is associated with a strong hierarchical subordination and less likely to ask question or critique those of higher hierarchical | **common understanding of norms, values and motivation among KT actors (CD2)**

Strong hierarchical subordination in organizational structure of companies that exists in various societies can be a barrier for KT (CD1) |
| | | | | | | | |

				Language skills in different cultures are main barrier for KT between foreign partners (CD2,3)
		positions and interactions between hierarchical levels tend to be more authoritative and paternalistic, as a result KT can be hindered.		a) The informant supports the (CD2) and (CD3) aspects in the literature that cultural differences can hinder KT due to misunderstanding including language
		won't just say, well, you know, idea here what you say but you're completely wrong, that will never fly. You need to think about this. And that was absolutely a mistake from my side. I was throwing out ideas. They were just doing what I told them. I expected to get some feedback as well. So, certain part of the world. You need to be very specific and down to detail what to do and when. And you will get exactly what you asked for. And nothing more, nothing less. That's come to work ethics then the culture side. With some cultures you need to be direct. So, in Scandinavia you go around and round and round until we come to the point. But, for instance, if you took people from Germany, from US, you need to be direct and it's okay. It won't be frowned upon if you are to direct. So, that's something also that you need to consider, and you shouldn't get insulted if they aren't yet"		The national cultural distance arises different barriers for the KT among international partners regardless the firm size such as (CD1) higher cost of entry and hinders the core transmitting the
Informant 7 (Large company)		Question: In your experience, when you have interacted with people from different countries, what success (hits) and mistakes did you have? *"I have interacted a lot with Japanese people when I worked, maybe I'll also a lot with German, where I worked with*		

Strong hierarchical subordination in organizational structure of companies that exists in various societies can be a barrier for KT (CD1)	barrier, which implies that language barrier can be reduced if organization transmits the knowledge in simple form. Informant also suggests that understanding of uniqueness of cultures through adjusting communication can facilitate KT activities. The aspect related to (1) is not discussed by informant. b) Informant provided examples that consistent with the literature regarding cultural aspects of China and Japan, where strong hierarchical subordination exists that affects KT activities.	competencies to the international markets; (CD2) facing operational challenges in terms of lacking common understanding of norms, values and motivation; (CD3) misunderstanding between foreign partners can limit sharing of critical organizational knowledge (Wijk et al., 2008)	*customers and operators. And I tried to joke a lot. And I've learned that doesn't fly at all. If you don't speak your native language to start with them in Japanese people kind of struggle with English. Maybe not so much for Swedes, Germans maybe as well. But then you also have the different cultural backgrounds. We don't really have the same like references and so on. So maybe be more, I shouldn't say formal, but maybe just a bit, a bit boring to increase the success rate. So I did try to joke and it kind of just went over their head and then it's just pointless. You make them feel a bit dumb and I don't feel right basically".*	
KT can be hindered in different cultures due to lacking common	The national cultural distance arises different barriers for the KT among international partners regardless the firm size. The informant suggests that different cultures can bring different operational challenges due to		Question: In your experience, when you have interacted with people from different countries, what success (hits) and mistakes did you have?	Informant 9 (Large company)

				"In the US, typically, there is one management agenda and won't matter much, everyone is kind of aligning to that, knowing silently that we won't be able to deliver on that time. So, but we deal with that later when we get there. In Scandinavian (countries), it's more like being open, telling up front as soon as possible as you know that you have a problem of some sort. You should share it (failure) immediately with your management. And they will, normally, depending on the person, try to help you what they can do to support you in solving it. *...you have to deliver something at someday...* *So, there's different as countries, whether with command or control...* *... But you need to learn how to maneuver between different cultures. So, it's not like my culture is going to make it fail or something.* *... And you can have different approaches. I mean, in organization needs to be transparency towards my counterparts. I would ask them for recommendation how to proceed with this. And if they are part of the problem, I will probably go into my organization to my level, go back through what we call the key accounts and have them*	such as (CD1) higher cost of entry and hinders the transmitting the core competencies to the international markets; (CD2) facing operational challenges in terms of lacking common understanding of norms, values and motivation; (CD3) misunderstanding between foreign partners can limit sharing of critical organizational knowledge (Wijk et al., 2008)	norms, values (CD2): Scandinavian counties prefer to know about possible problems in advance, while the US partners have higher risk-taking levels and no need to explicitly share the knowledge about possible failure. This implicitly means that the US partners have trust that failure will be solved by dedicated employees (key accounts from those culture) partner due to a common agenda (CD3). Moreover, the US partners prefer to have KT activities with organizational representative that have the top-level position at their organization, which can provide a successful partnership and KT activities. Aspect related to	**understanding of norms, values and motivation (CD2,3)** **Companies need to learn how to adapt in different cultures for better understanding (CD3)**

				address it at some other level. And since the key accounts are part of the culture in US so to speak, or in Europe or in Asia. They normally know how to handle it and they normally will advise you on how to move around in this. In Scandinavia, I would probably go more directly to the people. Maybe we can also sit, have a round table session. Which you cannot do in other cultures because you sometimes oppose cultures because, sometimes, the boss is the boss"		(CD1) is not mentioned. This is consistent with the literature.	
				Informan t 4 (SME)	Question: In your experience, when you have interacted with people from different countries, what success (hits) and mistakes did you have? *"I one mistake that I think about is that you often or at least me myself, I often think that the other person is just like me and that they have the same culture and the right thoughts about what's right and wrong and so on. That's a recurring mistake. I think. What success I think. Again, meeting in real life, it's good in business and otherwise. So, when I have been in, in business, travels and I think it's, it's good to meet the persons outside the business room. So, to do something fun. After the work time.*	The national cultural distance arises different barriers for the KT among international partners regardless the firm size such as (CD1) higher cost of entry and hinders the transmitting the core competencies to the international markets; (CD2) facing operational challenges in terms of lacking common understanding of norms, values and motivation; (CD3) misunderstanding between foreign partners can limit sharing of critical	Informant supports differences in cultures can lead fewer common understandings and each cultures have their own rules that can be barrier for KT activities (CD2,3). (CD1) was not mentioned The informant suggests that cultural distance for KT can be reduced by frequent interaction via training meetings and informal relations after formal	**KT can be hindered in different cultures due to lacking common understanding of norms, values and motivation (CD2,3)**

Informant	Interview excerpts (data)	Analysis	Literature connection	Emerging theme
	"…You never stop learning. And it is very different cultures even in Europe. So, like it's very different to work with companies in Finland and in Germany and so on. You will have different cultures and different rules. And I think it's really fun to get to know the differences as well. You can share the differences in meetings. And you both think it's fun that we are we are different. So that's, yeah, that's what I have learned"	…job meeting. Moreover, the informant suggests that informal relationships in terms of establishing social ties with the partner can also enhance KT and reduce cultural distance. a) The informant confirms that different cultures for KT can bring operational challenges (CD2) and suggest reducing cultural barriers for KT activities their organization is conducting cultural awareness trainings. b) The informant supports that in societies such as China, there is strong subordination exist and as a result this can serve as KT operational barriers (CD2,3), therefore the	organizational knowledge (Wijk et al., 2008) The national cultural distance arises different barriers for the KT among international partners regardless the firm size such as (CD1) higher cost of entry and hinders the transmitting the core competencies to the international markets; (CD2) facing operational challenges in terms of common lacking understanding of norms, values and motivation; (CD3) misunderstanding between foreign partners can limit sharing of critical organizational knowledge (Wijk et al., 2008)	**Training cultural awareness workshops facilitate common understanding of norms, values and motivation among KT actors (CD2)** **Strong hierarchical subordination in organizational structure of companies that exists in various societies can be a**
Informant 5 (SME)	Question: In your experience, when you have interacted with people from different countries, what success (hits) and mistakes did you have? a) *"I worked a lot with Sony Eriksson. I worked with China, Beijing and I attended a course when I went to a product leader, you need to know, you need to understand the culture and how different is it from Sweden to other countries I mean, we attended a course where they explain how Europe works, how Europe, Asia, and different kind of worlds. How you pronounce things, how you express yourself. So, you need to know how you as a leader or manager, what kind of role you have, probably Product Manager, how you express yourself, what kind of things*			

							barrier for KT (CD2)
		KT		*you request and how you pronounce it, how you express yourself in a meeting, and what do you want out of things"* *b) "…And, for example, I was working for a factory in China because we were producing mobile phone. And we are in Sweden, we talk to the people in the project manager and the team. And we had some opinions, want to change things. And then they are misunderstanding people, they had to leave because they thought they had done some mistake, and it wasn't a mistake. <u>We want to correct things. And in Sweden you can talk to people, but there they fire people right away because we had said something. I had a bad experience, but then we corrected it, but we got to learn that way. You have to be careful how you pronounce thing, how you say things, we wanted some small correction and then they misunderstand us completely.</u> I did a lot of experience there how to work with different culture in this way with China"*		informant suggests that it is important to know how you articulate the knowledge that you are going to transmit. This is in line with the literature.	

Informant 6 (SME)	Question: In your experience, when you have interacted with people from different countries, what success (hits) and mistakes did you have? a) *"So, I think when I work with international companies. Before starting to work with other people. I tried to understand their motivations and how they make decisions, how they get interested in things. So how they are motivated. So, this varies from different cultures. So, for instance, in Japan, when I work with them, I did this hierarchical structure to get attention from, from manager's managers. And then I found an individual that I worked with. And in American culture was quite similar, German as well. But they have different requirements and how you should have a dialogue with them. So, Germans and Japanese people, they are very structured and like structured information. We're, whereas American. They love stories. I just loved the success stories"* b) *"I mean, I'm sure that you have some Maslow equivalence of trust. So, you need to build before you can do anything. You need to have trust because that's the foundation of the*	The national cultural distance arises different barriers for the KT among international partners regardless the firm size such as (CD1) higher cost of entry and hinders the transmitting the core competencies to the international markets; (CD2) facing operational challenges in terms of lacking common understanding of norms, values and motivation; (CD3) misunderstanding between foreign partners can limit sharing of critical organizational knowledge (Wijk et al., 2008)	a) Informant agrees that motivation (CD2) and the process of decision making vary depending on cultures. The informant suggests that there is different communication culture are in place in different societies where (CD3), Germans (low hierarchical subordination) and Japanese (strong hierarchical subordination) have similarities in terms of more structured information. In contrast, the US partners prefer success stories that helps to better absorb the transferred knowledge. b) This is inconsistent with the literature. The informant suggests	**Lack of understanding the decision-making process in different cultures can hinder KT activities between partners.** **Trust is important aspect for KT activities between foreign partners**

					house you need to build. And that comes with every people that you interact with. And trust can be built in different ways. Many Swedes they build trust from a good interaction, qualities and good interpersonal qualities, good dialogue qualities. Whereas other cultures (Japan), they will *trust from while you asked my manager to tell me this and then I will do it. So, it differs very much because people in other cultures, they may build a trust on what their managers say that they should have trust in.* But as everything starts with trust"		that in low power distance societies such as Sweden are important to establish trust (CD2) for KT, while in cultures with higher power distance (Japan), trust to perform KT is built due to order from the higher-level managers. But Sweden belongs to lower PD. So, this implies that establishing trust as common motivation to transfer knowledge between foreign partners. Aspect (CD1) is not mentioned.	
				Informant 8 (SME)	Question: In your experience, when you have interacted with people from different countries, what success (hits) and mistakes did you have? a) *"I would say that my biggest learning is that to have respect for different cultures. Because how you work is usually very different in how you run your daily business. So I would say that's a general learning is to kind*	The national cultural distance arises different barriers for the KT among international partners regardless the firm size such as (CD1) higher cost of entry and hinders the transmitting the core competencies to the international markets;	The informant provided with the example from Indian partnership, where strong subordination exists as a result a communication by order can distort the projects since knowledge is not	**Lack of understanding the decision-making process in different cultures can hinder KT activities between partners.**

					of figure out the actual culture of the country that you're that you're working. And not just focusing on maybe doing things as you normally do and communicating. *b) ... I mean, but to try to figure out the culture of how you run things and how you take decisions. How you operate in that in that country. Or from all from a project, that project perspective to a more how the hierarchy is set up. If it's very, in some countries this more formal and in some countries it's more informal on how you take decisions.* *...I think that it's usually difficult to learn in advance. But I think you need to maybe go into the projects with an open mind in trying to figure out how to help to navigate, to set up the structure for the project in everything from how you take decisions to how you run meetings and how will you follow up on actions.* *...I've been working in a large project. It's almost 10 years ago now in an outsourcing project with the US and India. And there it was difficult because everything as soon as I spent a lot of time in India then and there it works so much that everything was fine when you are present. And then they started running the project their way when you*	(CD2) facing operational challenges in terms of lacking common understanding of norms, values and motivation; (CD3) misunderstanding between foreign partners can limit sharing of critical organizational knowledge (Wijk et al., 2008)	transferred due to absence of understanding how process goes and how decisions has been taken depending on formal or informal way in particular society. Implicitly, this suggests that due to lacking understanding of norms, values and motivation in different cultures can be a barrier for KT process between international allies (CD2,3). Aspect (CD1) is not mentioned.	

Dimensions/ Characteristics	Aspects that affect KT	Subcategory of the aspect	Primary data source	Quote from the informant	Relating literature	Conclusion/Analysis	Summary
				left. And then I was under the impression that we still had agreed to certain. But as soon as you stopped following up, they had their own way of taking decisions and deciding priorities, etc. And I'm not saying that their parties were wrong, but they just they didn't have a way of communicating that within the project"			
Relational characteristics	**Geographical distance/proximity**	**Geographical proximity**	Informant 1 (Large company)	Question: In a globalized context of operation, does physical proximity or distance affect your collaboration with other companies, especially small and medium sized partners? In what way? a) *"... I mean, of course, I think it always matters actually because it's always simpler to be where the customers are, or it is depending on what you're going to like discuss. But yeah, of course it matters. But in the in the world, we live in, we have to kind of adapt to not being side-by-side all the time. So that's why I think it's important to kind of come up with what helped both the partners and our company in that collaboration.*	Geographical distance negatively influences inter-organizational KT especially for tacit knowledge exchange but can be mediated in terms of (GD1) developing trust-based relations between distant partners (e.g., Capaldo & Petruzzelli, 2014); (GD2) organizational proximity in terms of common corporate culture and values (Korbi & Chouki, 2017); and (GD3) bilateral way of utilizing of communication	a) The informant argues that proximity to partners is important, but also, she mentioned that it is important to adapt in new reality (implicitly, pandemic time) and find solutions how to help each other (between partners) for better collaboration for KT activities. Aspect (GD1) was not explicitly described by the informant such as trust-based	**Adaptability capacity between partners is important for geographical distant partners** **Physical proximity is more prevalent than organizational proximity in terms of corporate culture and values**

		relationships but rather it is more important to have adaptability capacity for partners for better KT activities. Aspect (GD2) was not supported by the informant stating that large organizations have dedicated personnel to easily reach the partners in a specific location of the world for learning and exchanging knowledge with those partners. Implicitly, this implies that organizational proximity is still important to transfer tacit and complex knowledge.
… (2) I think so. I mean, we have a big part in the US, for instance, and I know that the people in our company, then we have like individuals that are very involved in different like the key accounts. I mean, I'm not talking about the key accounts' manager, So stuff like that. But all the bigger partners that we have are the big hotel groups, we have dedicated people to talk to them a lot and maybe try to see them and help them out with whatever it is. And I think that is very important to have at least a few people that know the customer and know what they want. And now they're using the systems. Because it's hard for us in Sweden to know exactly what America or China or India what needs they have, for instance. But I think we have to cope with it. …I think, for instance, the last year, during the pandemic. I think our company had an ease transition from being located or working in the office to the online working when we're doing mutually sits where it is. It has been a very easy transition because we kind of use it. We have like partners and companies in China. We have US, Australia. Yeah. I mean, all over the place, we are used to it. But, of course, it's better to be in-person.	technologies (Choi & Contractor, 2016)	(GD3) was supported by the informant who argues that in pandemic time the company could easily switch to online communication

				Willingness for interfirm KT between partners affect rather geographical proximity
				a) The informant suggests that proximity is not important for KT activities during the collaboration, but states that is more important to be motivated for geographical collaboration rather co-geographical presence. Implicitly, this means willingness is important aspect for interfirm KT. GD1 is not explicitly mentioned. The aspect (GD2) is not discussed by the informant, but the informant argues that geographical distance
			Geographical distance negatively influences inter-organizational KT especially for tacit knowledge exchange but can be mediated in terms of (GD1) developing trust-based relations between distant partners (e.g., Capaldo & Petruzzelli, 2014); (GD2) organizational proximity in terms of common corporate culture and values (Korbi & Chouki, 2017); and (GD3) bilateral way of utilizing of communication technologies (Choi & Contractor, 2016)	
	Informant 2 (Large company)	Question: In a globalized context of operation, does physical proximity or distance affect your collaboration with other companies, especially small and medium sized partners? In what way? a) *"I would say physical proximity. It doesn't affect collaboration edges. That it's a bit tricky if you work with Australia, the time differences…when I was in Canada that we had exactly 12 hours with China. So, eight in the morning was eight in the evening for them. Somebody would not be in the right time. Not in the regular day time, right? It makes it a bit challenging, but you need to, and of course, if everybody's in one building, then it makes it easier. I don't think it affects collaboration because collaboration means people are willing to do things. So, I don't think people are either will or unwilling to do things if they're far*		

				apart. It's just that it makes it more difficult to do it at the same time, but it doesn't mean they're unwilling to do it.		is more challenging due to time difference. GD3 is not explicitly mentioned.	
Relation al character istics	Geogr aphica l distan ce/pro ximity	Geogra phical distanc e	Informan t 3 (Large company)	Question:In a globalized context of operation, does physical proximity or distance affect your collaboration with other companies, especially small and medium sized partners? In what way? *a) One challenge absolutely is the time difference. So that is can be a drawback and also can make so that things take much longer than they should. So for instance, if we need to get a collaboration between US and China, It's hard. One of them needed to stay in the office after hours or someone on yet a sudden it to get up three in the morning to attend. And you don't want to do that much either. So that's a challenge. And then if you had a couple collaboration between R&D and the supplier. You have just two few hours overlap and then you need to wait until next day before you get feedback. And*	Geographical distance negatively influences inter-organizational KT especially for tacit knowledge exchange but can be mediated in terms of (GD1) developing trust-based relations between distant partners (e.g., Capaldo & Petruzzelli, 2014); (GD2) organizational proximity in terms of common corporate culture and values (Korbi & Chouki, 2017); and (GD3) bilateral way of utilizing of communication technologies (Choi & Contractor, 2016)	a) The informant suggests that geographical distance brings challenges due to time difference and supplier cost. The informant explicitly did not mention (GD1,2,3)	

				then you come, which often you that answer and then you wait till the next day. And that is also a challenge. And then of course, shipping. It's absolutely a challenge now. And we see, for instance, having a manufacturing site in China had their benefits before when it comes to cost. Today, that is maybe not the case. The situation has changed. So for instance, if you look at Eastern countries in Europe, you could have not, not much more expensive labor cost compared to China. But from a logistics point of view, it's much, much easier. You don't have lead time with which ship, right? And therefore, it can be very costly if you're moving heavy objects. Hardware wise. So that's also absolutely from that point of view, when it comes to medium-sized and small, have nothing too late to that aspect of it"			
			Informant 7 (Large company)	Question: In a globalized context of operation, does physical proximity or distance affect your collaboration with other companies, especially small and medium sized partners? In what way? a) *"If you have asked me this a year and a half ago, I would probably give it a different answer than today. Because we are really practice this global and*	Geographical distance negatively influences inter-organizational KT especially for tacit knowledge exchange but can be mediated in terms of (GD1) developing trust-based relations between distant partners (e.g., Capaldo & Petruzzelli,	a) This perspective is partially in line with the literature. The informant suggests that personal relationships are still very important for geographical distance partners, specifically when you establish	**Physical proximity in terms of face-to-face meetings with distant partners facilitates to build a trust for interfirm KT**

Theme	Informant	Quote	Description	Finding
		far away communication. And I think we have really improved and even become a bit tired of it as well. Using this kind of worn-out phrase, this is the new normal. So, I still see a great value in meeting people in person, definitely in the phase when you start to know them. And that's true for a mean, personal relationships, but also professional relationships. *...So travel might be required in the future, but follow-up meeting I think this kind of Teams meeting that we do now is very efficient and it's better for the environment as well"*	(GD2) 2014); organizational proximity in terms of common corporate culture and values (Korbi & Chouki, 2017); and (GD3) bilateral way of utilizing of communication technologies (Choi & Contractor, 2016) new relationship is essential to have face-to-face meetings that implicitly can be described as to establish trust for better transfer of the knowledge with new partners (GD1), while follow-up meetings (after establishing a trust) can be effectively performed via online communication tools that are efficient and environmentally friendly (GD3). GD2 was not mentioned by informant.	**Digital communication tools can enhance KT between distant organizations but as a complementary mechanism after establishing face-to-face meetings**
Geographical proximity	Informant 9 (Large company)	In a globalized context of operation, does physical proximity or distance affect your collaboration with other companies, especially small and medium sized partners? In what way? a) *"I think we have learned in the last year that we can solve most by actually teamwork and so on. And emails are good for nothing really. E-mails are just for exchanging the basic and then*	Geographical distance negatively influences inter-organizational KT especially for tacit knowledge exchange but can be mediated in terms of (GD1) developing trust-based relations between distant partners (e.g., Capaldo & Petruzzelli, a) The informant argues that proximity face-to-face and meetings can facilitate to transfer knowledge related the company's technology and solve products and emergent different situations. This	**Physical proximity in terms of face-to-face meetings with distant partners facilitates to build a trust for interfirm KT**

				you meet either on a meeting like this (online). But occasionally you read, we at least we have to meet the customer, we have to sit face to face with our major customers or our partners to mostly why not is to avoid an emergent situation going on. It is required from a technology point of view, actually to be able to do some real life testing of what we have develop. We would need to have be physically present, but that's because we just cannot connect these two body parts over the Atlantic, we have to actually be physically present. But physical proximity can affect our collaboration both from a technical point of view, actually testing out, doing, finding out the product. This is the customer see the problems that I have physically understand them" b) *"…And also that meeting the customers in order to sort out situation and have kind of a face-to-face talk. And very often that happens around two to three times a year. We tried to do this and around where the big exhibitions are, where the big market events are. But we also do it on ad hoc basis, when require five people and fly them out to part in Asia and talk to the major customers…"*	2014); (GD2) organizational proximity in terms of common corporate culture and values (Korbi & Chouki, 2017); and (GD3) bilateral way of utilizing of communication technologies (Choi & Contractor, 2016)	implies that tacit knowledge and related complexity can be better transmitted to other partners only via real physical interaction. Implicitly, this means that trust-based relationships are important aspect that can be built via physical proximity (GD1). This is perspective is inconsistent with the literature (GD1). Aspects (GD2,3) were not mentioned.	

				Physical proximity in terms of face-to-face meetings with distant partners facilitates to build a trust for interfirm KT between distant organizations.
				Digital communication tools can enhance KT between distant organizations.
			Geographical distance negatively influences inter-organizational KT especially for tacit knowledge exchange but can be mediated in terms of (GD1) developing trust-based relations between distant partners (e.g., Capaldo & Petruzzelli, 2014); (GD2) organizational proximity in terms of common corporate culture and values (Korbi & Chouki, 2017); and (GD3) bilateral way of utilizing of communication technologies (Choi & Contractor, 2016)	a) The informant supports that geographical proximity and face-to-face interaction facilitate KT activities between partners, while distance can be a barrier for knowledge sharing. Implicitly, this means that physical proximity allows building trustworthy relations between partners (GD1,2). b) Aspect (GD3) was confirmed by the informant that digital communication tools can enhance KT
Geographical proximity	Informant 4 (SME)	*… It is more, sometimes more difficult to get the perception of people and how they interact, body language and all these kinds of things. And see the premises which are about to buy when you are on a meeting like this, or if you have to trust other party that are close to them"*		

(In a globalized context of operation, does physical proximity or distance affect your collaboration with other companies, especially small and medium sized partners? In what way?)

a) *"I think it's harder to, to cooperate with a large distance. Because if you really want to solve something, it's easier if you can go and meet in person. So that's the thing that I think is hard and the culture in some way but, but not necessary. So the main thing I think is that if you or when you need to meet the distance can be a problem"*

Follow up question: Now in this pandemic time, how do you think this distance can be reduced for better knowledge sharing?)

"I think this digital improvement or usage that we have learned now this year is very good because it's so easy to

		between organizations.	a) The informant confirms that geographical distant partners need to establish face-to-face meetings for creating some relations. This implies that distance can be reduced via establishing personal relations to build trustworthy relations (GD1). b) This perspective is partially consistent with the literature. The informant supports that digital communication tools can enhance KT between (GD3)	**Physical proximity in terms of face-to-face meetings with distant partners facilitates to build a trust for interfirm KT** **Digital communication tools can enhance KT between distant organizations but as a complementary mechanism after establishing face-to-face meetings**
		Geographical distance negatively influences inter-organizational KT especially for tacit knowledge exchange but can be mediated in terms of (GD1) developing trust-based relations between distant partners (e.g., Capaldo & Petruzzelli, 2014); (GD2) organizational proximity in terms of common corporate culture and values (Korbi & Chouki, 2017); and (GD3) bilateral way of utilizing of communication technologies (Choi & Contractor, 2016)		
	Informant 5 (SME)	*get to decide about the meeting and defined. And it's easier to plan since you can adjust according to their schedule if they have it online in other words, to find an available slot in the schedule of the other person (online)"*	Question: In a globalized context of operation, does physical proximity or distance affect your collaboration with other companies, especially small and medium sized partners? In what way? a) *"We work with a company that we bought in Netherland. I haven't met them before. I think, If you get to know people that you meet them face to face one or two times. I mean, it will always be easier to communicate. You have you build a kind of relations. I always try to meet people for the first-time. That's why we went to China, for example, to meet people to be that they thought, then it was much easier than to have daily meetings where you think, you know, how they react. If they don't react, you know, how they pronounce thing, you understand more. I think that a good way to learn each other"*	

Theme	Informant	Transcript	Analysis	Findings
		organizations but as a complementary mechanism after establishing face meetings. b) (So we can summarize, when you don't have a relationship, if it is a distant partner, it's better to see and have a face-to-face meeting to establish some relationship. And then you can move to online version or remote version to decrease this, reduce this distance am I right?) I: Yeah.	a) Informant suggests that geographical proximity is important because it can facilitate to establish trust between partners for better learning capacity. This implies that to establish trust first it is needed to have physical proximity in terms of face-to-face interaction and then it can move to distant partnerships (GD1,2) b) This viewpoint is partially consistent with the literature. The informant suggests that digital communication tools	**Physical proximity in terms of face-to-face meetings with distant partners facilitates to build a trust for interfirm KT** **Digital communication tools can enhance KT between distant organizations but as a complementary mechanism after establishing face-to-face meetings**
Geographical proximity	Informant 6 (SME)	Question: In a globalized context of operation, does physical proximity or distance affect your collaboration with other companies, especially small and medium sized partners? In what way?) a) *"Physical proximity has an effect. When I worked in an international company. I always tend to build the trust part by being present with the person, so the person could get a grasp of what I am. Because when you do a team's discussions like we're doing now, I met you Aizhan, so I know more about you. But there are so much more in your communication that can be done over a physical meaning, the body language, the way we interact in social situations. That we become that come from the Savanna, right? We did not have video conferences on the Savanna. We worked as a group or as a group of people. And there were trust*	Geographical distance negatively influences inter-organizational KT especially for tacit knowledge exchange but can be mediated in terms of (GD1) developing trust-based relations between distant partners (e.g., Capaldo & Petruzzelli, 2014); (GD2) organizational proximity in terms of common corporate culture and values (Korbi & Chouki, 2017); and (GD3) bilateral way of utilizing of communication technologies (Choi & Contractor, 2016)	

				between these people. And that's due to many aspects of what we did on the Savanna. What people knew what they had competence in and what if they were strong, if they were good then in crafting. If they were good speakers, they were good in finding water or whatever. That's something you gain from physical proximity" (How do you think, for example, if you work here, but I know that you have businesses in the United State or a client from, let's say in other countries. How do you see how you can decrease this distance effect for collaboration or knowledge exchange, especially for large partners?) b) "I think I would have very hard to trust another people to full extent if I've only had one hour team's conference with that person at hand. So, I need to meet with the person to have full trust. So, I would say that yes, you need to meet at some point during your cooperation. Or you should have a longer session with some interactive social tools. Maybe you should show them around and your house, how you live. Maybe you should show your kids, or I mean to build deeper trust"		cannot fully substitute face-to-face interaction for KT due to difficulties to establish trustworthy relations, but the informant supports that trust can be built if online meetings can be done longer period and having other topics such as related to family (GD3).	

				Informant 8 (SME)	Question: In a globalized context of operation, does physical proximity or distance affect your collaboration with other companies, especially small and medium sized partners? In what way?) a) *"I would say that it doesn't affect <u>the collaboration with large organization so much because they're usually so widespread already from the beginning.</u> I think my experience is that the <u>proximity or physical distance isn't so important any longer. Even if it's nice to meet them from time to time, you still don't you usually don't have possibility to physical proximity to tool or parts of their organization.</u> So you kind of learn how to deal with that on a distance"* Follow up question: Say you had a partner that is maybe five minutes away with car compared to having partners in USA or Asia? How would just say that distance affects the collaboration?) b) *"I think that depends a bit on what type of organization it is. If it's just a company where you work with on a <u>more abstract level, with services or with general cooperation.</u> And, you don't have so much <u>physical on hand production,</u> for example. Then I don't*	Geographical distance negatively influences inter-organizational KT especially for tacit knowledge exchange but can be mediated in terms of (GD1) developing trust-based relations between distant partners (e.g., Capaldo & Petruzzelli, 2014); (GD2) organizational proximity in terms of common corporate culture and values (Korbi & Chouki, 2017); and (GD3) bilateral way of utilizing of communication technologies (Choi & Contractor, 2016)	The informant suggests physical proximity is not essential for KT activities and confirms that large organizations inherently have KT with distant partners due to widespread operations. The informant supports that physical distance can influence knowledge transfer in the case of manufacturing of physical products where tacit and complex knowledge inherently exists and by proximity can be easily transferred. However, for the service sector proximity does not have any importance. Informant did not specify aspects (GD1,2,3)	**Geographical distance/proximity does not influence interfirm KT**

think that I would have visited them so much more, just even if they were very close. It's more of important once you actually, work with physical products for example. We have our largest manufacturing sites with the supplier just about an hour away by car. And that's really good for us because we can always go and help them. If they have issues in production, then it is easier for us to go there and help them to do the search and trying to figure out what's wrong and how to fix it and collaborate with them in setting up the testing equipment, etc. But, so that has a positive a side effect in having a shorter physical distance, then for example, looking and moving some of that production to an facility in China. And then of course it's a directly becomes more difficult to cooperate and figure out things from afar when you're actually not there and can see things. But if it's just general corporations, just meetings and calls and documentations and systems, then it doesn't matter. From my perspective, the physical distance doesn't have so much impact"

Dimensions/ Characteristics	Aspects that affect KT	Subcategory of the aspect	Primary data source	Quote from the informant	Relating literature	Analysis	Summary
Relational characteristics	**Cultural distance**		Informant 1 (Large company)	Question: In your experience, when you have interacted with people from different countries, what success (hits) and mistakes did you have? What did you learn? a) *"I know that having like, online meetings… I mean, it's easy to misunderstand actually. I dont know if people are very good at English or not. But it's easy to misunderstand. But it's, it's really important that someone drives meetings and you keep track on what both of you are talking about. And if you're making decisions in the meeting…* *… I mean, it's on the other hand, if we are like in a meeting with them, with the customer in USA, for instance, you can't really go back and forth all the time either, that kind of cost a lot of money. But and that's what I mean if you had like online meetings, it's important that someone is meeting organizer and keep track on what topics you're supposed to discuss and also what you decide during these meetings…"*	The national cultural distance arises different barriers for the KT among international partners regardless the firm size such as (CD1) higher cost of entry and hinders the core transmitting the competencies to the international markets; (CD2) facing operational challenges in terms of lacking common understanding of norms, values and motivation; (CD3) misunderstanding between foreign partners can limit sharing of critical organizational knowledge (Wijk et al., 2008).	a) The informant supports the (CD2) and (CD3) aspects in the literature that cultural differences can hinder KT due to misunderstanding including language barrier. Therefore, and meetings interactions should be predefined and well organized to establish common understanding among partners. The aspect related to (CD1) is not specified.	**Language skills in different cultures are main barrier for KT between foreign partners (CD2,3).**

	Interview		
Informan t 2 (Large company)	Question: In your experience, when you have interacted with people from different countries, what success (hits) and mistakes did you have? a) *"So, when you communicate it's important that you remember who you are communicating with and that you adjust your communications with the people (from different culture). So when you communicate, for example, with colleagues in China, you need to make your communications very clear and not that you need to have. For example, short emails. Sentences, very simple. And like almost 1, 2, 3, 4, 5. If you have five questions, actually number them and they will answer. And you write your question and if it's a yes, no? And if it's a number you're looking for, you will write what number, you know, things like that, because you need to remember this. English is not their native language. And you need to make everything easy for them. And this will help you not be frustrated with their answer when, because sometimes people send a very long and complicated e-mail with like five questions that are embedded in the email. And people, one thing, and then they're like sending it back in. And this*	The national cultural distance arises different barriers for the KT among international partners regardless of the firm size such as (CD1) higher cost of entry and hinders the core transmitting the competencies to the international markets; (CD2) facing operational challenges in terms of lacking common understanding of norms, values and motivation; (CD3) misunderstanding between foreign partners can limit critical sharing of organizational knowledge (Wijk et al., 2008).	a) The informant supports the (CD2) and (CD3) aspects in the literature that cultural differences can hinder KT due to misunderstanding including language barrier, which implies that language barrier can be reduced if organization transmits the knowledge in simple form. Informant also suggests that understanding of uniqueness of cultures adjusting through communication can facilitate KT activities. The aspect related to (CD1) is not discussed by informant. b) Informant provided examples that consistent with the literature regarding cultural aspects of China and Japan, where strong hierarchical subordination exists that can lead to
			Language skills in different cultures are main barrier for KT between foreign partners (CD2,3) **Strong hierarchical subordination in organizationa l structure of companies that exists in various societies can be a barrier for KT (CD1)**

				Informant 3			
					is not how you communicate with people when it's not their native language. So you need to remember who you're talking to and make it super simple. <u>And I always had very successful communications with my colleagues in China because I made it very simple.</u> *<u>…So, these are things you need to remember, and different countries have different ways of communicating. And we work in an international situation… And people forget that they always think, you know, English is king. It's not true.</u>** Follow up question: Would say culture affects your communication? *b) "Well, I think you need to consider it in the same way, <u>I think the language and the culture as well</u>. So for example, <u>in Japan, they always CC their boss. So if somebody is late in sending you stuff, I do reporting and things like that. I will always be careful not to hound the person too much because I don't want them to get in trouble. I just need my stuff but I don't want them to be in trouble, you know. Yeah. So I think it's, all about cooperating with people"</u>*		misunderstanding and other operational challenges and eventually can hinder KT activities (CD2).	
				Informant 3	Question: In your experience, when you have interacted with people from	The national cultural distance arises different barriers for the	a) The informant confirms that different	**Training cultural**

| | | | | (Large company) | different countries, what success (hits) and mistakes did you have?
a) "_They (Human Resource unit) actually did courses regarding this topic_ for people that are not experienced to work with different cultures. Absolutely will hit some walls a few times. Because, I mean, we are different and you can absolutely start discussing what's wrong and right, _but you just need to accept that the cultures are different_"
b) "...So, I think actually this is one of the exciting parts of being a global company to be honest. Mistakes, so mistakes that I personally have encountered. So, for instance, we hired a consultancy company from India to develop a solution for us. So, they can provide you literally whatever expertise you want. I think they had like 5000 consultants in different areas. And when we started the project, what I'm used to combine an idea as a product manager, and _you want to concept; you have a vision. And then the details are done by the teams._ And this is where it was a bit different here because what my experiences is with people from India in general, they work a lot. They have really can take on a lot. _But you need to be very detailed in what_ | KT among international partners regardless the firm size such as (CD1) higher cost of entry and hinders the transmitting the core competencies to the international markets; (CD2) facing operational challenges in terms of lacking common understanding of norms, values and motivation; (CD3) misunderstanding between foreign partners can limit sharing of critical organizational knowledge (Wijk et al., 2008) | cultures can be the barrier for KT between organizations (CD2,3) as a result to reduce cultural barriers for KT activities their organization is conducting cultural awareness trainings that helps to increase common understanding, values, and motivation of KT actors. This is in line with the literature. Aspect to (CD1) was not mentioned.
b) The informant provided supportive information of the existing literature that different cultural operational challenges can appear in various cultures during KT activities (CD2,3): in societies with higher power distance, which is associated with a strong hierarchical subordination and less likely to ask question or | **awareness workshops facilitate common understanding of norms, values and motivation among KT actors (CD2)**

Strong hierarchical subordination in organizational structure of companies that exists in various societies can be a barrier for KT (CD1) |

			Language skills in different cultures are main barrier for KT
critique those of higher hierarchical positions and interactions between hierarchical levels tend to be more authoritative and paternalistic, as a result KT can be hindered.		The national cultural distance arises different barriers for the KT among international partners regardless the firm size such as (CD1) higher cost of entry and hinders the	a) The informant supports the (CD2) and (CD3) aspects in the literature that cultural differences can hinder KT due to
them to do. And if you want to get their ideas, you need to ask for them. They won't just say, well, you know, idea here what you say but you're completely wrong, that will never fly. You need to think about this. And that was absolutely a mistake from my side. I was throwing out ideas. They were just doing what I told them. I expected to get some feedback as well. So, certain part of the world. You need to be very specific and down to detail what to do and when. And you will get exactly what you asked for. And nothing more, nothing less. That's come to work ethics then the culture side. With some cultures you need to be direct. So, in Scandinavia you go around and round and round until we come to the point. But, for instance, if you took people from Germany, from US, you need to be direct and it's okay. It won't be frowned upon if you are to direct. So, that's something also that you need to consider, and you shouldn't get insulted if they aren't yet"	Informant 7 (Large company)	Question: In your experience, when you have interacted with people from different countries, what success (hits) and mistakes did you have?	

				"*I have interacted a lot with Japanese people when I worked, maybe I'll also a lot with German, where I worked with customers and operators. And I tried to joke a lot. And I've learned that doesn't fly at all. <u>If you don't speak your native language to start with them in Japanese people kind of struggle with English.</u> Maybe not so much for Swedes. Germans maybe as well. But then you also have the different cultural backgrounds. We don't really have the same like references and so on. So maybe be more, I shouldn't say formal, but maybe just a bit, a bit boring to increase the success rate. <u>So I did try to joke and it kind of just went over their head and then it's just pointless. You make them feel a bit dumb and I don't feel right basically</u>*".	transmitting the core competencies to the international markets; (CD2) facing operational challenges in terms of lacking common understanding of norms, values and motivation; (CD3) misunderstanding between foreign partners can limit sharing of critical organizational knowledge (Wijk et al., 2008)	misunderstanding including language barrier, which implies that language barrier can be reduced if organization transmits the knowledge in simple form. Informant also suggests that understanding of uniqueness of cultures through adjusting communication can facilitate KT activities. The aspect related to (1) is not discussed by informant. b) Informant provided examples that consistent with the literature regarding cultural aspects of China and Japan, where strong hierarchical subordination exists that affects KT activities.	**between foreign partners (CD2,3)** **Strong hierarchical subordination in organizationaI structure of companies that exists in various societies can be a barrier for KT (CD1)**
			Informant 9 (Large company)	Question: In your experience, when you have interacted with people from different countries, what success (hits) and mistakes did you have?	The national cultural distance arises different barriers for the KT among international partners regardless the firm size such as (CD1) higher cost	The informant suggests that different cultures can bring different operational challenges due to norms, values	**KT can be hindered in different cultures due to lacking**

				"In the US, typically, there is one management agenda and won't matter much, everyone is kind of aligning to that, knowing silently that we won't be able to deliver on that time. So, but we deal with that later when we get there. In Scandinavian (countries), it's more like being open, telling up front as soon as possible as you know that you have a problem of some sort. You should share it (failure) immediately with your management. And they will, normally, depending on the person, try to help you what they can do to support you in solving it. *...you have to deliver something at someday...* *So, there's different as countries, whether with command or control...* *... But you need to learn how to maneuver between different cultures. So, it's not like my culture is going to make it fail or something.* *... And you can have different approaches. I mean, in organization needs to be transparency towards my counterparts. I would ask them for recommendation how to proceed with this. And if they are part of the problem, I will probably go into my organization to my level, go back through what we call the key accounts and have them*	of entry and hinders the transmitting the core competencies to the international markets; (CD2) facing operational challenges in terms of lacking common understanding of norms, values and motivation; (CD3) misunderstanding between foreign partners can limit sharing of critical organizational knowledge (Wijk et al., 2008)	(CD2): Scandinavian counties prefer to know about possible problems in advance, while the US partners have higher risk-taking levels and no need to explicitly share the knowledge about possible failure. This implicitly means that the US partners have trust that failure will be solved by dedicated employees (key accounts from those culture) partner due to a common agenda (CD3). Moreover, the US partners prefer to have KT activities with organizational representative that have the top-level position at their organization, which can provide a successful partnership and KT activities. Aspect related to (CD1) is not mentioned. This is consistent with the literature.	**common understanding of norms, values and motivation (CD2,3)** **Companies need to learn how to adapt in different cultures for better understanding (CD3)**

			KT can be hindered in different cultures due to lacking common understandin g of norms, values and motivation (CD2,3)
		Informant supports differences in cultures can lead fewer common understandings and each cultures have their own rules that can be barrier for KT activities (CD2,3). (CD1) was not mentioned The informant suggests that cultural distance for KT can be reduced by frequent interaction via training meetings and informal relations after formal job meeting. Moreover, the	
	address it at some other level. And since the key accounts are part of the culture in US so to speak, or in Europe or in Asia. They normally know how to handle it and they normally will advise you on how to move around in this. In Scandinavia, I would probably go more directly to the people. Maybe we can also sit, have a round table session. Which you cannot do in other cultures because you sometimes oppose cultures because, sometimes, the boss is the boss"	The national cultural distance arises different barriers for the KT among international partners regardless the firm size such as (CD1) higher cost of entry and hinders the transmitting the core competencies to the international markets; (CD2) facing operational challenges in terms of lacking common understanding of norms, values and motivation; (CD3) misunderstanding between foreign partners can limit sharing of critical organizational knowledge (Wijk et al., 2008)	
Informan t 4 (SME)	Question: In your experience, when you have interacted with people from different countries, what success (hits) and mistakes did you have? *"I one mistake that I think about is that you often or at least me myself, I often think that the other person is just like me and that they have the same culture and the right thoughts about what's right and wrong and so on. That's a recurring mistake. I think. What success I think. Again, meeting in real life, it's good in business and otherwise. So, when I have been in, in business, travels and I think it's, it's good to meet the persons outside the business room. So, to do something fun. After the work time.*		

			informant suggests that informal relationships in terms of establishing social ties with the partner can also enhance KT and reduce cultural distance.	**Training cultural awareness workshops facilitate common understandin g of norms, values and motivation among KT actors (CD2)** **Strong hierarchical subordination in organizationa l structure of**
			The national cultural distance arises different barriers for the KT among international partners regardless the firm size such as (CD1) higher cost of entry and hinders the transmitting the core competencies to the international markets; (CD2) facing operational challenges in terms of lacking common understanding of norms, values and motivation; (CD3) misunderstanding between foreign partners can limit sharing of critical organizational knowledge (Wijk et al., 2008)	a) The informant confirms that different cultures for KT can bring operational challenges (CD2) and reducing cultural barriers for KT their activities their organization is conducting cultural awareness trainings. b) The informant supports that in societies such as China, there is strong subordination exist and as a result this can serve as KT operational barriers (CD2,3), therefore the informant
		...You never stop learning. And it is very different cultures even in Europe. So, like it's very different to work with companies in Finland and in Germany and so on. You will have different cultures and different rules. And I think it's really fun to get to know the differences as well. You can share the differences in meetings. And you both think it's fun that we are we are different. So that's, yeah, that's what I have learned"		
		Informan t 5 (SME)	Question: In your experience, when you have interacted with people from different countries, what success (hits) and mistakes did you have? a) *"I worked a lot with Sony Eriksson. I worked with China, Beijing and I attended a course when I went to a product leader, you need to know, you need to understand the culture and how different is it from Sweden to other countries I mean, we attended a course where they explain how Europe works, how Europe, Asia, and different kind of worlds. How you pronounce things, how you express yourself. So, you need to know how you as a leader or manager, what kind of role you have, probably Product Manager, what kind of express yourself, what kind of things*		

				you request and how you pronounce it, how you express yourself in a meeting, and what do you want out of things" *b) "…And, for example, I was working for a factory in China because we were producing mobile phone. And we are in Sweden, we talk to the people in the project manager and the team. And we had some opinions, want to change things. And then they are misunderstanding people, they had to leave because they thought they had done some mistake, and it wasn't a mistake. We want to correct things. And in Sweden you can talk to people, but there they fire people right away because we had said something. I had a bad experience, but then we corrected it, but we got to learn that way. You have to be careful how you pronounce thing, how you say things, we wanted some small correction and then they misunderstand us completely. I did a lot of experience there how to work with different culture in this way with China"*		suggests that it is important to know how you articulate the knowledge that you are going to transmit. This is in line with the literature.	**companies that exists in various societies can be a barrier for KT (CD2)**

				Lack of understanding the decision-making process in different cultures can hinder KT activities between partners. Trust is important aspect for KT activities between foreign partners
Informant 6 (SME)	Question: In your experience, when you have interacted with people from different countries, what success (hits) and mistakes did you have? a) *"So, I think when I work with international companies. Before starting to work with other people. I tried to understand their motivations and how they make decisions, how they get interested in things. So how they are motivated. So, this varies from different cultures. So, for instance, in Japan, when I work with them, I did this hierarchical structure to get attention from, from manager's managers. And then I found an individual that I worked with. And in American culture was quite similar, German as well. But they have different requirements and how you should have a dialogue with them. So, Germans and Japanese people, they are very structured and like structured information. We're, whereas American. They love stories. I just loved the success stories"* b) *"I mean, I'm sure that you have some Maslow equivalence of trust. So, you need to build before you can do anything. You need to have trust because that's the foundation of the*	The national cultural distance arises different barriers for the KT among international partners regardless the firm size such as (CD1) higher cost of entry and hinders the transmitting the core competencies to the international markets; (CD2) facing operational challenges in terms of lacking common understanding of norms, values and motivation; (CD3) misunderstanding between foreign partners can limit critical sharing of organizational knowledge (Wijk et al., 2008)	a) Informant agrees that motivation (CD2) and the process of decision making vary depending on cultures. The informant suggests that there is different communication culture are in place in different societies (CD3), where Germans (low hierarchical subordination) and Japanese (strong hierarchical subordination) have similarities in terms of more structured information. In contrast, the US partners prefer success stories that helps to better absorb the transferred knowledge. b) This is inconsistent with the literature. The informant suggests that in low power distance societies such as Sweden are important	

		to establish trust (CD2) for KT, while in cultures with higher power distance (Japan), trust to perform KT is built due to order from the higher-level managers. But Sweden belongs to lower PD. So, this implies that establishing trust as common motivation to transfer knowledge between foreign partners. Aspect (CD1) is not mentioned.	**Lack of understanding the decision-making process in different cultures can hinder KT activities between partners.**
	house you need to build. And that comes with every people that you interact with. And trust can be built in different ways. Many Swedes they build trust from a good interaction, qualities and good interpersonal qualities, good dialogue qualities. Whereas other cultures (Japan), they will trust from while you asked my manager to tell me this and then I will do it. So, it differs very much because people in other cultures, they may build a trust on what their managers say that they should have trust in. But as everything starts with trust"		
Informant 8 (SME)	Question: In your experience, when you have interacted with people from different countries, what success (hits) and mistakes did you have? *a) "I would say that my biggest learning is that to have respect for different cultures. Because how you work is usually very different in how you run your daily business. So I would say that's a general learning is to kind of figure out the actual culture of the country that you're that you're working. And not just focusing on maybe doing*	The national cultural distance arises different barriers for the KT among international partners regardless the firm size such as (CD1) higher cost of entry and hinders the transmitting the core competencies to the international markets; (CD2) facing operational challenges in terms of lacking common understanding of norms, values and motivation; (CD3) misunderstanding between	

				things as you normally do and communicating. *b) ... I mean, but to try to figure out the culture of how you run things and how you take decisions. How you operate in that in that country. Or from all from a project, that project perspective to a more how the hierarchy is set up. If it's very, in some countries this more formal and in some countries it's more informal on how you take decisions.* *...I think that it's usually difficult to learn in advance. But I think you need to maybe go into the projects with an open mind in trying to figure out how to help to navigate, to set up the structure for the project in everything from how you take decisions to how you run meetings and how will you follow up on actions.* *...I've been working in a large project. It's almost 10 years ago now in an outsourcing project with the US and India. And there it was difficult because everything as soon as I spent a lot of time in India then and there it works so much that everything was fine when you are present. And then they started running the project their way when you left. And then I was under the impression that we still had agreed to certain. But as soon as you stopped*	foreign partners can limit sharing of critical organizational knowledge (Wijk et al., 2008)	decisions has been taken depending on formal or informal way in particular society. Implicitly, this suggests that due to lacking understanding of norms, values and motivation in different cultures can be a barrier for KT process between international allies (CD2,3). Aspect (CD1) is not mentioned.	

Dimensions/ Characteristics	Aspects that affect KT	Subcategory of the aspect	Primary data source	Quote from the informant	Relating literature	Conclusion/Analysis	Summary
				following up, they had their own way of taking decisions and deciding priorities, etc. And I'm not saying that their parties were wrong, but they just they didn't have a way of communicating that within the project"			
Relational characteristics	**Geographical distance/proximity**	**Geographical proximity**	Informant 1 (Large company)	Question: In a globalized context of operation, does physical proximity or distance affect your collaboration with other companies, especially small and medium sized partners? In what way? a) *"… I mean, of course, I think it always matters actually because it's always simpler to be where the customers are, or it is depending on what you're going to like discuss. But yeah, of course it matters. But in the in the world, we live in, we have to kind of adapt to not being side-by-side all the time. So that's why I think it's important to kind of come up with what helped both the partners and our company in that collaboration.*	Geographical distance negatively influences inter-organizational KT especially for tacit knowledge exchange but can be mediated in terms of (GD1) developing trust-based relations between distant partners (e.g., Capaldo & Petruzzelli, 2014); (GD2) organizational proximity in terms of common corporate culture and values (Korbi & Chouki, 2017); and (GD3) bilateral way of utilizing of communication technologies (Choi & Contractor, 2016)	a) The informant argues that proximity to partners is important, but also, she mentioned that it is important to adapt in new reality (implicitly, pandemic time) and find solutions how to help each other (between partners) for better collaboration for KT activities. Aspect (GD1) was not explicitly described by the informant such as trust-based relationships but rather	**Adaptability capacity between partners is important for geographical distant partners** **Physical proximity is more prevalent than organizational proximity in terms of**

				... *(2) I think so. I mean, we have a big part in the US, for instance, and I know that the people in our company, then we have like individuals that are very involved in different like the key accounts. I mean, I'm not talking about the key accounts' manager, So stuff like that. But all the bigger partners that we have are the big hotel groups, we have dedicated people to talk to them a lot and maybe try to see them and help them out with whatever it is. And I think that is very important to have at least a few people that know the customer and know what they want. And now they're using the systems. Because it's hard for us in Sweden to know exactly what America or China or India what needs they have, for instance. But I think we have to cope with it. ...I think, for instance, the last year, during the pandemic. I think our company had an ease transition from being located or working in the office to the online working when we're doing mutually sits where it is. It has been a very easy transition because we kind of use it. We have like partners and companies in China. We have US, Australia. Yeah. I mean, all over the place, we are used to it. But, of course, it's better to be in-person.*		it is more important to have adaptability capacity for partners for better KT activities. Aspect (GD2) was not supported by the informant stating that large organizations have dedicated personnel to easily reach the partners in a specific location of the world for learning and exchanging knowledge with those partners. Implicitly, this implies that organizational proximity is still important to transfer tacit and complex knowledge. (GD3) was supported by the informant who argues that in pandemic time the company could easily switch to online communication (implicitly, by heavily applying modern interaction technologies) but still	**corporate culture and values**

			Willingness for interfirm KT between partners affect more rather geographical proximity
		argues that face-to-face communication is better for KT activities since it is still difficult to know exactly foreign partners needs.	a) The informant suggests that proximity is not important for KT activities during the collaboration, but states that is more important to be motivated for collaboration rather geographical co-geographical presence. Implicitly, this means willingness is important aspect for interfirm KT. GD1 is not explicitly mentioned. The aspect (GD2) is not discussed by the informant, but the informant argues that geographical distance is more challenging due to time difference.
		Geographical distance negatively influences inter-organizational KT especially for tacit knowledge exchange but can be mediated in terms of (GD1) developing trust-based relations between distant partners (e.g., Capaldo & Petruzzelli, 2014); (GD2) organizational proximity in terms of common corporate culture and values (Korbi & Chouki, 2017); and (GD3) bilateral way of utilizing of communication technologies (Choi & Contractor, 2016)	
	Informant 2 (Large company)	Question: In a globalized context of operation, does physical proximity or distance affect your collaboration with other companies, especially small and medium sized partners? In what way? a) *"I would say physical proximity. It doesn't affect collaboration edges. That it's a bit tricky if you work with Australia, the time differences…when I was in Canada that we had exactly 12 hours with China. So, eight in the morning was eight in the evening for them. Somebody would not be in the right time. Not in the regular day time, right? It makes it a bit challenging, but you need to, and of course, if everybody's in one building, then it makes it easier. I don't think it affects collaboration because collaboration means people are willing to do things. So, I don't think people are either will or unwilling to do things if they're far*	

				apart. It's just that it makes it more difficult to do it at the same time, but it doesn't mean they're unwilling to do it.		GD3 is not explicitly mentioned.	
Relation al character istics	Geogr aphica l distan ce/pro ximity	Geogra phical distanc e	Informan t 3 (Large company)	Question:In a globalized context of operation, does physical proximity or distance affect your collaboration with other companies, especially small and medium sized partners? In what way? a) *One challenge absolutely is the time difference.* So that is can be a drawback and also can make so that things take much longer than they should. So for instance, if we need to get a collaboration between US and China, It's hard. One of them needed to stay in the office after hours or someone on yet a sudden it to get up three in the morning to attend. And you don't want to do that much either. So that's a challenge. *And then if you had a couple collaboration between R&D and the supplier. You have just two few hours overlap and then you need to wait until next day before you get feedback. And*	Geographical distance negatively influences inter-organizational KT especially for tacit knowledge exchange but can be mediated in terms of (GD1) developing trust-based relations between distant partners (e.g., Capaldo & Petruzzelli, 2014); (GD2) organizational proximity in terms of common corporate culture and values (Korbi & Chouki, 2017); and (GD3) bilateral way of utilizing of communication technologies (Choi & Contractor, 2016)	a) The informant suggests that geographical distance brings challenges due to time difference and supplier cost. The informant explicitly did not mention (GD1,2,3)	

363

					then you come, which often you that answer and then you wait till the next day. And that is also a challenge. And then of course, shipping. It's absolutely a challenge now. And we see, for instance, having a manufacturing site in China had their benefits before when it comes to cost. Today, that is maybe not the case. The situation has changed. So for instance, if you look at Eastern countries in Europe, you could have not, not much more expensive labor cost compared to China. But from a logistics point of view, it's much, much easier. You don't have lead time with which ship, right? And therefore, it can be very costly if you're moving heavy objects. Hardware wise. So that's also absolutely from that point of view, when it comes to medium-sized and small, have nothing too late to that aspect of it"		
			Informant 7 (Large company)	Question: In a globalized context of operation, does physical proximity or distance affect your collaboration with other companies, especially small and medium sized partners? In what way? *a) "If you have asked me this a year and a half ago, I would probably give it a different answer than today. Because we are really practice this global and*	Geographical distance negatively influences inter-organizational KT especially for tacit knowledge exchange but can be mediated in terms of (GD1) developing trust-based relations between distant partners (e.g., Capaldo & Petruzzelli, 2014); (GD2) organizational proximity in	a) This perspective is partially in line with the literature. The informant suggests that personal relationships are still very important for geographical distance partners, specifically when you establish new	**Physical proximity in terms of face-to-face meetings with distant partners facilitates to build a trust**

Dimension	Informant	Data extract	Interpretation	Key finding
		…far away communication. And I think we have really improved and even become a bit tired of it as well. Using this kind of worn-out phrase, this is the new normal. So, I still see a great value in meeting people in person, definitely in the phase when you start to know them. And that's true for a mean, personal relationships, but also professional relationships. …So travel might be required in the future, but follow-up meeting I think this kind of Teams meeting that we do now is very efficient and it's better for the environment as well	…terms of common corporate culture and values (Korbi & Chouki, 2017); and (GD3) bilateral way of utilizing of communication technologies (Choi & Contractor, 2016) relationship is essential to have face-to-face meetings that implicitly can be described as to establish trust for better transfer of the knowledge with new partners (GD1), while follow-up meetings (after establishing a trust) can be effectively performed via online communication tools that are efficient and environmentally friendly (GD3). GD2 was not mentioned by informant.	**for interfirm KT** **Digital communication tools can enhance KT between distant organizations but as a complementary mechanism after establishing face-to-face meetings**
Geographical proximity	Informant 9 (Large company)	In a globalized context of operation, does physical proximity or distance affect your collaboration with other companies, especially small and medium sized partners? In what way? a) *"I think we have learned in the last year that we can solve most by actually teamwork and so on. And emails are good for nothing really. E-mails are just for exchanging the basic and then you meet either on a meeting like this (online). But occasionally you read, we*	Geographical distance negatively influences inter-organizational KT especially for tacit knowledge exchange but can be mediated in terms of (GD1) developing trust-based relations between distant partners (e.g., Capaldo & Petruzzelli, 2014); (GD2) organizational proximity in terms of common corporate culture and values (Korbi & a) The informant argues that proximity and face-to-face meetings can facilitate to transfer knowledge related the company's technology and products and solve different emergent situations. This implies that tacit knowledge and related complexity can be better	**Physical proximity in terms of face-to-face meetings with distant partners facilitates to build a trust for interfirm KT**

	transmitted to other partners only via real physical interaction. Implicitly, this means that trust-based relationships are important aspect that can be built via physical proximity (GD1). This is perspective is inconsistent with the literature (GD1). Aspects (GD2,3) were not mentioned.
	Chouki, 2017); and (GD3) bilateral way of utilizing of communication technologies (Choi & Contractor, 2016)
at least we have to meet the customer, we have to sit face to face with our major customers or our partners to mostly why not is to avoid an emergent situation going on. It is required from a technology point of view, actually to be able to do some real life testing of what we have develop. We would need to have be physically present, but that's because we just cannot connect these two body parts over the Atlantic, we have to actually be physically present. But physical proximity can affect our collaboration both from a technical point of view, actually testing out, doing, finding out the product. This is the customer see the problems that I have physically understand them b) *"...And also that meeting the customers in order to sort out situation and have kind of a face-to-face talk. And very often that happens around two to three times a year. We tried to do this and around where the big exhibitions are, where the big market events are. But we also do it on ad hoc basis, when require five people and fly them out to part in Asia and talk to the major customers..."* *... It is more, sometimes more difficult to get the perception of people and how they interact, body language and all*	

Geographical proximity	Informant 4 (SME)	*these kinds of things. And see the premises which are about to buy when you are on a meeting like this, or if you have to trust other party that are close to them"* (In a globalized context of operation, does physical proximity or distance affect your collaboration with other companies, especially small and medium sized partners? In what way?) a) *"I think it's harder to, to cooperate with a large distance. Because if you really want to solve something, it's easier if you can go and meet in person. So that's the thing that I think is hard and the culture in some way but, but not necessary. So the main thing I think is that if you or when you need to meet the distance can be a problem"* Follow up question: Now in this pandemic time, how do you think this distance can be reduced for better knowledge sharing?) *"I think this digital improvement or usage that we have learned now this year is very good because it's so easy to*	Geographical distance negatively influences inter-organizational KT especially for tacit knowledge exchange but can be mediated in terms of (GD1) developing trust-based relations between distant partners (e.g., Capaldo & Petruzzelli, 2014); (GD2) organizational proximity in terms of common corporate culture and values (Korbi & Chouki, 2017); and (GD3) bilateral way of utilizing of communication technologies (Choi & Contractor, 2016)	a) The informant supports that geographical proximity and face-to-face interaction facilitate KT activities between partners, while distance can be a barrier for knowledge sharing. Implicitly, this means that physical proximity allows building trustworthy relations between partners. (GD1,2). b) Aspect (GD3) was confirmed by the informant that digital communication tools can enhance KT	**Physical proximity in terms of face-to-face meetings with distant partners facilitates to build a trust for interfirm KT** **Digital communicati on tools can enhance KT between distant organizations** .

				get to decide about the meeting and defined. And it's easier to plan since you can adjust according to their schedule if they have it online in other words, to find an available slot in the schedule of the other person (online)"		between distant organizations.	
Informant 5 (SME)	Question: In a globalized context of operation, does physical proximity or distance affect your collaboration with other companies, especially small and medium sized partners? In what way? a) *"We work with a company that we bought in Netherland. I haven't met them before. I think, <u>If you get to know people that you meet them face to face one or two times. I mean, it will always be easier to communicate.</u> <u>You have you build a kind of relations. I always try to meet people for the first-time.</u> That's why we went to China, for example, to meet people to be that they thought, <u>then it was much easier than to have daily meetings where you think, you know, how they react.</u> If they don't react, you know, how they pronounce thing, you understand more. I think that a good way to learn each other"*	Geographical distance negatively influences inter-organizational KT especially for tacit knowledge exchange but can be mediated in terms of (GD1) developing trust-based relations between distant partners (e.g., Capaldo & Petruzzelli, 2014); (GD2) organizational proximity in terms of common corporate culture and values (Korbi & Chouki, 2017); and (GD3) bilateral way of utilizing of communication technologies (Choi & Contractor, 2016)	a) The informant confirms that geographical distant partners need to establish face-to-face meetings for creating some relations. This implies that distance can be reduced via establishing personal relations to build trustworthy relations (GD1). b) This perspective is partially consistent with the literature. The informant supports that digital communication tools can enhance KT (GD3) between organizations but as a	**Physical proximity in terms of face-to-face meetings with distant partners facilitates to build a trust for interfirm KT** **Digital communication tools can enhance KT between distant organizations but as a complementary mechanism**

		b) (So we can summarize, when you don't have a relationship, if it is a distant partner, it's better to see and have a face-to-face meeting to establish some relationship. And then you can move to online version or remote version to decrease this, reduce this distance am I right?) I: Yeah.		complementary after mechanism establishing face-to-face face meetings.	**after establishing face-to-face meetings**
Geographical proximity	Informant 6 (SME)	Question: In a globalized context of operation, does physical proximity or distance affect your collaboration with other companies, especially small and medium sized partners? In what way?) a) *"Physical proximity has an effect. When I worked in an international company. I always tend to build the trust part by being present with the person, so the person could get a grasp of what I am. Because when you do a team's discussions like we're doing now, I met you Aizhan, so I know more about you. But there are so much more in your communication that can be done over a physical meaning, the body language, the way we interact in social situations. That we become that come from the Savanna, right? We did not have video conferences on the Savanna. We worked as a group or as a group of people. And there were trust*	Geographical distance negatively influences inter-organizational KT especially for tacit knowledge exchange but can be mediated in terms of (GD1) developing trust-based relations between distant partners (e.g., Capaldo & Petruzzelli, 2014); (GD2) organizational proximity in terms of common corporate culture and values (Korbi & Chouki, 2017); and (GD3) bilateral way of utilizing of communication technologies (Choi & Contractor, 2016)	a) Informant suggests that geographical proximity is important because it can facilitate trust to establish trust between partners for better learning capacity. This implies that to establish trust first it is needed to have physical proximity in terms of face-to-face interaction and then it can move to distant partnerships (GD1,2) b) This viewpoint is partially consistent with the literature. The informant suggests that digital communication tools cannot fully substitute face-to-face	**Physical proximity in terms of face-to-face meetings with distant partners facilitates to build a trust for interfirm KT** **Digital communication tools can enhance KT between distant organizations but as a complementary mechanism**

				between these people. And that's due to many aspects of what we did on the Savanna. What people knew what they had competence in and what if they were strong, if they were good then in crafting. If they were good speakers, they were good in finding water or whatever. That's something you gain from physical proximity" (How do you think, for example, if you work here, but I know that you have businesses in the United State or a client from, let's say in other countries. How do you see how you can decrease this distance effect for collaboration or knowledge exchange, especially for large partners?) *b) "I think I would have very hard to trust another people to full extent if I've only had one hour team's conference with that person at hand. So, I need to meet with the person to have full trust. So, I would say that yes, you need to meet at some point during your cooperation. Or you should have a longer session with some interactive social tools. Maybe you should show them around and your house, how you live. Maybe you should show your kids, or I mean to build deeper trust"*		interaction for KT due to difficulties to establish trustworthy relations, but the informant supports that trust can be built if online meetings can be done longer period and having other topics such as related to family (GD3).	**after establishing face-to-face meetings**

							Geographical distance/prox imity does not influence interfirm KT
				Informan t 8 (SME)	Question: In a globalized context of operation, does physical proximity or distance affect your collaboration with other companies, especially small and medium sized partners? In what way?) a) *"I would say that it doesn't affect the collaboration with large organization so much because they're usually so widespread already from the beginning. I think my experience is that the proximity or physical distance isn't so important any longer. Even if it's nice to meet them from time to time, you still don't you usually don't have possibility to physical proximity to tool or parts of their organization. So you kind of learn how to deal with that on a distance"* Follow up question: Say you had a partner that is maybe five minutes away with car compared to having partners in USA or Asia? How would just say that distance affects the collaboration?) b) *"I think that depends a bit on what type of organization it is. If it's just a company where you work with on a more abstract level, with services or with general cooperation. And, you don't have so much physical on hand production, for example. Then I don't*	Geographical distance negatively influences inter-organizational KT especially for tacit knowledge exchange but can be mediated in terms of (GD1) developing trust-based relations between distant partners (e.g., Capaldo & Petruzzelli, 2014); (GD2) organizational proximity in terms of common corporate culture and values (Korbi & Chouki, 2017); and (GD3) bilateral way of utilizing of communication technologies (Choi & Contractor, 2016)	The informant suggests physical proximity is not essential for KT activities and confirms that large organizations inherently have KT with distant partners due to widespread operations. The informant supports that physical distance can influence knowledge transfer in the case of manufacturing of physical products where tacit and complex knowledge inherently exists and by proximity can be easily transferred. However, for the service sector proximity does not have any importance. Informant did not specify aspects (GD1,2,3)	

think that I would have visited them so much more, just even if they were very close. *It's more of important once you actually, work with physical products* for example. We have our largest manufacturing sites *with the supplier just about an hour away by car. And that's really good for us because we can always go and help them*. If they have issues in production, then *it is easier for us to go there and help them to do the search and trying to figure out what's wrong and how to fix it and collaborate with them in setting up the testing equipment*, etc. But, so that has a positive a side effect in having a shorter physical distance, then for example, *looking and moving some of that production to an facility in China. And then of course it's a directly becomes more difficult to cooperate and figure out things from afar when you're actually not there and can see things*. But if it's just general corporations, just meetings and calls and documentations and systems, then it doesn't matter. *From my perspective, the physical distance doesn't have so much impact"*

Appendix 4 – The coding of secondary data

Dimensions/ Characteristics	Aspects that affect KT	Subcategory of the aspect	Secondary data source	Paragraph from the secondary data	Relating literature	Conclusion/Analysis	Summary
Characteristics of the relationship	Social ties	**(SOCT1)** How to establish new relationships and **(SOCT2)** How to maintain relationships	(LC) Document: External communications guideline 2020	*"The Disclosure Policy: (**SOCT1+2**) Who may act as a spokesperson for the (large company)"* *"Communication to media and other stakeholders shall always be planned and carried out in line with the (**1.SOCT1**) Group's External Disclosure Policy and Crisis Management Policy"*	Strong social ties acts as a facilitator for knowledge transfer (Kang & Sauk Hau, 2014) and Liu et al. (2015) argues that there are three steps to establishing a high quality relationship, (**SOCT1**) first step involves to locate a partner firm and identify its key contacts, (**SOCT2**) secondly to develop and maintain the relationship established, last stage involves dispatching of knowledge from the source along with knowledge	The secondary data state that who may act as (**SOCT1+2**) a spokesperson for the LC, this indicates that the LC utilize a role of socialization to establish and maintain social ties. The secondary data state that to engage in external communication there are strict guidelines to follow, (**1.SOCT1**) this implicitly implies that the LC follows very formal means to establish social ties.	Social ties are established and maintained by a role of socialization in the large company (**SOCT1+2**) The LC establish social ties by formal means with SMEs (**1.SOCT1**)
			(SME)	-			

Characteristics of the relationship	Social ties	(SOCT3) Which social tie does the organization prefer, informal or formal relations	(LC)	-	Social ties according to Tangaraja et al. (2016) is a relationship based concept, its either conducted directly or indirectly	
					absorption of the recipient, which is how the knowledge is exchanged.	
			(SME)	-	between two parties and differentiated by (SOCT3) informal or formal ways.	

Dimensions / Characteristics	Aspects that affect KT	Subcategory of the aspect	Secondary data source	Paragraph from the secondary data	Relating literature	Conclusion/Analysis	Summary
Organizational characteristics of the recipient and the source	Absorptive capacity	(ABS1) Implementation of acquired knowledge (ABS2) Diffusion of	(LC) Document: Avenue News Global Insights	*"We are launching the Insights Database in an effort to increase the sharing of knowledge between divisions and business units,"* says Klas Rydstrand from the Global Innovation Management team. *"The Gateway process and generation planning process are just two*	The literature of absorptive capacity agrees that it consist of aspects such as (ABS1) intra-firm knowledge transmission	The secondary data shows that the LC/MNC (ABS1) implements acquired external knowledge into their system called "Insights Database", (ABS2) which is also how the	LCs rely on implementing acquired knowledge from individuals to the organizationa

		knowledge, internally	Database makes knowledge-sharing easier	*examples where insights are critical to make fact-based decisions.*" The **(ABS1+2)** Insights Database's content is categorized in four groups: **Customer insights** – Insights from Voice of Customer (VOC) research conducted by ASSA ABLOY and its entities. Discover what existing and potential customers say about us and our products and based on this, what business opportunities can be identified. **Market insights** – Marketing intelligence including internal market studies, external market reports, and more. **Technology insights** – Knowledge and research related to critical technologies such as sustainable energy solutions, wireless connectivity, software and identification. **News** – Handy updates on competitors, partners, customers, tradeshows, standards and more.	because after acquiring knowledge it needs to be **(ABS2)** diffused within the firm, which is how they are these aspects are interrelated (De Luca & Cano Rubio, 2019).	knowledge is diffused in the organization. This is explicitly consistent with the literature.	l level in systematic data bases, where it is then diffused in the organization (**ABS1+2**).

			(LC) Docume nt: Avenue Identify knowled ge gaps	*"If we work systematically to identify and close knowledge gaps our decision making is improved, resulting in solutions that delight customer, excellent product quality and efficent projects. (**ABS2**) A project kick-off meeting is an excellent opportunity to do some brainstorming and leverage the team members' experiences to build a good initial set of knowledge gaps. The team can immediately build the tracking, how and knowledge capture systems to manage the gaps."*		The secondary data shows a method of (**ABS2**) disseminating knowledge in the organisation, this is consistent with the literature.	LCs diffuse acquired knowledge amongst the employees with meetings and projects (**ABS2**)
			(SME) The official website of the SME.	The official website of the SME contains a Knowledge Bank, (**ABS1+2**) where various knowledge is accessible regarding services, manuals, recordings, products, customer articles and so on.		The secondary (**ABS1+2**) illustrates that the SME disseminates knowledge via their knowledge bank, this is consistent with the literature.	SMEs rely on implementing acquired knowledge from individuals to the organizationa l level in systematic data bases, where it is then diffused in the organization (**ABS1+2**).

Dimensions/ Characteristics	Aspects that affect KT	Subcategory of the aspect	Secondary data source	Paragraph from the secondary data	Relating literature	Conclusion/Analysis	Summary
Organizational characteristics of the recipient and the source	Motivation	Which facilitate knowledge exchange with partners more **(MOT1a)** intrinsic or **(MOT1b)** extrinsic motivation. **(MOT2a)** Intrinsic and **(MOT2b)** extrinsic motivation from the organization.	(LC) Document: Avenue News Global – Motivation is the key to individual and collective growth	*"For Mike Ong, Opening Solutions Country Director for the Philippines, growth is a crucial factor in achieving your goals, and motivation is the key to that growth. "You must keep motivating, despite the challenges," he says."* *"Last year was a year of limitations. Interactions with customers and each other, business conversations, and even fulfilling our overall objectives became more difficult than ever. For Mike's team, employee morale was low, and anxiety was high. Covid cases on the team meant providing strong emotional support to the sick and the healthy. These limitations had the potential to devastate Mike's team. "The Philippines has always been a growth engine for the SEA region, but 2020 was a bad year." Instead, (MOT1a) Mike motivated his team through those challenges and is now looking forward to driving initiatives put in place to not only bounce*	Martín Cruz et al. (2009) state that motivation consists of **(MOT1a+2a)** intrinsic and **(MOT1b+2b)** extrinsic motives, where the former takes place inside an individual, affected by per example pleasant work environment, atmosphere of mutual respect, accomplishment, self-respect, prestige, personal values and involvement. As for the latter, extrinsic motivation according to the authors is external motives which affects the individual, by per example monetary rewards that comes in direct or indirect	The secondary data state that employees are motivated by the challenges ahead, this implies that **(MOT1a)** intrinsic motivation in the form of accomplishment is a driving factor for the organization. This is consistent with the literature.	LCs are more incentivises by intrinsic motivation in the form of accomplishment to acquire new knowledge from partners and sharing knowledge with them **(MOT1a)**.

Source	Raw data	Interpretation	Finding
	ways, whereas the former relates to wages, incentives or bonuses and the latter relates to time not working, job education, health benefits or allowances.	The secondary data states that the organization (**MOT2b**) offer extrinsic motivation in the form of job education. This is consistent with the literature.	LCs offer direct extrinsic motivation in the form of job education to the employees to acquire and share knowledge (**MOT2b**)
(LC) Document: External communications guideline 2020	*forward, but to grow. "It's about the team working together. If each member grows, together we grow."* "*All spokespersons must participate (**MOT2b**) in a formal media training*"		
(SME) https://www.phoniro.com/sv/kunskapsbanken/modellkommuner-for-aldreomsorgens-digitalisering/	*(Translated from Swedish)* "*(**MOT2a**) The 1st June 2020, Swedens municipalities and regions organization (SKR) appointed ten municipalities that shall act as models for digitalization of elderly care. These role models will together with SKR support other municipalities with knowledge regarding digital services and welfare technology. Seven of these municipalities are customers to us (the SME) and we are really proud over this. Please, take inspiration and tips from these municipalities about how your municipality can strengthen the digitalization of elderly care. If you wish to contact*	The SME offer (**MOT2a**) intrinsic motivation in the form of accomplishment, this is consistent with the literature. As the publiced article shows the accomplishments of the SME, they are now referred by a governmental owned organization for future advancements within their field, this shows that the SME offer direct intrinsic	SMEs offer direct intrinsic motivation in the form of accomplishment to acquire and share knowledge with partners (**MOT2a**).

| | | | | any of these municipalities or if you have any questions, feel free to contact us (the SME) and we'll help you." | | motivation to their employees. | |

Dimensions/ Characteristics	Aspects that affect KT	Subcategory of the aspect	Secondary data source	Paragraph from the secondary data	Relating literature	Conclusion/Analysis	Summary
Characteristics of knowledge	Knowledge characteristics	Tacit knowledge / Explicit knowledge	(LC) Document: Identify knowledge gaps	*"If we work systematically (**KNOTAC**) to identify and close knowledge gaps our decision making is improved, resulting in solutions that delight customer, excellent product quality and efficient projects. A project kick-off meeting is an excellent opportunity to do some brainstorming and leverage the team members' experiences to build a good initial set of knowledge gaps. The team can immediately build the tracking, flow and knowledge capture systems to manage the gaps."*	Wijk et al. (2008) claims that knowledge ambiguity is defined by the uncertainty of the knowledge, (**KNOAMB**) what underlying components and sources are in play (content-dependency), along with how they interact with each other (context-dependency).	The secondary data shows that the LC/MNC is systematically attempting to close knowledge gaps, which implies that they are increasing their understanding of said knowledge, this is consistent with the literature.	When tacitness is high it becomes easier as a source LC/MNC to transfer knowledge to a recipient. (**KNOTAC**)
			(LC) Document: Innovation culture and	*"The elements of reusable knowledge ensure that the person receiving the knowledge has all the pieces to decide whether or not the knowledge applies and how far it can be trusted, (**KNOAMB**) understood, believable,*	Complexity consider how the knowledge that is transferred	The secondary data shows that the LC/MNC is systematically attempting to close reduce ambiguity by	The LC/MNC attempts to reduce ambiguity by codifying their tacit knowledge.

| | | | knowledg e | *actionable and generalized. If any of these elements are missing, the knowledge won't be reused to its full potential. The company culture needs to support the effort it takes to generate reusable knowledge and reward those who actively seek the benefits of knowledge reuse. This can be achieved through policies, processes and performance expectations that reward people for capturing and using reusable knowledge to save the company's time, money and capacity.* ” | (**KNOCOM**) is affected by different variations as it is comprehended by different skills or experiences of organizations (Stock & Tatikonda, 2000)… complexity can be measured by how many different variables are involved (content and context dependency) in a certain situation (Milagres & Burcharth, 2019). Kalling, (2003), state that tacitness exists for both tacit and explicit knowledge, whereas tacitness is affected by (**KNOTAC**) the individuals understanding of the knowledge they are handling… to measure tacitness in knowledge to some | codifying knowledge, which is consistent with the literature. | |
| | | | (SME) | - | | | |

				extent we will analyse the individuals understanding of the context at hand (context dependency).		

Dimensions/ Characteristics	Aspects that affect KT	Subcategory of the aspect	Secondary source data	Paragraph from the secondary data	Relating literature	Conclusion/Analysis
Relational characteristics	**Knowledge governance**	**Formal knowledge governance (FKG)**	**Docume nt:** "A3 Thinking" instruction to report The Knowledge Brief A3 Report template **Source:** News/ The internal website of the	*"The Knowledge Brief A3 is a one-page report about some area of knowledge that the author wishes to share.* *Knowledge Briefs require the most flexible format because the topic could cover so many different things. It could be about test procedures, market forecasting methods, platform architecture models and many others. To make a good Knowledge Brief, the author needs the flexibility to use the space on the page in whatever way will best suit the subject matter. The organization may want to develop a lightweight template for a Knowledge Brief, just to give the author a place to start. The best way to*	Knowledge governance mechanisms can take two modes such as formal and informal: (FKG1) the formal governance mechanisms consist of management policies and procedures, organizational structure, incentive schemes, information systems, and other control and coordination systems, whereas (IKG2) informal	This document shows that company has formal mechanisms to share knowledge in form of specific report template that can help to the knowledge source to transfer its knowledge to other actors. This is consistent with the literature.

			large company", "Avenue", posting date 14/05/20 20	*learn how to write a Knowledge Brief is to simply make the attempt and get feedback on the draft, and then refine it. It gets easier with experience"*	(also referred to as relational) includes corporate culture, trust, social networks and communities (Foss, 2007).	Secondary data shows that the company uses project planning tools as formal procedures that helps to acquire knowledge and information from the customer and share within teams to transform it into clear goals and having common understanding before starting any projects.
Relational characteristics	**Knowledge governance**	**Formal knowledge governance (FKG)**	*Docume nt: Lean project planning How to plan and execute projects in a lean way* **Source:** News/ The internal website of the large company, "Avenue "	*A workshop to clarify what the project is supposed to deliver. It is done in order to transform voice of the customer data into actionable, prioritized product design requirements that maximize customer value, market acceptance and profit. It also ensures that all participants and stakeholders agree on what is part of the delivery of the project, and what is not. And it needs to be clarified and agreed before project planning can begin, otherwise the team doesn't know what to plan for.*	Knowledge governance mechanisms can take two modes such as formal and informal: (FKG1) the formal governance mechanisms consist of management policies and procedures, organizational structure, incentive schemes, information systems, and other control and coordination systems, whereas informal (IKG2) (also referred to as relational) includes corporate culture,	

382

<table>
<tr>
<td></td>
<td>trust, social networks and communities (Foss, 2007).</td>
<td></td>
<td></td>
<td></td>
<td>This knowledge base section is aimed to share the knowledge with partners of SME. In this section, you can find case studies, procedures, white papers, instructions and manuals that can describe to the partners the key formal procedures in KT activities at the company (FKG1)</td>
</tr>
<tr>
<td rowspan="2">Relational characteristics</td>
<td rowspan="2">Knowledge governance</td>
<td rowspan="2">Formal knowledge governance (FKG)</td>
<td>Document: Knowledge base section

Source: the web page of SME</td>
<td>"Welcome to the Knowledge base, a site where you will find lots of inspiration and knowledge in digitalisation and e-health"</td>
<td>Knowledge governance mechanisms can take two modes such as formal and informal: (FKG1) the formal governance mechanisms consist of management policies and procedures, organizational structure, incentive schemes, information systems, and other control and coordination systems, whereas informal (IKG2) (also referred to as relational) includes corporate culture, trust, social networks and communities (Foss, 2007).</td>
</tr>
<tr>
<td></td>
<td></td>
<td></td>
</tr>
<tr>
<td>Relational</td>
<td>Knowledge</td>
<td>Formal knowle</td>
<td>Document:</td>
<td>Instructions and manuals that are aimed to transfer knowledge about different</td>
<td>Knowledge governance

Service and product instructions and manuals are formal knowledge governance mechanisms</td>
</tr>
</table>

characte ristics	gover nance	dge govern ance (FKG)	Knowled ge base section **Source:** the web page of the large organizat ion	products and services to the customers and suppliers of the large organization	mechanisms can take two modes such as formal and informal: (FKG1) the formal governance mechanisms consist of management policies and procedures, organizational structure, incentive schemes, information systems, and other control and coordination systems, whereas (IKG2) informal (also referred to as relational) includes corporate culture, trust, social networks and communities (Foss, 2007).	that facilitate KT activities to the partners of the large organization
Relation al characte ristics	**Geog raphi cal proxi mity/ distan ce**		**1) Docume nt:** The Global Product Manage ment training	1) The Global Product Management training program has been instrumental in building a common knowledge base and capabilities within product management and it has now been rolled out to over 200 participants globally. In the video, one of the participants suggests that this training helps to	Geographical distance negatively influences inter-organizational KT especially for tacit knowledge exchange but can be mediated in terms of (GD1)	1) This secondary data is consistent with the literature. The company establish training programs to reach common knowledge base that facilitate common corporate culture and values for different geographical distant parties (GD2). 2) MS Teams as digital tool applied by companies for digital communication

			program (video) **Source:** News/ The internal website of <u>the large company</u>, "Avenue", posting date 22/08/2019 2) Microsoft Office 365 licenses for digital meeting and video streaming	*establish common knowledge for participants from different geographical regions.* *2) Both companies use MS Teams as tool for digital communication between partners*	developing trust-based relations between distant partners (e.g., Capaldo & Petruzzelli, 2014); (GD2) organizational proximity in terms of common corporate culture and values (Korbi & Chouki, 2017); and (GD3) bilateral way of utilizing of communication technologies (Choi & Contractor, 2016)	

Organizational characteristics	Disseminative capacity	Motivation to teach	Document:			
Organizational characteristics	Disseminative capacity	Motivation to teach	Document: Customer insights Source: Corporate library portal/ The internal website of the large company, "Avenue", posting date 17/01/2020	*"Growth through customer relevance" is one of the four strategic objectives defined by the company. That means Voice of the Customer (VOC) constitutes one of the foundations for the successful development of the Group as a global leader.* <u>*VOC is an in-depth process of capturing customers' expectations, preferences and aversions. In-depth means understanding why a customer has a specific preference, not just determining or knowing about it. VOC researchers need to understand drivers, barriers and pain points, keep an open mind and replace assumptions with understanding.*</u>	The following aspects drives motivation to teach by knowledge source: (MTDC1)Direct interaction and mutual experience sharing (Squire et al., 2009). (MTDC2) to build a mutual trust between partners (Minbaeva et al., 2018)	mutual experience for knowledge sharing pushes the source to transfer knowledge to its external partners but the main purpose is to achieve better business performance due to creating value for customers.
Organizational characteristics	Disseminative capacity	Capacity to teach	Document: Knowledge base from the company website (SMEs) Source: Knowled	*"Senior Housing Experts in Electronic Locking Solutions. …Our locks are installed to over 100 000 senior's doors globally. Additionally, our innovative technology evolves with the organization's needs, increasing efficiencies, enhancing security, and providing seamless maintenance"*	The following aspects influence ability to teach of the knowledge source: (ATDC1) attractiveness of source companies in terms of competence and knowledge holding value (Pérez-	This document shows how the case company from SME demonstrates its expertise and value of their innovative products to its partners via a knowledge-sharing base page at their website. This knowledge base might facilitate KT between partners through ability to codify of knowledge by SME (ATDC2).

	ge base section from the company website (SMEs)	Nordtvedt et al., 2008) (ATDC2) the degree of the source firm's ability to codify knowledge by developing common glossaries that can facilitate mutual understanding (Schulze et al., 2014) (ATDC3) the degree of partner's expertise and competence (e.g Schulze et al., 2014)	
Trust	Docume nt: Template s of Non-Disclosu re Agreeme nt (NDA) **Source:** The internal website of <u>the large company</u>	To assess interfirm trust for KT in terms of fulfillment of commitments agreed between partners; honest negotiation; and avoidance of receiving excessive benefits of ally organizations (Cummings & Bromiley, 1996).	For the company NDAs are a formal document that passed the formal procedures to formalize agreements that facilitate knowledge transfer.
		Group Legal has developed two general NDA templates - one <u>mutual and one unilateral.</u> *The mutual agreement is intended to be used if both the large company and the counterparty will provide confidential information. The unilateral agreement is intended to be used if only the large organization will provide confidential information (to the partner).* *Each of the NDA template is available in a version governed by Swedish laws and a version governed by the laws of the State of New York.* *A NDA guideline is available.*	

				, "Avenue "		
	Trust		**Docume nt:** Template of Non-Disclosu re Agreeme nt (NDA) **Source:** company provided the template by email	From NDA template (translated from Swedish): *The undersigned is aware that during my collaboration with Company. I will receive technical, scientific, financial and strategically important information regarding the company. I am also aware that Company considers such information to be valuable trade secrets, the value of which depends on it not being disclosed or disclosed. I undertake that both during the time the cooperation / assignment relationship is ongoing and for 5 years thereafter not for any unauthorized disclosure or disclosure of such information. The above does not apply to such information as:* • *on receipt was demonstrably publicly known or who subsequently became publicly known without breach of this Agreement;* • *at the reception I was demonstrably already in possession of* • *I have legally demonstrably obtained from a third party without a duty of confidentiality*	To assess interfirm trust for KT in terms of fulfillment of commitments agreed between partners; honest negotiation; and avoidance of receiving excessive benefits of ally organizations (Cummings & Bromiley, 1996).	For the company NDAs are a formal document that passed the formal procedures to formalize agreements that facilitate knowledge transfer.

				• developed independently of company's data. *If and to the extent that I need to discuss confidential information in order to fulfill my tasks in collaboration with the company, I undertake to ensure that the persons concerned are authorized to take part in such information and bound to keep it secret to the same extent as myself according to this confidentiality link.* *I am obliged to store and handle all documentation concerning such information in a secure manner, inaccessible to unauthorized access. I undertake, at the request of the company, to return it and not to keep any copies thereof.* *Should I violate this confidentiality agreement, I am aware that I may be liable for the damage that the company may suffer as a result.*		
	Cultur al distan ce		Docume nt: English Opens Doors - Improve your English Skills	*English opens doors for you in the company. The English language course is an interactive and web-based program designed to help improve your communication skills in English.* *English Opens Doors is targeted towards employees who need to improve their English skills to become more efficient in their jobs. Based on the Language policy and the competency*	The national cultural distance arises different barriers for the KT among international partners regardless the firm size such as (CD1) higher cost of entry and hinders the transmitting the core	The secondary data supports that language skills can be barrier for communication (implicitly knowledge transfer) for their job responsibilities

			Source: The internal website of <u>the</u> <u>large</u> <u>company</u> , "Avenue "	*profiles (see documents below), you can see if you are eligible for training. The program assesses your current level, and adapts the training to your level. You can also choose to deep dive into specific business areas to improve your communication skills within e.g. finance. If you are interested in knowing more, or sign up for the course,* **contact your HR manager**. *Find more details on the program offering and setup in the* **documents below**. *For beginners, the online portal is available in: English, Portuguese, Spanish, French, Dutch, Italian, Korean, Simplified Chinese, Traditional Chinese, Russian, Indonesian, Thai and Turkish.*	competencies to the international markets; (CD2) facing operational challenges in terms of lacking common understanding of norms, values and motivation; (CD3) misunderstanding between foreign partners can limit sharing of critical organizational knowledge (Wijk et al., 2008)